# Occupational Health and Safety Handbook

# Occupational Health and Safety Handbook

Edited by Alfred Scott

Clanrye International
www.clanryeinternational.com

Clanrye International,
750 Third Avenue, 9th Floor,
New York, NY 10017, USA



ISBN: 978-1-63240-695-8

**Cataloging-in-Publication Data**

Occupational health and safety handbook / edited by Alfred Scott.
p. cm.
Includes bibliographical references and index.
ISBN 978-1-63240-695-8
1. Industrial hygiene. 2. Industrial safety. I. Scott, Alfred.
HD7261 .O23 2018
658.382--dc23

For information on all Clanrye International publications visit our website at www.clanryeinternational.com

CLANRYE INTERNATIONAL

Printed in China.

# Contents

# Preface

Occupational health and safety management deals with the security and well-being of persons at their worksite. The main aim of this subject is to promote a place of work which is safe and has a good working environment. The hazards that can occur in a workplace are categorized into physical hazards, biological hazards and psychological hazards. One of the ways of protecting oneself from these hazards is personal protective equipment which are clothing or equipment used in safeguarding the wearer's body from any damage. Different approaches, evaluations, methodologies and advanced studies on occupational health and safety have been included in this book. It will provide comprehensive knowledge to the readers.

Various studies have approached the subject by analyzing it with a single perspective, but the present book provides diverse methodologies and techniques to address this field. This book contains theories and applications needed for understanding the subject from different perspectives. The aim is to keep the readers informed about the progresses in the field; therefore, the contributions were carefully examined to compile novel researches by specialists from across the globe.

Indeed, the job of the editor is the most crucial and challenging in compiling all chapters into a single book. In the end, I would extend my sincere thanks to the chapter authors for their profound work. I am also thankful for the support provided by my family and colleagues during the compilation of this book.

**Editor**

1

# Assessment and Risk Management of Potential Hazards by Failure Modes and Effect Analysis (FMEA) Method in Yazd Steel Complex

**Mehrzad Ebrahemzadih[1,2], G. H. Halvani[3], Behzad Shahmoradi[2], Omid Giahi[1,2*]**

[1]Department of Occupational Health, Faculty of Health, Kurdistan University of Medical Sciences, Sanandaj, Iran
[2]Kurdistan Environmental Health Research Center, Kurdistan University of Medical Sciences, Sanandaj, Iran
[3]Departments of Occupational Health, Faculty of Health, Shahid Sadoghi University of Medical Sciences, Yazd, Iran
Email: *Omidgi71@yahoo.com

## Abstract

**Background: Failure mode and effect analysis (FMEA) is a widely used quality improvement and risk assessment tool in manufacturing. The aim of this study is to assess potential hazards by failure modes and effect analysis (FMEA) method in Yazd Steel Complex. Methods: In this descriptive study, we evaluated the risks in different parts of the complex by using FMEA method and by using FMEA Worksheets (PFMEA) derived from the standard (MIL_STD-882). Failure modes and the various components and effects as using quantitative score to the risk priority (RPN) were obtained. PFMEA worksheets were completed and, we reevaluated the weaknesses part of the system. Activities related to each from the different parts of Yazd Steel Complex by using the scores risk priority (RPN) were evaluated. Then the results obtained by using SPSS software were performed by evaluation and analysis. Results: The findings showed that the steel maker lime unite and steel making ingot casting achieved the highest of RPN before and after corrective actions measures (490, 168) and environmental health unite and roll styles unite achieved the lowest of RPN before and after corrective actions measures (28, 20). Conclusions: The results show that the FMEA technique can identify a higher number of hazards than any other technique. The important point is that selection of an appropriate technique plays an important role in identifying a higher number of hazards.**

## Keywords

**Failure Modes and Effects Analysis (FMEA), Risk Management, Steel, Risk**

*Corresponding author.

## 1. Introduction

In the past few decades, many actions and research works have been undertaken to prevent potential accidents and to promote safety in the chemical processes; the output of these measures is the systematic management of safety in these processes. One of the key elements of safety management systems is to identify hazards, to assess risks and their control, which helps security professionals to investigate the ability of rational decision to reduce the risk and severity of accidents and their consequences [1].

Failure mode and effects analysis (FMEA) is commonly defined as "a systematic process for identifying potential design and process failures before they occur, with the intent to eliminate them or minimize the risk associated with them". The FMEA technique was first reported in the 1920s but its use has only been significantly documented since the early 1960s. It was developed in the USA in the 1960s by national aeronautics space agency (NASA) as a means of addressing a way to improve the reliability of military equipment [2].

Despite the importance of risk assessment as a scientific basis for national and international activities, this category has different meanings among people and experts in various disciplines, and is often controversial and incorrect interpretations. Diversity and distribution of scientific topics, definition of risk assessment, the distinction between risk assessment and risk management, *et al.* are among the reasons that have offered different interpretations of the concept of risk assessment. Based on objectives, data, resources, and other factors for risk assessment, methods and definitions are introduced. Hazard identification and safety evaluation have several methods, including safety audits patrols and inspections of HAZOP, JSA, and FMEA. An OSHA guideline is one of the quality tools that every manager needs to be familiar with [3].

Every day in the United States, a large number of workers are injured or killed at work. Considering work place conditions, providing proper work procedure, and training all workers are among suitable and useful strategies to prevent disease incidence, injury and trauma [4]. Risk management includes techniques to identify, assess and control risk, as other tools are multifarious. Generally it can be divided into two categories: qualitative and quantitative; qualitative methods are usually applied to small companies with a low number of activities, whereas quantitative methods are usually used for large organizations with high activity. One of the few ways to identify, assess and control risks is FMEA method [5].

In 1950, FMEA method was established by engineers for reliability and safety assessment in military systems. It was quickly spread so that it was used to evaluate safety of Concorde and Airbus aircraft in America and France respectively. This method was also developed in nuclear safety after the Maryland disaster [6]. Failure mode and effects analysis (FMEA) is a comprehensive engineering technique that manufacturers are able to improve the quality, reliability, and safety of their products through applying this technique [7]. In particular, FMEA technique is used to identify, define, and eliminate known and potential failures, problems and errors in the products, programs, systems and services before they reach the customer.

FMEA is a systematic method, which is applied because of the following reasons:

1) To identify and prioritize potential failure modes in a system, product, process, or service;

2) To define and run measures in order to eliminate or reduce the incidence of potential failure modes;

3) To record analysis results in order to provide a comprehensive reference for solving future issues and problems [8].

FMEA could be described as a structured method to find and identify failure modes in a system, object, or an activity and calculation of the failure effects on upper steps [9] [10].

According to the 2011 disaster, which killed 14 and injured 5 people and the importance of steel industry as well as the preservation of labor force, we conducted a comprehensive and targeted study.

## 2. Materials and Methods

This is a descriptive study conducted in Yazd Steel Complex. In this study, we evaluated the risks in different parts of the complex using FMEA method. Using FMEA Worksheets derived from the standard (MIL_STD-882), we obtained failure modes along with its different components and failure effects in quantity using risk priority number (RPN) equation and PFMEA worksheets were completed. Finally, we reevaluated the weaknesses of the system. Activities related to each of the different parts of Yazd Steel Complex were studied using RPN equation. The result of the evaluation and analysis was performed using SPSS software.

**-System description**

After visiting and studying different parts of Yazd Steel Complex, risks involved in the steel complex were examined in detail and likelihood, severity, and frequency of exposure were quantitatively studied. Moreover, the risk of each activity was obtained and in another part, the control measures were expressed. Later, we again stated the likelihood, severity, and frequency of exposure and risk reduced. At the end of this study, we would present tables of the risk probability, risk severity, frequency of exposure, and risk number obtained before control measures and after control measures separately. **Figure 1** shows the FMEA process.

**-RPN methodology**

Decision making using RPN scoring and crisis level.

The RPN is a mathematical product of the severity, the occurrence and the detection. The number is used to identify the most critical failure mode, leading to corrective action [12].

RPN scoring is based on the fact that numbers with higher risk priority have priority for analyzing and resource allocation aimed at improvement, and the team should work on failure modes having a higher RPN. RPN is obtained by multiplying three factors of intensity, possibility of occurrence, and detection possibility [13]. It is calculated using Equation (1).

$$\text{RPN} = \text{Severity} \times \text{Occurrenc} \times \text{Detection} \quad (1)$$

RPN to evaluate the risk level of a component or process.

The RPN is obtained by finding the multiplication of three factors, which are the severity of the failure (S), the probability of occurrence (O) and the probability of detection (D). In this project, we used risk criterion number to determine the level of acceptable and unacceptable risk in RPN. Risk criterion is an index for separating acceptable and unacceptable risks in the system studied. A failure that its RPN number is greater than the risk criterion is considered as unacceptable risk and a failure that its RPN is lower than the risk criterion is called acceptable risk. This index is varied based on the laws and regulations of each organization and its ability to pay for needed projects costs.

**-Determination of the severity rate**

**Severity** is a rating corresponding to the seriousness of an effect of a potential failure mode. Severity or seriousness of the risk is considered just in case of "the effect"; reducing the risk severity is possible only through changing the process and the manner of performing activities.

**-Determination of the occurrence rate**

**Occurrence** is ranked according to the failure probability, which represents the relative number of failures anticipated during the design life of the item. The effects of a failure mode are normally described by the effects on the user of the product or as they would be seen by the user.

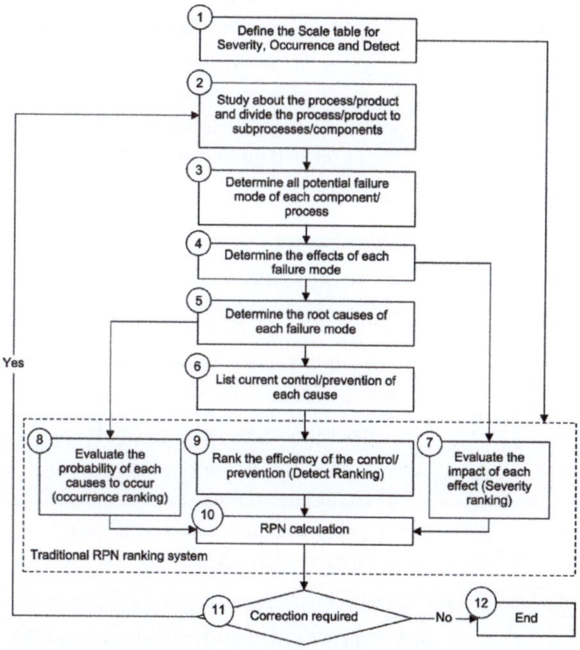


**Figure 1.** The traditional FMEA procedure [11].

**-Risk detection probability rate**

**Detection possibility** is an assessment of the ability existing to identify a cause/mechanism of a risk occurrence. In other words, detection possibility is a rating corresponding to the likelihood that the detection methods or current controls will detect the potential failure mode before the product is released for production for design, or for process before it leaves the production facility. Assessing control process of standards, requirements and laws of labor and how to apply them to achieve this number are very useful.

In the RPN methodology the parameters used to determine the "criticality" of an item failure mode are, the severity of its failure effects, its frequency of occurrence, and the likelihood that subsequent testing of the design will detect that the potential failure mode actually occurs. **Tables 1-3** show the qualitative scales commonly used for the severity, the occurrence and the detect ability indexes [14].

**Table 1.** Severity guidelines for design FMEA (1 - 10 qualitative scale) [14].

| Effect | Rank | Criteria |
|---|---|---|
| No | 1 | No effect |
| Very slight | 2 | Customer not annoyed |
| Slight | 3 | Customer slight annoyed |
| Minor | 4 | Customer experiences minor nuisance |
| Moderate | 5 | Customer experiences some dissatisfaction |
| Significant | 6 | Customer experiences discomfort |
| Major | 7 | Customer dissatisfied |
| Extreme | 8 | Customer very dissatisfied |
| Serious | 9 | Potential hazardous effect |
| Hazardous | 10 | Hazardous effect |

**Table 2.** Occurrence guidelines for design FMEA (1 - 10 qualitative scale) [14].

| Effect | Rank | Criteria |
|---|---|---|
| Almost never | 1 | Failure unlikely. History shows no failure |
| Remote | 2 | Rare number of failures likely |
| Very slight | 3 | Very few failures likely |
| Slight | 4 | Few failures likely |
| Low | 5 | Occasional number of failures likely |
| Medium | 6 | Medium number of failures likely |
| Moderately high | 7 | Moderately high number of failures likely |
| High | 8 | High number of failures likely |
| Very high | 9 | Very high number of failures likely |
| Almost certain | 10 | Failure almost certain |

**Table 3.** Detectability guidelines for design FMEA (1 - 10 qualitative scale) [14].

| Effect | Rank | Criteria |
|---|---|---|
| Almost certain | 1 | Proven detection methods available in concept stage |
| Very high | 2 | Proven computer analysis available in early design stage |
| High | 3 | Simulation and/or modeling in early stage |
| Moderately high | 4 | Tests on early prototype system elements |
| Medium | 5 | Tests on preproduction system components |
| Low | 6 | Tests on similar system components |
| Slight | 7 | Tests on product with prototypes and system components installed |
| Very slight | 8 | Proving durability tests on products with system components installed |
| Remote | 9 | Only unproven or unreliable technique(s) available |
| Almost impossible | 10 | No known techniques available |

In this research work, we considered both risks with high RPN and low RPN having one or two abovementioned factors, *i.e.* while determining risk criterion and decision making for considering a failure within the acceptable or unacceptable risk domain, the team attention was not only towards RPN values but also each of the tree failure factors were assessed. For this purpose, a criterion was defined as the level of crisis. The crisis is a criterion that expresses the importance of a potential/actual risk in the system studied. In addition, it is used to measure the level of crisis in the system. Crisis grade is composed of normal, semi-critical, and critical levels which are explained in detail below:

Level 1: Normal level in which all of the three factors of RPN (especially the severity and probability of occurrence) have values less than 5. Or RPN number is very low and does not require corrective and preventive actions (however, according to the concerned engineer, the corrective/preventive action could be presented (usually RPN < 70)).

Level 2: Semi-critical level in which at least a factor of three factors of RPN (especially the severity and probability of occurrence) has a value greater than 5 but RPN is relatively low. In this case, corrective/preventive action is essential (typically 70 < RPN < 140).

Level 3: Critical level in which at least two factors of the three-factor of RPN have high values or RPN number is too high. Since this level has been considered for high RPN, it is obvious and clear that it has a corrective/preventative action (usually RPN > 140) [15].

## 3. Results

**Table 4** and **Table 5** summarize the results of implementing FMEA in Yazd Steel Complex. These tables tabu-

**Tables 4.** The sample FMEA work sheet in the steel making (unit of lime).

| FAILURE MODE AND EFFECTS ANALYSIS<br>Work sheet item: steel making (unit of lime)<br>Core team: Mehrzad Ebrahemzadih, G.H. Halvani, B. Shahmoradi | | | | | | | | | | | | |
|---|---|---|---|---|---|---|---|---|---|---|---|---|
| **Process Function** | **Potential Failure Mode** | **Potential Effect(s) of Failure** | **S** | **O** | **D** | **RPN** | **Recommended Action(s)** | **S RE** | **O RE** | **D RE** | **RPN RE** |
| 1 Furnace operator | Dust | Lung damages | 7 | 5 | 5 | 175 | The use of respiratory mask | 2 | 5 | 5 | 50 |
| | Noise | Hearing damage | 7 | 4 | 5 | 140 | The use of ear plugs | 2 | 5 | 5 | 50 |
| 2 Raw ore charging into the furnace | The fall basket Netscape | Death | 10 | 5 | 6 | 300 | Creating safeguards | 3 | 5 | 6 | 90 |
| | Rock fall | Fracture of lumbar | 6 | 5 | 4 | 120 | Creating safeguards + training | 3 | 4 | 4 | 48 |
| 3 Lime discharge | Dust | Lung damages | 7 | 8 | 5 | 280 | The use of respiratory mask | 3 | 7 | 5 | 105 |
| 4 Driving loaders | Dust | Lung damages | 6 | 5 | 6 | 180 | The use of respiratory mask | 2 | 5 | 5 | 50 |
| | Noise | Hearing damage | 7 | 6 | 4 | 168 | The use of ear plugs | 3 | 5 | 3 | 30 |

**Tables 5.** The sample FMEA work sheet in the steel making (IC unit-ingot casting).

| FAILURE MODE AND EFFECTS ANALYSIS<br>Work sheet item: steel making (IC unit-ingot casting)<br>Core team: Mehrzad Ebrahemzadih, G.H. Halvani, B. Shahmoradi | | | | | | | | | | | |
|---|---|---|---|---|---|---|---|---|---|---|---|
| **Process Function** | **Potential Failure Mode** | **Potential Effect(s) of Failure** | **S** | **O** | **D** | **RPN** | **Recommended Action(s)** | **S RE** | **O RE** | **D RE** | **RPN RE** |
| 1 Filling the sand blasting tank | The fall sand sacking | Death | 10 | 5 | 3 | 150 | Alarm by crane | 10 | 5 | 2 | 100 |
| 2 Casting | Infrared light | Cataract | 7 | 5 | 8 | 280 | The use of face shield | 7 | 3 | 6 | 126 |
| | High temperature | Burn | 4 | 6 | 9 | 216 | The use of helmet and gloves fireproof and body shield | 2 | 6 | 8 | 96 |
| 3 Checking the furnace burners | Biting | Death | 10 | 5 | 4 | 200 | The use of detector | 8 | 4 | 2 | 64 |
| 4 Work in steel making | Dust | Lung damages | 7 | 5 | 6 | 180 | The use of respiratory mask | 3 | 4 | 5 | 60 |
| | Noise | Hearing damage | 7 | 6 | 6 | 252 | The use of ear plugs | 3 | 5 | 5 | 75 |

late the risks and their causes, severity and occurrence possibility of risks in two parts of steel making (unit of lime) and ingot casting in order to show the way of filling risk assessment sheets.

**Graph 1** and **Graph 2** show the risks involved in the systems before and after corrective actions in different units of Yazd Steel Complex.

## 4. Discussion

After determining the level of risk associated with different parts of the steel industry, the following results related to the MAX and MIN, mean and SD, RPN before and after corrective actions was defined with regard to the job in the process function (**Table 6**).

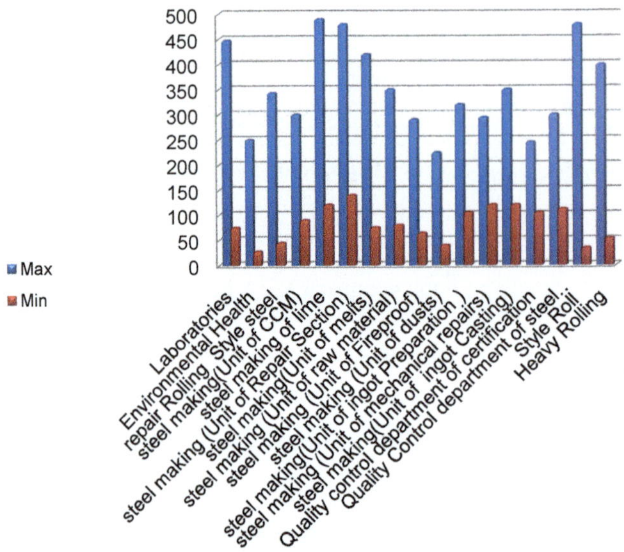


**Graph 1.** Risks evaluation before corrective actions in different units of Yazd Steel Complex.

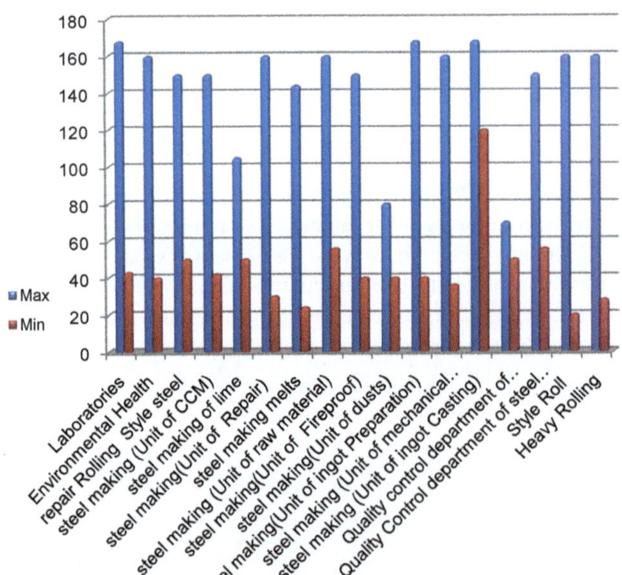


**Graph 2.** Risks evaluation after corrective actions in different units of Yazd Steel Complex.

**Table 6.** Summary of results risk assessment by FMEA method.

| Parameters RPN | Process Function | MAX | Process Function | MIN | Mean ± SD | |
|---|---|---|---|---|---|---|
| | | | | | MIN | MAX |
| Before corrective actions | Steel making of lime | 490 | Department environmental health | 28 | 352 ± 86.58 | 82.82 ± 34.61 |
| After corrective actions | Steel making ingot casting | 168 | Style roll | 20 | 144.88 ± 30.09 | 45 ± 21.96 |

Although initially it seemed that there is no big and much risks in these sections and the subset employees persist this problem, our results indicated that there are many unknown hazards with a higher risk.

A part of the forms of hazard identification, harmful factors, and risk assessment at steel industry (lime unit).

Equipment failures analysis using FMEA provides the possibility to identify various factors having the potential to create conditions for the possibility of an accident or stop operating phases [16]. Because no preventive and control measures are conducted on them and finally one day they could cause an accident.

It is noteworthy that most of the known risks threaten the system while they could be eliminated or controlled easily and by spending the minimum cost. However, the accidents resulted from them could impose huge losses on the system [17]. FMEA is applied in the industry at all stages of an industrial project from the manufacturing stage to the production stage to improve product quality and productivity [18]. Given that the traditional approach to the design industry without considering the risk of accidents has caused many deficiency and failures in past, nowadays, in implementing industrial plans, especially in the early stages of design, FMEA technique has limited and minimized failures through identifying and assessing risks involved [19].

OSHA believes that the process of risk analysis should be at best as a team work with expertise in engineering and process operations. Moreover, it should involve at least one person from the workers, who has expertise and knowledge in the field of the method used to identify risks [20].

In a study, Rezvani showed that among the possible existing risks in milk company, noise from the production line equipment including tetra packing equipment and basket washing machine had the highest relative frequency (64%) followed by hazards such as inhaling NaOH and acid fumes (32%), and burns from acid and NaOH (32%). In this study, the number of employees studied was 28 and the number of hazards identified was 280 risks [21].

In another study in oil refining company, Asadi reported that the risk of falling from a height with a relative frequency of 12% and the risk of slipping with a relative frequency of 10% led to accidents. The total number of identified risks was 4250 cases [22]. Benjamin showed that accidents and disasters could be significantly prevented if during the production process of the system, FMEA could be launched and carefully managed at the beginning of the process [23].

In another study conducted by Ebrahimzadeh *et al.* at Shiraz Refinery, Iran, it was found that activities with low RPN have more priority than the activities with higher RPN in terms of severity. In addition, scoring high RPN in some activities such as handling and transportation of objects and milling, one can apply appropriate control measures to the level of acceptable risk indicating usefulness and effectiveness of FMEA method [24].

Naturally, not only organization culture but also community culture must booster safety. Economic condition of the organization is also an influencing factor on the risk assessment. Organizations in which productivity is low and revenue is not satisfactory, is not able to pay for safety. This, in turn, reduces the level of safety in the organization and would result in more accidents or disasters. These disaster themselves will bring about direct losses and reducing product quality and quantity. This means lower productivity and profits, and thus reducing the organization funding on safety. Organizational factors have also remarkable influences on the development or progression or system safety development [25].

## 5. Conclusion

In order to prevent potential accidents and improve safety in industrial processes, systematic management of safety is essential in these processes. It seems that the implementation of a documentation system for recording equipment deficiencies and events can be basic information needed to assess the subsequent safety ideally. In addition, performing preventive maintenance can reduce the possibility of the equipment defects and their consequences. Our results showed that compared with other methods of risk assessment, FMEA can identify more risks and an important point is that choosing an appropriate method plays a crucial role in identifying more risks.

## References


[1] Mirghiasi, Z. (2009) Hazard Identification and Risk Analysis. B.Sc. Graduation Thesis, Abadan Petrochemical Unit 300, PUT.

[2] Johnson, K.G. and Khan, M.K. (2003) A Study into the Use of the Process Failure Mode and Effects Analysis (PFMEA) in the Automotive Industry in the UK. *Journal of Materials Processing Technology*, **139**, 348-356. http://dx.doi.org/10.1016/S0924-0136(03)00542-9

[3] Joy, J. (2004) Occupational Safety Risk Management in Australian Mining. *Occupational Medicine*, **54**, 311-315.

[4] US Department of Labor, Occupational Safety and Health Administration (2001) Job Hazard Analysis. OSHA Publication.

[5] Russomanno, D.J., Bonnell, R.D. and Bowles, J.B. (1993) Functional Reasoning in a Failure Modes and Effects Analysis (FMEA) Expert-System. *Proceedings of the Annual Reliability and Maintainability Symposium*, Atlanta, 26-28 January 1993, 339-347.

[6] Chen, H.C. (1996) Failure Modes and Effects Analysis Training Manual. Personal Communication, Hen Technology Inc., USA.

[7] Huang, G.Q., Nie, M. and Mak, K.L. (1999) Web-Based Failure Mode and Effect Analysis. *Computers & Industrial Engineering*, **37**, 177-180. http://dx.doi.org/10.1016/S0360-8352(99)00049-2

[8] Wirth, R., Berthold, B., kramer, A. and Gerhard, P. (1996) Knowledge-Based Support of System Analysis for Analysis of Failure Modes and Effects. *Engineering Applied Artificial Intelligence*, **9**, 219-229. http://dx.doi.org/10.1016/0952-1976(96)00014-0

[9] Rozenfeld, O., Sacks, R. and Rosenfeld, Y. (2009) CHASTE—Construction Hazard Analysis with Spatial and Temporal Exposure. *Construction Management & Economics*, **27**, 625-638. http://dx.doi.org/10.1080/01446190903002771

[10] SAE (2002) Potential Failure Mode and Effects Analysis in Design (Design FMEA), Potential Failure Mode and Effects in Manufacturing and Assembly Processes (Process FMEA), and Potential Failure Mode and Effects Analysis for Machinery (Machinery FMEA).

[11] Tay, K.M. and Lim, C.P. (2006) Fuzzy FMEA with a Guided Rules Reduction System for Prioritization of Failures. *International Journal of Quality & Reliability Management*, **23**, 1047-1066. http://dx.doi.org/10.1108/02656710610688202

[12] Palady, P. (1995) Failure Modes and Effects Analysis: Predicting & Preventing Problems before They Occur. PT Publications Inc., Florida.

[13] International Electro Technical Commission (1991) IEC 60812: Analysis Techniques for System Reliability-Procedure for Failure Mode and Effects Analysis (FMEA).

[14] Narayanagounder, S. and Gurusami, K. (2009) A New Approach for Prioritization of Failure Modes in Design FMEA Using ANOVA. *World Academy of Science, Engineering and Technology*, **3**, 477-484.

[15] USA Department of Defense (1980) Military Standard, Procedure for Performing a Failure Mode, Effects and Criticaly Analysis. Department of Defense, Washington DC, 27-33.

[16] Wirth, R., Berthold, B., Kramer, A. and Peter, G. (1996) Knowledge-Based Support of System Analysis for the Analysis of Failure Modes and Effects. *Engineering Applications of Artificial Intelligence*, **9**, 219-229.

[17] Elmqvist, J. and Nadjm-Tehrani, S. (2008) Formal Support for Quantitative Analysis of Residual Risks in Safety-Critical Systems. *Proceedings of HASE* 2008, Nanjing, 3-5 December 2008, 154-164.

[18] Ale, B.J.M., Bellamy, L.J., Baksteen, H., Damen, M., Goossens, L.H.J., Hale, A.R., Mud, M., Oh, J., Papazoglou, I.A. and Whiston, J.Y. (2008) Accidents in the Construction Industry in the Netherlands: An Analysis of Accident Reports Using Storybuilder. *Reliability Engineering and System Safety*, **93**, 1523-1533. http://dx.doi.org/10.1016/j.ress.2007.09.004

[19] FAA (Federal Aviation Administration) (2000) Lesson 5: Failure Modes and Effects Analysis. In: *System Safety Handbook*, 2.

[20] USA Department of Labor, Occupational Safety and Health Administration (OSHA) (2000) Process Safety Management, 10.

[21] Rezvani, Z. and Gholami, M. (2006) Identifying Analysis of Occupational Hazards in a Milk Company. 1*st National Conference of Safety Engineering and Management*, Tehran.

[22] Asadi, A. (2004) The Investigation Hazards Make Accidents in an Oil Refining Company. First Symposium of Occupational Health and Safety in Oil Refining Companies.

[23] Benjamin, P.C., Menzel, C.C., Mayer, R.J., Fillion, F., Futrell, M.T., Dewitte, P.S. and Lingineni, M. (1994) Informa-

tion Integration for Concurrent Engineering, IDEF5 Method Report. http://www.idef.com/pdf/idef3_fn.pdf

[24] Ebrahimzadeh, M., Halvani, G.H., Mortazavi, M. and Soltani, R. (2011) Assessment of Potential Hazards by Failure Modes and Effect Analysis (FMEA) Method in Shiraz Oil Refinery. *Occupational Medicine Journal*, **3**, 16-23.

[25] Darligton, J.F. and Booker, J.D. (2006) Development of a Design Technique for the Identification of Fatigue Initiating Features. *Engineering Failure Analysis*, **13**, 809-876.

# 2

# Microcontroller Based Home Security System with GSM Technology

**Raqibull Hasan, Mohammad Monirujjaman Khan, Asaduzzaman Ashek, Israt Jahan Rumpa**

Department of Electrical and Electronic Engineering, Primeasia University, Dhaka, Bangladesh
Email: monirkhan.qmul@gmail.com

## Abstract

**In this paper, design and implement of a microcontroller based home security system with GSM technology have been presented and analyzed. Two microcontrollers with other peripheral devices which include Light Emitting Diode (LED), Liquid Crystal Display (LCD), Buzzer and Global System for Mobile Communication (GSM) Module are responsible for reliable operation of the proposed security system. In addition, a mobile phone is interfaced with microcontroller through a Bluetooth device in order to control the system. Moreover, a manual keypad is another way to lock or unlock the system. A Compiler Code Vision AVR is used to design a program that controls the system along with maintaining all security functions. The designed program is applied in Proteus Software for simulation. At last, the results of practical circuit show the proper functions and also verify the reliable security within reasonable cost.**

## Keywords

**Home, Security, System, Control, Microcontroller, Bluetooth, GSM**

## 1. Introduction

Home automation is a process for improving the quality of resident's life by facilitating a flexible, comfortable and secure environment [1]. Home security system is the most prominent feature for home automation. Traditional techniques of alarm based security have gained much popularity in past decades. Nowadays, embedded system is designed to provide security due to tremendous improvement in microcontroller unit and widespread applications of GSM technology. In literatures, researchers suggested a number of security systems based on new technologies like GSM, GPRS (General Packet Radio Service), Internet, USN (Ubiquitous Sensors Network) and implemented through FPGA (Field Programmable Gate Arrays), ASICs (Application Specific Integrated Circuit), DSP (Digital Signal Processing), and MCU (Microcontroller Unit) [2]. In [3] [4], home automation system has been explained using FPGA, GSM, Internet and Speech Recognition. In this system, the home gateway is internet which requires personal computer (PC). However, it is difficult to manage PC and keep it ON all the time which also consumes more power. The system presented in [5] is an internet-based intelligent system for home power management aiming to reduce energy consumption. This system also uses internet cloud as a home gateway having the same limitations as described earlier. A java equipped mobile based home automation system is described in [6]. Although the research proposes an embedded home server but it still requires internet connectivity for GPRS. In [7], authors proposed a zigbee based home network configuration which controlled all home appliances through zigbee-infrared combination and zigbee power adapter. Lee *et al.* in [8] proposed a wireless network protocol providing a bidirectional communication channel between a gateway and the control device, highlighting the significance of wireless sensors network in controlling home appliances. In [9], PIC18F452 microcontroller based home security system has been designed without GSM technology. In this paper, ATMEGA16 microcontroller based home security system with GSM technology has been described. In addition, Bluetooth application has been used to control the system. Consequently, the proposed system provides reliable security within reasonable cost and also removes the circuit complexity.

In this work, firstly proposed system will be described in section 2 which includes software design process and implementation of practical circuit. Moreover, Section 3 will present the results along with discussing the functional sequence of practical circuit. Finally, section 4 will summarize the whole research.

## 2. Proposed System

General block diagram of the proposed microcontroller based home security system has been shown in **Figure 1**. It consists of three sections. Those are input, controller and output sections. At first the controller section remains at waiting state for receiving a signal from input section which includes Bluetooth, 4 × 4 Keypad, Pressure sensors and a mobile phone. In input section, a mobile phone through a Bluetooth device has been used for interfacing with microcontroller to lock or unlock the door of the system. Consequently, a manual keypad performs the similar function without communication with Bluetooth device. On the other hand, pressure sensors have been placed both at door and window to protect the system.

If anyone tries to unlock the system without authority, then those sensors will activate and controller will receive a suitable pulse. According to the pulse of input devices, the controller section takes decision and activates the output section which includes LCD display, GSM Module, Servo Motor, LED and Buzzer. However, the ac-

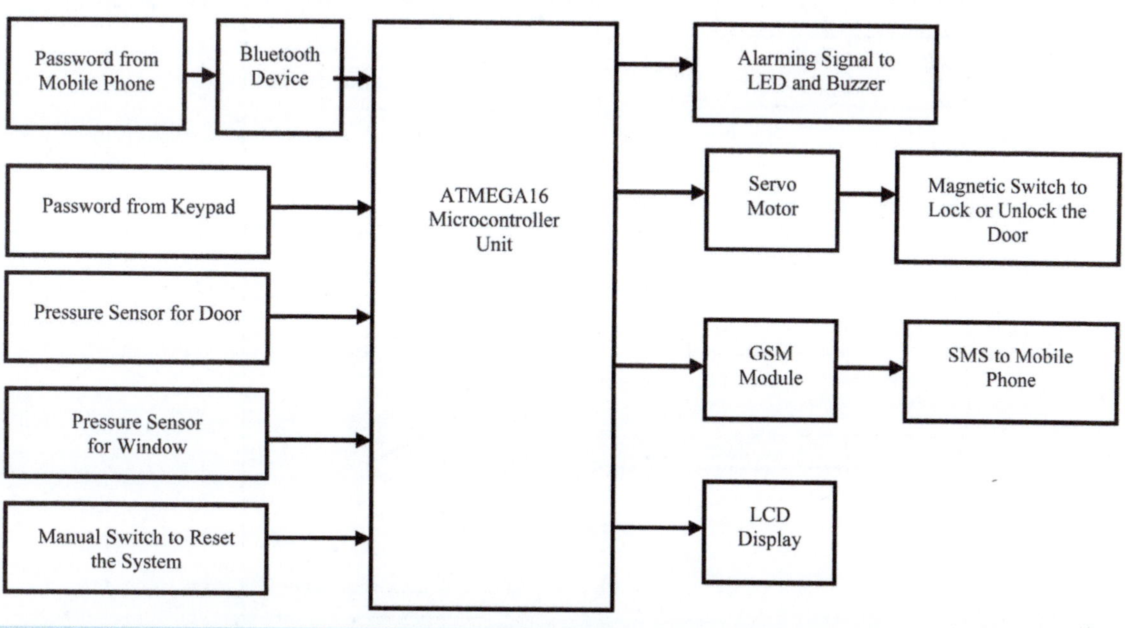


**Figure 1.** General block diagram of proposed home security system.

tivation of output device depends on the activity of input section that indicates user unlock the system either authoritative or without authoritative way. For authoritative way, controller sends a pulse signal to the servo motor that is also connected with a magnetic switch placed with a door. After 90 degree rotation of servo motor activates the magnetic switch which is used to lock or unlock the system. At the same time, a message is displayed at LCD. Conversely, at without authoritative way controller sends signal to the LED, Buzzer and GSM Module simultaneously. LED has been placed surrounding the home which indicates something wrong events that have occurred within the system. Consequently a Buzzer produces sound signal and GSM module throws a message towards the owner's mobile phone. Those activities are easily aware both the owner and security guard. As a result they will detect the mal operation within the system. At last a reset switch is used to regain the whole system at normal condition.

The proposed system includes two parts such as: software and hardware. At software part, a program has been designed in Code Vision AVR Software according to the functional sequence of block diagram. In accordance to software instruction a suitable circuit diagram has been designated for hardware implementation.

## 2.1. Software Design

The program flow chart of the proposed home security system has been shown in **Figure 2** which explains the

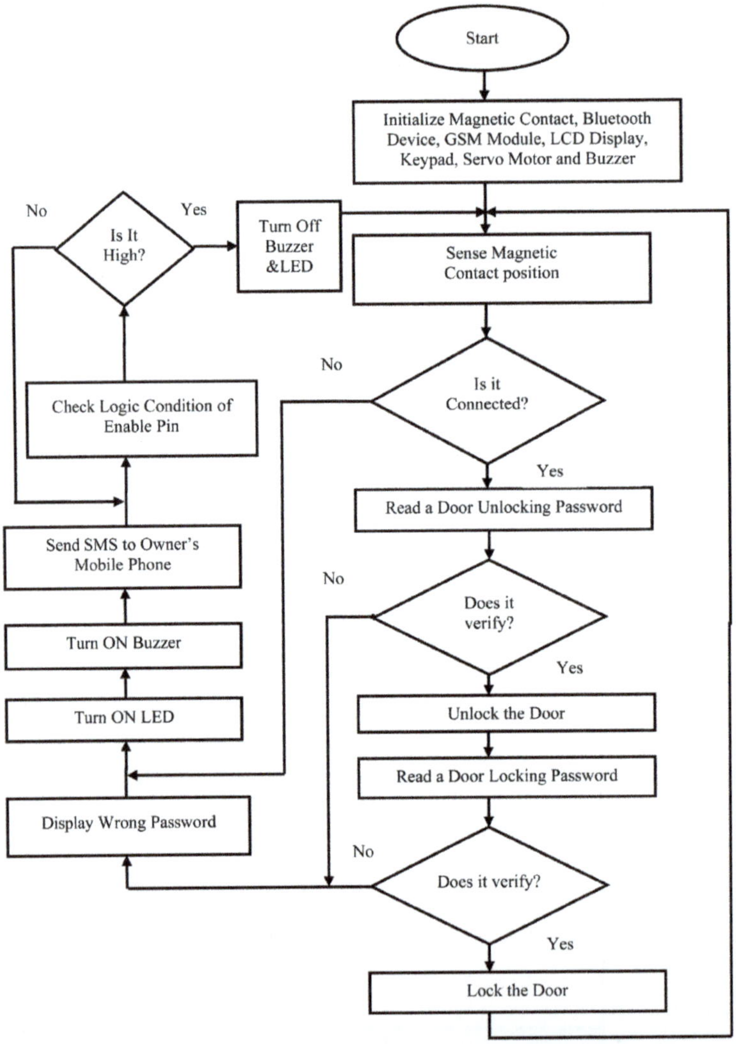


**Figure 2.** Program flow chart of the proposed home security system.

whole software design procedures. Software Code Vision AVR is used as a compiler for designing the program in C-Language. The proper instructions perform the desired function sequentially according to the flow chart. At first, the program initializes all peripheral devices (Magnetic Contact, Bluetooth Device, GSM Module, LCD Display, Keypad, Servo Motor and Buzzer). This is the precondition for interfacing between microcontroller and peripheral devices. Then the program examines the magnetic contact's position either it is connected or separated. If it is connected with door or window, then the program will give the authority to enter a password that will lock the door. Otherwise program instructions have been mentioned for mal-operation resulting it sends signal to the LED, Buzzer and GSM Module for their individual function sequentially. On the contrary, after receiving a password program verifies the password with the door unlocking password. If the result is affirmative, the program will give the authority to unlock the door. Otherwise mal-operation indicates in LCD by displaying an error message and also the previous functions remain unchanged. However, program remains at waiting state for receiving a door locking password after getting the permission to access the system. Similarly program verifies the password resulting it locks the door. Otherwise it sustains the mal-operation until reset pin is high.

However, the instructions in program flow chart have been represented in **Figure 2** which maintains the following procedure for the proper operation of proposed home security system.

1) Start;

2) Initialize Magnetic Contact, Bluetooth, GSM module, LCD display, Keypad, Servo Motor and Buzzer;

3) Sense the magnetic switch position;

4) If the magnetic switch is connected, the system will remain at waiting state for receiving a password. Otherwise, the system will turn on the alarming signal and send SMS to the owner's mobile;

5) Receive a password from keypad due to unlock the system;

6) Verify the password with the unlocking password which is mentioned in program. If the input password is exactly equal to the system password, a pulse signal will send to the servo motor which will pressurize the door resulting the system will be unlocked. Otherwise, LCD display will show wrong password and the system will follow the similar procedure which is mentioned in step 4 for the false condition, and the system will jump to step 7.

7) Receive a password from keypad due to lock the system. After opening the door, the system will remain at waiting state for receiving a number which is designated for locking the door.

8) Verify the password with the locking password which is mentioned in program. If the input password is exactly equal to the system password for closing the door, the system will be locked and return to step 3. Otherwise, the system will operate at false mode and follow the similar procedure mentioning in step 5.

9) Check the logic condition of enable pin. If it remains at low state, the system will sustain the mal-operation and remain at waiting state until it will obtain high logic condition. On the contrary, the system will start again from step 2 for the high logic condition of enable pin.

## 2.2. Hardware Design

The implementation of practical circuit has been designated using Proteus Software as shown in **Figure 3**. Moreover, the circuit is also simulated using this software before practical implementation. The result of simulation ensures that the circuit works at proper way. The practical implementation of simulated circuit has been presented in **Figure 4**. In this circuit diagram two microcontrollers are the main components which are used for controlling other devices (Magnetic Contact, Bluetooth Device, GSM Module, LCD Display, Keypad, Servo Motor and Buzzer). Among the controlling devices, Magnetic Contact is used as a sensor for determining the present status of the system. This device will operate when the position of magnetic contacts are separated due to pressure from external environment. Bluetooth technology is a networking protocol that allows two or more devices to connect with each other through a Personal Area Network (PAN). In this part, Bluetooth device is used as a medium for the communication between microcontroller and mobile phone. A Servo Motor consists of three major parts: a motor, control board, and potentiometer (variable resistor) connected to the output shaft. The motor utilizes a set of gears to rotate the potentiometer and the output shaft at the same time. The potentiometer controls the angle of the servo motor and allows the control circuitry to monitor the current angle of the servo motor. If the control circuit detects that the angle is not correct, it will turn the servo motor to the correct direction until the angle is correct. In this system, the servo motor is used to control an angular motion between 0 and 90 degrees. Moreover, GSM is a standard developed by the European Telecommunications Standards Institute

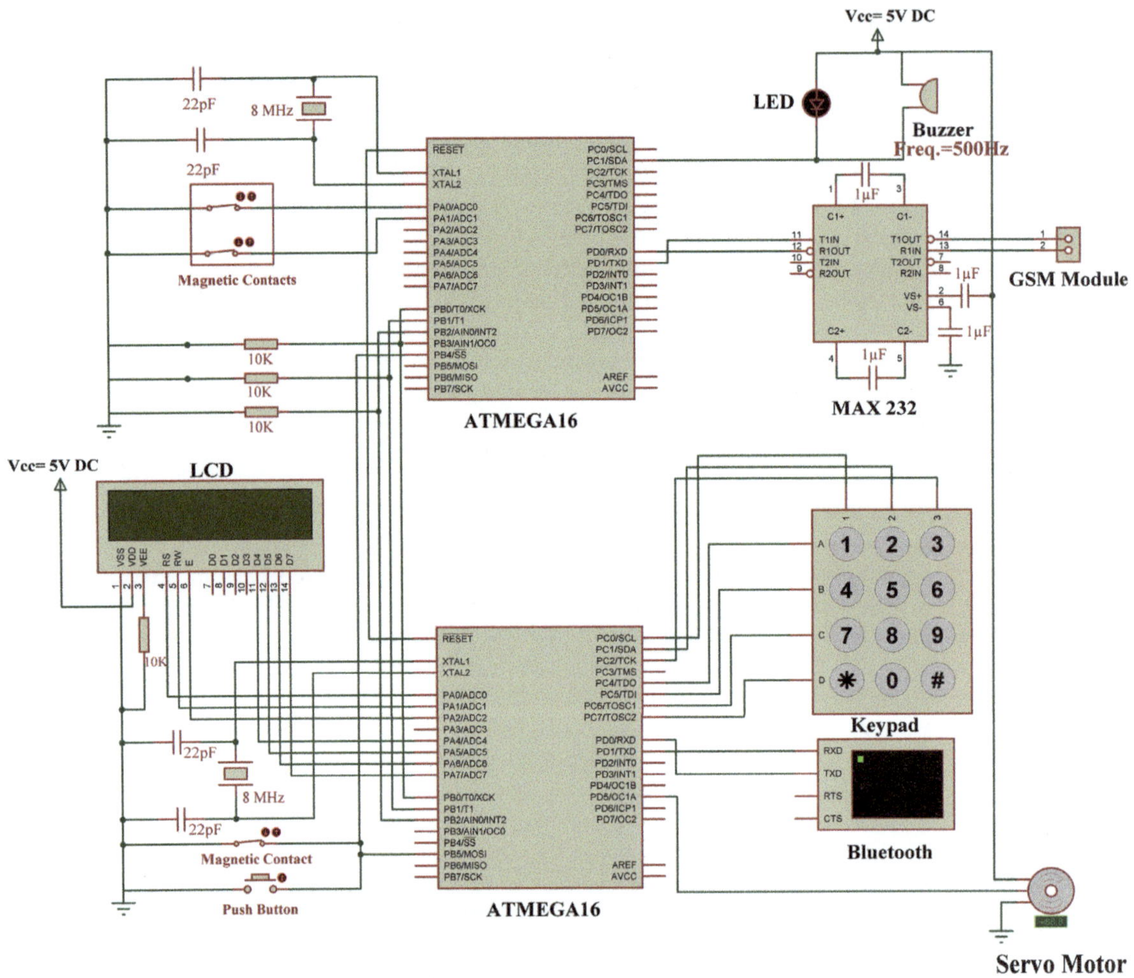


**Figure 3.** Schematic diagram of the proposed home security system.

(ETSI) to describe protocols for second generation (2G) digital cellular networks used by mobile phones. So, GSM Module makes a strong wireless communication network between owner and security system.

However, the system receives signal from input devices (Magnetic Contact, Bluetooth Device, Keypad) and sends signal to the output devices (LCD Display, Servo Motor, LED and Buzzer) in order to maintain security.

## 3. Results

In this section, firstly **Figure 5** & **Figure 6** describe the overall procedures in order to lock or unlock the system using Keypad or Mobile Phone. Then the result from mal-operation is also presented in **Figure 7**.

**Figure 5(a)** shows a practical implementation of home security system which is receiving a password from physical environment to unlock the door. This function is accomplished by interfacing with keypad and micro-controller. Moreover, the password is displayed at LCD and the system remains at waiting state until password is conformed from user end.

**Figure 5(b)** shows that the system has been unlocked by the rotation of servo motor and separation of magnetic contacts after verifying the password. At the same time, a conformation message "DOOR OPEN" is displayed at LCD. This indicates someone is accessing the system in proper way. However, the system again returns at initial mode for receiving another password in order to lock the system. This password is also displayed at LCD until the password is verified from controller section.

In **Figure 5(c)**, the system has become locked after the verification of door locking password. To perform this

**Figure 4.** Hardware setup for home security system.

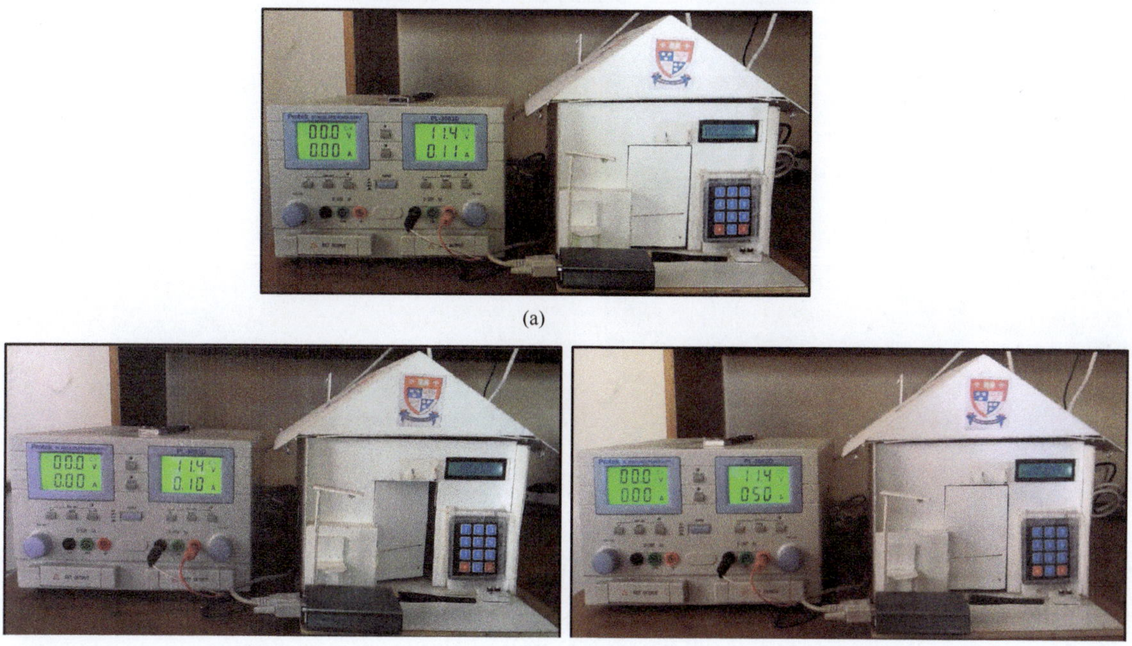

(a) (b) (c)

**Figure 5.** Interfacing between Keypad and Microcontroller resulting the system (a) receives password, (b) open and (c) close the door.

function, microcontroller sends a signal to the servo motor and magnetic contacts so that servo motor rotates at reverse order resulting it pressurizes the door along with connecting the magnetic contacts. Moreover, the system also ensures the functions by displaying a conformation message "DOOR CLOSED" at LCD. This message also indicates that the system has been locked without any mal-operation.

In **Figure 6**, the android application has been implemented through a Bluetooth device in order to control the system. In this work, an android mobile phone is used for accessing the system without help of Keypad. At first, mobile phone sends a signal to the Bluetooth device through a wireless medium. Then the transmitting signal is received by microcontroller from Bluetooth device which is directly connected with microcontroller. Using mobile keypad if anyone enters a wrong password, the system will activate the security mode. On the contrary, after getting an appropriate password system ensured the proper function by displaying a conformation message at LCD. Password system from mobile phone has been accomplished by designing mobile apps for android application.

**Figure 6.** Implementation of Android application.

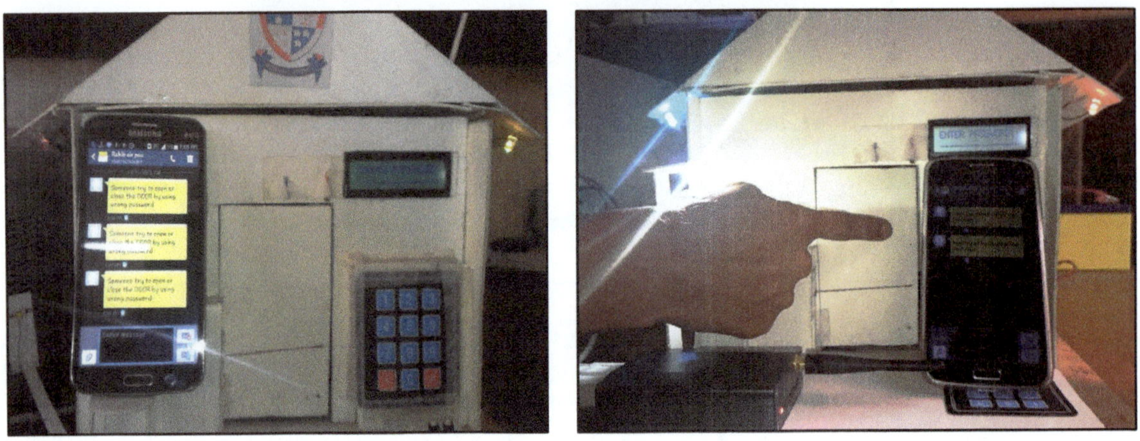

(b) (c)

**Figure 7.** Mal operation introduced by (a) wrong password and (b) breaking door.

**Figure 7(a)** shows the practical implementation of mal-operation which has been occurred by pressing wrong password. As a result the system activates all the security system. This includes an alarming message to owner mobile phone, turn on the Buzzer and LED. The system sustains the hazardous condition until the whole system is become reset.

In **Figure 7(b)**, another mal-operation is introduced by breaking the door without authentication. In this case, microcontroller activates GSM Module, LED and Buzzer. Consequently, the system follows the similar procedure described in previous wrong condition in order to maintain security. Moreover, the system performs the similar security functions in order to protect windows.

## 4. Conclusion

This paper presents design and implementation of a smart home security system based on microcontroller along with GSM for user friendly application. The system is intelligent enough to monitor the secure environment. In addition, the user is informed about the security breach through GSM network that provides a special opportunity whenever the user stays at far away from home. However, Android application is the most stunning feature in order to control the system through a Bluetooth device. Moreover, the system provides the reliable operation within reasonable cost and removes the system complexity. In this work, traditional burglar alarm mode, LED lights and LCD are the promising features used to ensure reliability. The whole system is implemented on a practical home security system which requires considerable effort to install it. Consequently, the system is also applicable for commercial purposes due to versatile ways of security and controllability.

## References

[1] Rana, G.M.S.M., Khan, A.A.M., Hoque, M.N. and Mitul, A.F. (2013) Design and Implementation of a GSM Based Remote Home Security and Appliance Control System. *Proceedings of the* 2*nd International Conference on Advances in Electrical Engineering*, Dhaka, 19-21 December 2013, 291-295.

[2] Ahmad, A.W., Jan, N., Iqbal, S. and Lee, C. (2011) Implementation of ZigBee—GSM Based Home Security Monitoring and Remote Control System. *IEEE* 54*th International Midwest Symposium on Circuits and Systems*, Seoul, 7-10 August 2011, 1-4. http://dx.doi.org/10.1109/mwscas.2011.6026611

[3] El-Medany, W.M. and El-Sabry, M.R. (2008) GSM-Based Remote Sensing and Control System using FPGA. *Proceedings of International Conference on Computer and Communication Engineering*, Kuala Lumpur, 13-15 May 2008, 1093-1097.

[4] Yuksekkaya, B., Kayalar, A.A., Tosun, M.B., Ozcan, M.K. and Alkar, A.Z. (2006) A GSM, Internet and Speech Controlled Wireless Interactive Home Automation System. *IEEE Transactions on Consumer Electronics*, **52**, 837-843. http://dx.doi.org/10.1109/TCE.2006.1706478

[5] Golzar, M.G. and Tajozzakerin, H.R. (2010) A New Intelligent Remote Control System for Home Automation and Reduce Energy Consumption. 4*th Asia International Conference on Mathematical/Analytical Modelling and Computer Simulation*, Kota Kinabalu, 26-28 May 2010, 174-180.

[6] Van Der Werff, M., Gui, X. and Xu, W.L. (2005) A Mobile-Based Home Automation System. *Proceedings of the* 2*nd International Conference on Mobile Technology, Applications and Systems*, Guangzhou, 15-17 November 2005, 1-5.

[7] Hwang, I.-K., Lee, D.-S. and Baek, J.-W. (2009) Home Network Configuring Scheme for All Electric Appliances using ZigBee-Based Integrated Remote Controller. *IEEE Transactions on Consumer Electronics*, **55**, 1300-1307. http://dx.doi.org/10.1109/TCE.2009.5277992

[8] Lee, H.-B., Park, J.-L., Park, S.-W., Chung, T.-Y. and Moon, J.-H. (2010) Interactive Remote Control of Legacy Home Appliances through a Virtually Wired Sensor Network. *IEEE Transactions on Consumer Electronics*, **56**, 2241-2248. http://dx.doi.org/10.1109/TCE.2010.5681096

[9] Islam, M.S. (2014) Home Security System Based on PIC18F452 Microcontroller. *Proceedings of* 2014 *IEEE International Conference on Electro/Information Technology*, Milwaukee, 5-7 June 2014, 202-205. http://dx.doi.org/10.1109/EIT.2014.6871762

# 3

# Skin Problems among Users of the Urine-Based Fertiliser in Ouagadougou Periurban Areas, Burkina Faso

**Joseph M. Makaya[1*], Serge Aho[2], Joseph Wethé[3], Dayéri Dianou[4], Nicolas Barro[1], Alfred S. Traoré[1]**

[1]Research Center for Biological, Food and Nutritional Sciences, Research and Training Unit, Life and Earth Sciences, University of Ouagadougou, Ouagadougou, Burkina Faso
[2]Epidemiology and Infection Control Unit, Public Hospital of Dijon, Dijon, France
[3]Pan African Agency, Water and Sanitation for Africa, Ouagadougou, Burkina Faso
[4]National Center for Scientific and Technological Researches, Institute for Health Sciences Research, Ouagadougou, Burkina Faso
Email: *makaya.joseph7@gmail.com

## Abstract

**The great challenge for the sustainable use of excreta (urine, faeces) in agriculture is to increase the benefits of these products as resources and decrease the negative effects on human health. The risk of gastrointestinal diseases associated with the use of human excreta as natural fertilisers is well established, while information on skin problems remains largely anecdotal. The objective of this study was to evaluate the prevalence of skin problems among people involved in the handling of the urine used as fertiliser along a productive sanitation system in Ouagadougou periurban areas. A questionnaire was used for each targeted worker to collect data on sociodemographic characteristics, conditions of work and reported health symptoms such as skin burning, itching, eye irritation and paronychia. The exposure measurements were essentially based on field observations. A total of 435 people were interviewed, including 45 workers in urine storage sites, 209 farmers using urine-based fertiliser and 181 control farmers. More than 35% of site workers reported skin symptoms. For farmers using urine-based fertiliser and their controls, 17.2% and 26.0% reported skin symptoms, respectively. Overall, the associations between skin symptoms and the handling of urine used as fertiliser were almost significant (*P* < 0.06). Also, the study showed that these skin symptoms reported by the urine handlers are associated with conditions**

*Corresponding author.

**of work. Effective preventive measures such as wearing suitable protective equipment and practicing good personal hygiene should be emphasized. This topic needs to be further examined using longitudinal studies.**

## Keywords

**Human Urine, Fertiliser, Exposure, Hazardous Agents, Skin Problems**

## 1. Introduction

In order to decrease reliance on chemical fertilisers, human excreta (urine, faeces) are used in agriculture around the world [1]-[3]. In Burkina Faso, urine-based fertiliser was used firstly by truck farmers in Ouagadougou periurban areas, along productive sanitation systems [4]. The main steps in the management of a sanitation system are: urine collection from the eco-toilets (Urine Diversion Dry Toilet), transportation and storage (on a site) before the use as liquid fertiliser. The quality of the urine from these systems was previously evaluated [3] [4].

Human health risks associated with the use of wastewater and excreta in agriculture have been well shown [5]. Although the risk for certain gastrointestinal diseases has been established, the information on other potential human health risks is still largely anecdotal. Studies in many countries with farmers using wastewater suggested a high prevalence of the skin ailments [6]-[8]. In general, causes of these skin problems remained unknown [8] [9].

Also, farmers and workers involved in handling activities (e.g. cleaners) often complain about skin problems caused by various irritants and allergens, including sun [10], biological agents such as bacteria [11] [12] and chemicals such as pesticides and heavy metals [13]-[15]. Some soaps and detergents can damage the skin [16] [17].

The skin is a considerable interface between man and his environment [18]; it is an important portal of entry for hazardous agents and a vulnerable target tissue as well [19] [20]. Colonized by a normal resident flora, the skin is frequently contaminated by pathogens. Staphylococci and enterococci are the most common skin bacterial contaminants [12] [21] [22]. About 20% of the normal population carries continuously *Staphylococcus aureus* on the body. This chronic carriage is a risk factor for skin infections [23] [24]. Paronychia a skin infection of the fingers, occurs especially among manual workers. Staphylococci are the main causative agents, but can also be an Enterococcus or gram-negative bacteria [21].

In Burkina Faso, the incidence of skin diseases associated with exposure to wastewater and agricultural inputs (fertilisers and pesticides) is not known. This incidence is probably high for the following reasons: 1) The presence in irrigated water of some hazardous agents [25]-[27]; 2) The large amounts of chemical inputs frequently handled by farmworkers; 3) The use of prohibited pesticides in the agricultural sector [27]; and 4) The non-compliance with protective measures on handling some products in the fields [28].

Ammonia is used to produce chemical fertilisers. It also helps to preserve stored fruits [29]. However, this gas can cause severe skin irritation or burns due to its high solubility in water. Therefore, skin irritation can occur after sweating if the skin is in contact with ammonia, because of ammonia's caustic effect [29] [30]. Irritation due to ammonia can also affect eyes, causing watering and conjunctival ulceration. Glaucoma can occur after a long-term exposure to ammonia [29] [31].

This study was conducted among handlers of the urine used as fertiliser in Ouagadougou periurban areas, Burkina Faso. The main objective was to evaluate the prevalence of skin problems reported by people involved in the urine handling along a productive sanitation system.

## 2. Materials and Methods

### 2.1. Research Setting

The study was conducted in Ouagadougou periurban areas, where a large scale project in productive sanitation was implemented. For a period of three years (2006-2009), the project has equipped around 1000 households with eco-toilets (Urine Diversion Dry Toilets, UDDT). In each sector covered by the project, a productive sanitation system was set up, in which excreta is collected separately via UDDT and are transported and stored on a site before the use as fertilisers in the field. Urine was stored in jerry cans (20 L) and tanks (1000 L).

Each filled container was closed for at least 30 days. On average, 600 jerry cans and 15 tanks per site were used to manage the urine stocks. Our study was limited to two sites and conducted for a period of 11 months (February 2011-January 2012). Parameters of interest (pH, ammonia, heavy metals, *E. coli*, *S. aureus*, *Enterococcus sp.*) in urine from the studied sanitation system have been previously measured [3].

### 2.2. Study Population

Two groups of workers along productive sanitation systems were targeted: 1) Handlers of the urine in storage sites, in which the main tasks were collection and transportation of the urine from households to storage sites, and the decanting, cleaning and maintenance of containers; and 2) Farmers using urine-based fertiliser. A control group was integrated in this study, including non-urine using farmers, working in the same conditions as the others. All participants in this study were volunteers.

### 2.3. Data Collection

The survey on skin problems among targeted groups was preceded by the field visits in order to report the work conditions of people. To do this, forms were filled out at each visit, indicating observations concerning the hygiene practices in urine storage sites, the quality of irrigation water, and the protective measures taken during the urine handling. Data from field visits guided us to formulate relevant items for a questionnaire, which were individually administered at the workplace. The interviews were conducted face-to-face with investigators. The questionnaire was comprised of three parts: 1) The first part consists of items to collect sociodemographic characteristics of respondents; 2) The second part includes items relating to the environment and work conditions. These are the questions to find out what activities the person did, frequency of the urine handling, and about protective equipment used when working; and 3) The third part, which includes items relating to recent and/or current skin problems self-reported for the duration of work. Skin burning, itching, eye irritation and paronychia were the health symptoms examined. The location of these symptoms was also required to respondents.

### 2.4. Data Analysis

Statistical analysis of data was performed using the STATA V12.1 (Stata Corp 4905 Lakeway Drive, Texas, USA) software. The variables were analyzed by Chi-square test or Fisher's exact test. A *P*-value less than 0.05 was considered statistically significant.

## 3. Results

### 3.1. Description of the Study Population

In total, 435 people responded to the questionnaire, including forty-five (45) site urine handlers, two hundred and nine (209) farmers using urine-based fertiliser and one hundred eighty-one (181) as control farmers. The sociodemographic characteristics of the respondents are presented in **Table 1**.

The average age of respondents was about 38 years. Male interviews predominated, with 60% and 80% for site workers and farmers respectively. The majority of those interviewed people had a low level of education (primary or uneducated). Among field workers, truck farmers were the most numerous (97%), followed by flower growers (3%). For the duration of work, 53% of site urine handlers had more than 3 years and 20% had one year or less. About 87% of farmers had more than 3 years in working.

### 3.2. Skin Problems among Targeted Groups

More than 35% of site urine handlers have reported skin symptoms (**Table 2**), including skin burning (8.8%), itching (22.2%) and eye irritation (4.4%). No case of paronychia has been reported. Overall, the association between reported skin symptoms and the handling of urine on storage site was significant ($P < 0.05$).

In farmers, handling and non handling (control) the urine, the percentage of reported skin symptoms were 17.2% and 26.0% respectively (**Table 2**). These health symptoms reported by farmers handling the urine and their controls included, skin burning (4.8% vs. 7.7%), itching (6.7% vs. 4.4%) and paronychia (5.7% vs. 13.8%).

**Table 1.** Main sociodemographic characteristics of all respondents.

| Personal characteristics | Site urine handlers n = 45 | Urine handling farmers n = 209 | Control farmers n = 181 |
|---|---|---|---|
| Age (year) | | | |
| Mean[δ] (SD) | 37.1 (12.6) | 39.9 (9.0) | 38.6 (11.9) |
| Age groups | | | |
| <21 n (%) | 5 (11.1) | 7 (3.3) | 22 (12.1) |
| 21 - 49 n (%) | 30 (66.6) | 163 (77.9) | 115 (63.5) |
| >49 n (%) | 10 (22.2) | 39 (18.6) | 44 (24.3) |
| Sex | | | |
| Male n (%) | 27 (60) | 193 (92.3) | 127 (70.1) |
| Female n (%) | 18 (40) | 16 (7.6) | 54 (29.8) |
| Educational level | | | |
| High school n (%) | 0 (0) | 1 (0.5) | 0 (0) |
| Secondary school n (%) | 10 (22.2) | 25 (11.9) | 17 (9.4) |
| Primary school n (%) | 20 (44.4) | 73 (34.9) | 29 (16.0) |
| Uneducated n (%) | 15 (33.3) | 110 (52.6) | 135 (74.5) |
| Farming type | | | |
| Horticultural n (%) | / | 206 (98.5) | 173 (95.5) |
| Ornamental n (%) | / | 3 (1.4) | 8 (4.5) |
| Duration of work (year) | | | |
| <1 n (%) | 9 (20) | 6 (2.8) | 6 (3.3) |
| 1 - 3 n (%) | 12 (26.6) | 11 (5.3) | 25 (13.8) |
| >3 n (%) | 24 (53.3) | 192 (91.8) | 150 (82.8) |

[δ]Values given as mean (SD).

**Table 2.** Percentage of skin problems reported by respondents.

| Symptoms | Site urine handlers n = 45 | Urine handling farmers n = 209 | Control farmers n = 181 |
|---|---|---|---|
| | n (%) | n (%) | n (%) |
| Overall | 16 (35.5) | 36 (17.2) | 47 (26.0) |
| Burning | 4 (8.8) | 10 (4.8) | 14 (7.7) |
| Itching | 10 (22.2) | 14 (6.7) | 8 (4.4) |
| Paronychia | 0 (0) | 12 (5.7) | 25 (13.8) |
| Eye irritation | 2 (4.4) | 0 (0) | 0 (0) |

## 3.3. Occurrence of Skin Symptoms and Work Conditions

In **Table 3**, it is shown the incidence rates of skin symptoms related to work conditions in handlers of the urine on storage site. Sixteen (16) workers (n = 45) reported skin problems “due to the urine exposure”. In less than one year in working, 50% and 30% of them respectively had complained about skin burning and itching. Those who reported eye irritation had more than 3 years in the work.

**Table 3.** Incidence rate of skin symptoms reported by site urine handlers.

| Variables | Skin burning | Itching | Eye irritation |
|---|---|---|---|
| | % | % | % |
| Duration of work (year)† | | | |
| <1 | 50 | 30 | 0 |
| 1 - 3 | 25 | 40 | 0 |
| >3 | 25 | 30 | 100 |
| Frequency in urine handling† | | | |
| ≥Twice per week | 75 | 60 | 100 |
| ≤Twice per week | 25 | 40 | 0 |
| Personal protective equipment | | | |
| Yes | 100 | 90 | 100 |
| No | 0 | 10 | 0 |
| Location of skin symptoms | | | |
| Hand | 37.5 | 25 | 0 |
| Forearm | 37.5 | 50 | 0 |
| Foot | 25 | 0 | 0 |
| Other | 0 | 25 | 100 |

†$P = 0.06$.

By handling the urine on storage site, with a frequency of more than twice per week, had given incidence rates of 75%, 60% and 100% respectively for skin burning, itching and eye irritation (**Table 3**). Surprisingly, we observed that almost all workers on the urine storage site and who reported skin symptoms had personal protective equipment consisting of a coat with short sleeves, a gas mask, boots and gloves. This data confirms the observations made in the field in that there was a lack of interest in wearing this protective equipment, the poor state and the unsuitability of these protective materials. Skin symptoms mentioned were reported more than once in the concerned site workers. Itching and skin burning were most often located on forearms, hands and feet.

In **Table 4**, the incidence rates of skin symptoms reported by farmers using urine-based fertiliser are presented.

In farmers handling urine-based fertiliser which reported skin symptoms (26 among 209 in total), about 85% had at least two years in the work. For these farmers, the frequency of handling urine averaged once per week. More than half of farmers (59%) did not use a personal protective equipment (**Table 4**). Skin burning and itching were experienced by all reporting farmers (100%). For paronychia, 83.3% of reporting people were truck farmers and 16.7%, flower growers. Overall, the skin burning was most often located on hands and feet, while itching occurred on forearms. Except for paronychia, the other skin symptoms were experienced more than once by almost all reporting farmers. Observations in the field have shown that the farmers' existing protective equipment was not worn automatically during the urine handling. The reasons for this disinterest on wearing personal protective equipment have not been clearly given. Also, in the opinion of almost all farmworkers interviewed, "the quality of the water used for irrigation was good, since it was extracted from wells and dams". Taking into account all of these skin symptoms that were reported, the associations found between work factors, particularly the frequency of the urine handling and the duration of work (**Table 3** and **Table 4**) were almost significant ($P =$ 0.06).

## 4. Discussion

In this study, we evaluated the possible links between skin problems and the handling of urine among people

**Table 4.** Incidence rate of skin symptoms reported by farmers using urine-based fertilizer.

| Variables | Skin burning | Itching | Paronychia |
|---|---|---|---|
| | % | % | % |
| Duration of work (year)† | | | |
| <1 | 10 | 0 | 8.3 |
| 1 - 3 | 30 | 40 | 25 |
| >3 | 60 | 60 | 66.7 |
| Frequency in urine handling† | | | |
| ≥Twice per week | 0 | 0 | 0 |
| ≤Twice per week | 100 | 100 | 100 |
| Personal protective equipment | | | |
| Yes | 50 | 29 | 41.7 |
| No | 50 | 71 | 58.3 |
| Farming type | | | |
| Horticultural | 100 | 100 | 83.3 |
| Ornamental | 0 | 0 | 16.7 |
| Location of skin symptoms | | | |
| Hand | 36.3 | 35.7 | 100 |
| Forearm | 27.2 | 50 | 0 |
| Foot | 36.3 | 14.3 | 0 |
| Other | 0 | 0 | 0 |

†$P = 0.06$.

involved within productive sanitation systems in Ouagadougou periurban areas. The health symptoms examined were essentially skin burning, itching, eye irritation and paronychia, which can be caused by various agents potentially existing in the work environment. Exposure measurements were based mainly on observations made at any moment by the urine handlers on the workplace (storage site and farm field). We also considered as indicators of risk, data from a previous study on the characterization of chemical (pH, ammonia, heavy metals) and bacteriological (*E. coli*, *S. aureus*, *Enterococcus sp.*) parameters in the urine of the studied sanitation system [3].

Overall, the results of this study indicate skin symptoms reported by respondents were associated with the work they do. The prevalence of skin problems in the urine site urine handlers was almost the double of that found in farmers handling urine-based fertiliser (**Table 2**). Yet, according to field observations, the wearing of the personal protective equipment while handling urine was spontaneous among site workers compared to farmers. The handling of the untreated urine (before storage) may have more risk of disease transmission [32]. Based on this assumption, the high prevalence of skin symptoms in handlers of urine on storage sites (compared to farmers) was uncovered. Wearing personal protective equipment did not effectively prevent skin symptoms. This data confirms the observations made at the field regarding protective materials (poor state and unsuitable).

We observed that farmers not handling urine-based fertiliser also reported skin symptoms. Further investigations on the risks faced by farmers in the fields are needed. It is likely that the causes of skin diseases among farmworkers are many. However, some studies suggested that exposure to chemical inputs, e.g. pesticides and fertilisers [14] [33] and irrigation water [8] [34] were the main causes of skin problems self-reported among farmers. Wastewater often used in urban agriculture potentially contains chemical (e.g. heavy metals, pesticide residues) and biological agents (e.g. bacteria, viruses, enzymes) which, in contact with skin can affect it. Farmers interviewed in this study extracted their irrigation water from wells and dams. Regular skin contact with water can have a risk of disease because this irrigated water probably contains chemical and biological contaminants from soil or various releases [25] [27] [35]. However, the wearing of personal protective equipment during the handling of chemical inputs and wastewater could minimize the risk of some skin symptoms.

In addition, we observed significant associations between examined skin symptoms and work conditions, particularly the frequency of the urine handling and the duration of work (**Table 3** and **Table 4**). Our field observations showed that all urine handlers complained of strong smells due to ammonia, while the so-called "gas mask" was worn. We also found that there could be phases in the urine handling that are more at risk than others. For example among site urine handlers, exposure risks are high by transporting and decanting urine. In farm, workers using urine-based fertiliser, mix preparation (dilution) are probably the crucial exposure phase. During these all activities, parts of the body not covered such as face, hands and forearms were in frequent contact with aerosols from urine.

Previous studies have shown that the handling of products with high alkalinity (pH > 12), exposed people to the risk of skin burning [30] [36]. However, pH values of the handled urine by targeted actors were less than 9 [3]; therefore, the alkaline effect of urine on the skin of handlers is probably negligible. Also, the concentrations of chromium, lead, copper, cadmium and nickel in the urine were low [3]. So, the implication of heavy metals effects explaining the skin symptoms experienced by the urine handlers can be excluded, as much as the affected body parts (hands, forearms, eyes) were not directly "dipped" in urine during the handling. Moreover, skin irritation due to ammonia can be considered, because of its high concentration in the handled urine and its possible caustic effect on a sweated skin. Based on the above data, it seems clear that the hot and semi-arid climate favoring excessive sweating can be a worsening factor of ammonia toxicity, that also was confirmed by authors [30].

Other associations found in this study are difficult to explain. This is the case of the connection between the occurrence of a paronychia and exposure to the urine-based fertiliser, in the sense that the causative agent is often carried by the subject himself [12] [21]. In this situation, it is necessary that good hygiene practices, including washing the body after work be maintained to prevent some skin diseases.

Few limits to this study can be mentioned. First, respondents were asked to recall the experienced skin symptoms, since the beginning of their work. It is possible that subjective responses have been provided by the surveyed population. In order to minimize these errors due to oversight, Vallejos and *et al.* [14] asked to farmworkers handling pesticides to recall their skin symptoms reported in the previous seven days before the interview. But, we think that approach to be too limiting the period for recalling health symptoms, potentially hiding some evidence related to exposures. Second, skin problems examined in this study may result from many causes and cannot be solely attributed to the urine exposure. As proof, farmers not handling urine-based fertiliser reported the same skin symptoms that handlers. Third, the method of data collection used do not reflect precise measurements as the severity and the duration of the reported skin symptoms, which are important parameters for assessing effects associated with exposure to hazardous substances. In order to lessen the mentioned limits, we think that conducting longitudinal studies could be necessary, because they analyze in long-term targeted groups. Unfortunately, in the context of the current study, it was difficult to constitute cohorts among the urine handlers since their work was unstable and precarious.

Having a better understanding of factors associated with the occurrence of skin problems can help in efforts to prevent and to treat [14] [37]. Thus, one of the strengths of this study was to provide information on the prevalence of skin symptoms in people exposed to the urine used as fertiliser. The results of this first investigation may call out the people exposed to products possessing a potential health risk (e.g. chemical inputs, excreta-based fertilisers) to reinforce personal protective measures. Wearing a suitable protective equipment including coat with long sleeves, gloves covering forearms, boots and goggles, should certainly reduce the risk of skin symptoms here investigated.

## 5. Conclusion

This study suggests the existence of associations between health symptoms such as skin burning, itching, eye irritation and the handling of the urine used as fertiliser. The harmful effect of ammonia on the skin seems predominant compared to other measurable parameters in the handled urine. However, skin symptoms examined in this study could not be exclusively attributed to the urine exposure; other factors inherent to the workplace can be implicated. Based on the data obtained, it is necessary that people involved in the handling of the urine used as fertiliser, have to maintain personal good hygiene practices and wear protective equipment while working to reduce the health risks. In order to make our data more incisive and better explore the topic, additional studies are needed. For example, longitudinal studies by monitoring regularly some parameters (e.g. diagnosis, severity and duration of health symptoms) within cohorts, should better assess the effects of ailments due to some exposure. Thus, data from such studies can guide health policy makers, especially in the agricultural sector where the effects associated with exposure to chemical inputs are not well established.

## Acknowledgements

The authors acknowledge the collaboration of the managers of excreta storage sites along productive sanitation systems in Ouagadougou. Furthermore, we wish to give our special thanks to Dr. Marylise Mbakop for the help in the preparation of this study. Finally, we are grateful to the International Foundation for Science (IFS) for the funding source.

## References

[1] Mnkeni, P.N.S., Kutu, F.R., Muchaonyerwa, P. and Austin, L.M. (2008) Evaluation of Human Urine as a Source of Nutrients for Selected Vegetables and Maize under Tunnel House Conditions in the Eastern Cape, South Africa. *Waste Management & Research*, **26**, 132-139. http://dx.doi.org/10.1177/0734242X07079179

[2] Richert, A., Gensch, R., Jönsson, H., Dagerskog, L., Stenström, T. and Bonzi, M. (2010) Food Security and Productivesanitation: Practical Guideline on the Use of Urine in Crop Production. *Sustainable Sanitation Practice*, **3**, 31-33.

[3] Makaya, J.M., Savadogo, A., Somda, M.K., Bour, J.B., Barro, N. and Traoré, A.S. (2014) Quality of Human Urine Used as Fertilizer: Case of an Ecological Sanitation System in Ouagadougou Peri-Urban Areas—Burkina Faso. *Journal of Environmental Protection*, **5**, 467-474. http://dx.doi.org/10.4236/jep.2014.56049

[4] Dagerskog, L. and Bonzi, M. (2010) Opening Minds and Closing Loops—Productive Sanitation Initiatives in Burkina Faso and Niger. *Sustainable Sanitation Practice*, **3**, 4-11.

[5] World Health Organization (WHO) (2006) Guidelines for the Safe Use of Wastewater, Excreta and Greywater, Vol. 4, Wastewater Use in Agriculture. World Health Organization, Geneva, 196.

[6] Paulsen, E., Skov, P.S. and Andersen, K.E. (1998) Immediate Skin and Mucosal Symptoms from Pot Plants and Vegetables in Gardeners and Greenhouse Workers. *Contact Dermatitis*, **39**, 166-170. http://dx.doi.org/10.1111/j.1600-0536.1998.tb05888.x

[7] Douwes, J., Mannetje, A. and Heederik, D. (2000) Work-Related Symptoms in Sewage Treatment Workers. *Annals of Agricultural and Environmental Medicine*, **8**, 39-45.

[8] Trang, D.T., van der Hoek, W., Tuan, N.D., Cam, P.D., Viet, V.H., Luu, D.D., Konradsen, F. and Dalsgaard, A. (2007) Skin Disease among Farmers Using Wastewater in Rice Cultivation in Nam Dinh, Vietnam. *Tropical Medicine and International Health*, **12**, 51-58. http://dx.doi.org/10.1111/j.1365-3156.2007.01941.x

[9] van der Hoek, W., Anh, V.T., Cam, P.D., Vicheth, C. and Dalsgaard, A. (2005) Skin Diseases among People Using Urban Wastewater in Phnom Penh. *UA Magazine*, **14**, 30-31.

[10] Schenker, M.B., Orenstein, M.R. and Samuels, S.J. (2002) Use of Protective Equipment among California Farmers. *American Journal of Industrial Medicine*, **42**, 455-464. http://dx.doi.org/10.1002/ajim.10134

[11] Mahe, A., Cissé, I.A., Faye, O., N'Diaye, H.T. and Niamba, P. (1998) Skin Diseases in Bamako (Mali). *International Journal of Dermatology*, **37**, 673-676. http://dx.doi.org/10.1046/j.1365-4362.1998.00454.x

[12] Lorette, G., Beaulieu, P.H., Guihard, W., Lemaître, M., Jarlier, N., Bismuth, R., Duru, G. and Janin, N. (2003) Skin Infections in Communities: Bacteria Involved and Sensibility to Antibiotics. *Annales de Dermatologie et de Venereologie*, **130**, 723-731.

[13] Guo, Y.L., Wang, B.J., Lee, C.C. and Wang, J.D. (1996) Prevalence of Dermatoses and Skin Sensitisation Associated with Use of Pesticides in Fruit Farmers of Southern Taiwan. *Occupational Environ Medicine*, **53**, 427-431. http://dx.doi.org/10.1136/oem.53.6.427

[14] Vallejos, Q.M., Schulz, M.R., Quandt, S.A., Feldman, S.R., Galvan, L., Verma, A., Fleischer, A.B., Rapp, S.R. and Arcury, T.A. (2008) Self Report of Skin Problems among Farmworkers in North Carolina. *American Journal of Industrial Medicine*, **51**, 204-212. http://dx.doi.org/10.1002/ajim.20550

[15] Horiuchi, N., Oguchi, S., Nagami, H. and Nishigaki, Y. (2008) Pesticide-Related Dermatitis in Saku Districk, Japan, 1975-2000. *International Journal of Occupational and Environmental Health*, **14**, 25-34. http://dx.doi.org/10.1179/oeh.2008.14.1.25

[16] Pedersen, L.K., Johansen, J.D., Held, E. and Agner, T. (2004) Augmentation of Skin Response by Exposure to a Combination of Allergens and Irritants—A Review. *Contact Dermatitis*, **50**, 265-273.

http://dx.doi.org/10.1111/j.0105-1873.2004.00342.x

[17] Lynde, C.B., Obadia, M., Liss, G.M., Ribeiro, M., Holness, D.L. and Tarlo, S.M. (2009) Cutaneous and Respiratory Symptoms among Professional Cleaners. *Occupational Medicine*, **59**, 249-254. http://dx.doi.org/10.1093/occmed/kqp051

[18] Suskind, R.R. (1977) Environment and the Skin. *Environmental Health Perspectives*, **20**, 27-37. http://dx.doi.org/10.1289/ehp.772027

[19] Thyssen, J.P., Johansen, J.D., Linneberg, A. and Menné, T. (2010) The Epidemiology of Hand Eczema in the General Population-Prevalence and Main Findings. *Contact Dermatitis*, **62**, 75-87. http://dx.doi.org/10.1111/j.1600-0536.2009.01669.x

[20] Smedley, J., Williams, S., Peel, P. and Pedersen, K. (2012) Management of Occupational Dermatitis in Healthcare Workers: A Systematic Review. *Occupational and Environmental Medicine*, **69**, 276-279. http://dx.doi.org/10.1136/oemed-2011-100315

[21] Larson, E.L., Hughe, C.A. and Pyrek, J.D. (1998) Changes in Bacterial Flora Associated with Skin Damage on Hands of Healthcare Personnel. *American Journal of Infection Control*, **26**, 513-521. http://dx.doi.org/10.1016/S0196-6553(98)70025-2

[22] Gong, J.Q., Lin, L. and Lin, T. (2006) Skin Colonization by *Staphylococcus aureus* in Patients with Eczema and Atopic Dermatitis and Relevant Combined Topical Therapy: A Double Blind Multicentre Randomized Controlled Trial. *British Journal of Dermatology*, **155**, 680-687. http://dx.doi.org/10.1111/j.1365-2133.2006.07410.x

[23] Harbarth, S., Liassine, N., Dharan, S., Herrault, P., Auckenthaler, R. and Pittet, D. (2000) Risk Factors for Persistent Carriage of Methicillin-Resistant *Staphylococcus aureus*. *Clinical Infectious Diseases*, **3**, 1380-1385. http://dx.doi.org/10.1086/317484

[24] Wertheim, H.F.L., Melles, D.C., Vos, M.C., van Leeuwen, W., van Belkum, A. and Verbrugh, H.A. (2005) The Role of Nasal Carriage in *Staphylococcus aureus* Infections. *The Lancet Infectious Diseases*, **5**, 751-762. http://dx.doi.org/10.1016/S1473-3099(05)70295-4

[25] Sou, Y.M. (2009) Recycling Wastewater for Irrigation: Fertilizing Potential, Health Risks and Impacts on Soil Quality. Ph.D. Dissertation, EPF Lausanne, Lausanne.

[26] Dianou, D., Savadogo, B., Zongo, D., Zougouri, T., Poda, J.N., Bado, H. and Rosillon, F. (2011) Surface Waters Quality of the Sourou Valley: The Case of Mouhoun, Sourou, Debe and Gana Rivers in Burkina Faso. *International Journal of Biological and Chemical Sciences*, **5**, 1571-1589.

[27] Nitiema, L.W., Savadogo, B., Zongo, D., Kaboré, A., Poda, J.N., Traoré, A.S. and Dianou, D. (2013) Microbial Quality of Wastewater Used in Urban Truck Farming and Health Risks Issues in Developing Countries: Case Study of Ouagadougou in Burkina Faso. *Journal of Environmental Protection*, **4**, 575-584. http://dx.doi.org/10.4236/jep.2013.46067

[28] Bassolé, D. and Ouédraogo, L. (2007) Problem of the Use of Pesticides in Crop Conservation and Truck Farming in Urban and Peri-Urban Areas, Burkina Faso. International Fertilizer Development Center, IFDC, 51 p.

[29] Agency for Toxic Substances and Disease Registry, ATSDR (2004) Toxicological Profiles for Ammonia. US Department of Health and Human Services, Public Health Services, Atlanta. http://www.atsdr.cdc.gov/toxpro2.html

[30] Blanco, L.E., Aragon, A., Lundberg, I. and Lidén, C. (2005) Determinants of Dermal Exposure among Nicaraguan Subsistence Farmers during Pesticide Applications with Backpack Sprayers. *Annals of Occupational Hygiene*, **49**, 17-24.

[31] Beare, J.D., Wilson, R.S. and Marsh, R.J. (1988) Ammonia Burns of the Eye: Old Weapon in New Hands. *British Medicine Journal*, **296**, 590. http://dx.doi.org/10.1136/bmj.296.6622.590

[32] Jönsson, H. and Vinnerås, B. (2007) Experiences and Suggestions for Collection Systems for Source-Separated Urine and Faeces. *Water Science and Technology*, **56**, 71-76. http://dx.doi.org/10.2166/wst.2007.558

[33] Strong, L.L., Thompson, B., Coronado, G.D., Griffith, W.C., Vigoren, E.M. and Islas, I. (2004) Health Symptoms and Exposure to Organophosphate Pesticides in Farmworkers. *American Journal of Industrial Medicine*, **46**, 599-606. http://dx.doi.org/10.1002/ajim.20095

[34] Shrutakirthi, S.D., Suraj, D., Sugandhi, R., Girija, R. and Sreekumaran, N. (2005) Dermatoses among Paddy Field Workers—A Descriptive, Cross-Sectional Pilot Study. *Indian Journal of Dermatology, Venereology and Leprology*, **71**, 254-258. http://dx.doi.org/10.4103/0378-6323.16617

[35] Cissé, G. (1997) Health Impact of Wastewater Utilization in Urban Agriculture: Case of Truck Farming in Ouagadougou. Ph.D. Dissertation, EPF Lausanne, Lausanne.

[36] Géraut, C. and Frimat, P. (1993) Dermatoses in Cement Users. *Archives des Maladies Professionnelles*, **54**, 312-314.

[37] Susitaival, P., Flyvholm, M.A. and Meding, B. (2003) Nordic Occupational Skin Questionnaire (NOSQ-2002): A New Tool for Surveying Occupational Skin Diseases and Exposure. *Contact Dermatitis*, **49**, 70-76. http://dx.doi.org/10.1111/j.0105-1873.2003.00159.x

# 4

# The Analysis of Process Accidents Due to Risks in the Petrochemical Industries—The Case Study of Radiation Intensity Determination Proportional to Distance from Tank Level

**Mehrzad Ebrahemzadih[1,2], Afshin Maleki[1], Ebrahim Darvishi[1,2], Mehdi Meimanat Abadi[3*], Saeed Dehestaniathar[1]**

[1]Kurdistan Environmental Health Research Center, Kurdistan University of Medical Sciences, Sanandaj, Iran
[2]Department of Occupational Safety & Health, School of Health, Kurdistan University of Medical Sciences, Sanandaj, Iran
[3]Children's Cancer Charity, Kurdistan University of Medical Sciences, Sanandaj, Iran
Email: *mehdi_meimanat@yahoo.com

## Abstract

**Today, one of the basic methods in engaging the decision making of senior managers for securing the organizations regarding assessing risks that the industries face, is the analysis of outcomes of processing accidents. The development of process industries is always associated with an increased risk of accidents. Therefore, paying attention to this case is a priority for the safety units in the oil refineries and petrochemical plants. The aim of this study was to investigate safe distance between the storage tanks of gasoline tanks in the petrochemicals of South Pars Region; comparisons to the American petroleum institute were studied. The maximum of length of flame and the ratio of length to radius of flame were measured 20 m and 2.6 m respectively. If occurrence of leakage and pool fire in one of these tanks, the maximum intensity of thermal radiation was equal to 13.7kw/m$^2$. Hence, there is a possibility to control of fire at a distance of 15 meters from the tanks to heat radiation due to a pool fire, thereby, domino effect does not occur in the event of fire.. The fire is controllable but in case of leakages aggregation and creation of flammable ponds around these tanks, the gathered radiator intensities have immediate explosion potential in the tanks. So, it is proposed, besides increasing the distance between the storage tanks, other measures like quality and safety inspection planning of the connections and controlling instruments of the tanks, inspection of the fire lighting equipment like hydrants, monitors, fire hose reels are done and performance of reaction maneuvers in emergency conditions can take place**

*Corresponding author.

**and be documented and planned.**


## Keywords

**Process Accidents, Pool Fire, Fire Safety**


## 1. Introduction

The process industries have production, storage and use of flammable chemicals material units. Therefore, these industrials are most apt and prepared industries for facing the risks and explosions [1]. Always fire and explosions have enormous implications for human resources and financial capital. For example, in 1997, the explosion of pressure pipeline in the French Vitzan Petrohemica occurred because of leakage of liquid gas and killed 57 members of the industry staff [2]. Also, in 1992, release of light Nephatan from the pressurized tank resulted in the explosion in the Bambi refinery and 35 experts were killed [3]. According to risk matrixes, the risks with consequence of death are the unacceptable and critical risks [4] [5]. The lack of safety inspections has basic role in such accidents. So, the planning of safety inspection programs, risk assessment detonation and fire and determine preventive measures based safety is essential in the different units of the refinery. The items of inspection programs and risk assessment detonation and fire must be including identifying the warehouses of flammable materials, value and operating parameters of material, leakage paths of material, identifying places and storage locations for flammable materials, value and their operating parameters, possible leakage paths, identifying scenarios possible and consequences of accident, identifying of preventive measures fire and explosion [6]. Although the processing facilities and equipment are used for productivity, profitability and also wealth production, but they have important (main) potential for damaging human being, property and the environment. Since the number of refineries and their products increased thus exposure to hazards is also higher for workers. For example, the accident of Felix borough in England in 1984 left 28 dead and 26 injured [7]. Another case of chemical accidents occurred in December 1984 in Bhopal, India [8]. Also, because of leakage and firing of oil (petroleum), from gas platform of BP company in Mexico gulf, 11 people died and the damage amounted to more than 40 billion $ and caused 22 percent decrease in shares and 63 percent decrease in profits [9]. In order to prevent the accidents, assessment and identification of the accidents and risks in the refinery units are very important. For multiple injuries resulting from accidents in processing industries are illness cases and cause injury and death, damages to property, environmental pollutions, decreasing production and profits, loss of reputation and credibility in the organization. In the following, the common scenarios relating to fire has been investigated in oil and gas industries.

### 1.1. Eruptive Fire

This kind of fire usually consists of flammable materials that are exited from a pressurized source and there have been source of ignition in the environment. The speed of flammables effects greatly on the kind of fire, because initially the exit speed must be high enough to be sent to the air into the jet, and secondly the flame will be constant at a point that the flame speed will be equal to the local velocity of the gas mixture. When the exit speed of flammables increases, the air increases (in the mixture) too and as a result fuel consideration decreases in the jet, the flame distances more from the fuel exit and finally when the flame is far enough, the fuel density is lower than the low limit of flammability and the fire (flame) extinguishes. In the modeling of eruptive fire, the thermal radiation is the only factor that is studied as the sample for the damage of this type of hazard. The aim of eruptive fire modeling was assessing to the jet diameter and length and the rate of thermal radiation of burning jet at any point. It must be noted that in the modeling diagrams, the limit of radiation rate due to fire that causes severe injuries to humans is considered as $37 Kw/m^2$ [4] [10].

### 1.2. Abrupt Fire

Abrupt fire is a non-explosive combustion due to gas cloud of flammable material releases in the air. In other words, if the gas is uniformly dispersed in the environment and its flame is without acceleration, the fire is non-explosive. But if the flammable gas cloud non-uniformity be concentrated in a tank, then be exposed with a spark, it will cause explosive in the gas cloud [11] [12].

### 1.3. Pool Fire

It is possible to release a flammable liquid from equipment into the environment. Around the releasing source, a liquid pond may be formed and gradually evaporated. In case that the vapors of the pool reach to the ignition source, fire breaks out and forms pools that are flaring. This kind of fire is called pool fire, it must be noticed that the resulted vapors of pools reach to the ignition source and the resulted flames return back to the pool then caused fire. The outcome of a pool fire is assessed by the rate of radiation [11] [12]. The consequences due to the thermal radiations are different from burns to death. The domino effects of thermal radiations on equipment installations of the refinery industries are clearly evident. In the case, the outcomes caused by radiation internally, the exposure time played basic roles. The English standard BSS908, 1990, has been presented for the effects of the thermal radiation dose (**Table 1**).

## 2. Work Description Method

The type and the system characteristics must be described; also the work limits that consist installations and activities and conditions must be studied, identified and defined. So, in this study, the pool fire, and the effects of thermal radiations of the gasoline storage tank were studied in the utility unit in one of the South Pars petrochemicals in Iran. Also, distances between storage tanks have been studied according to the American petroleum institute standard (API's Standards).

**Table 2** shows the data (information) relating the damage caused by thermal radiation on the equipment and installation of processing industries.

### Description of Work Methods

In the following, the said tank specifications and the considered scenarios for this aim is described in details (**Table 3**).

Using the following formula [14], the flame length of the pool time is calculated.

$$\frac{L}{D} = 42\left[\frac{M}{r_a\sqrt{gD}}\right]$$

**Table 1.** Thermal rate and its effect on the human body [13] [14].

| Thermal Rate (KJ/m²) | Radiation Intensity (Kw/m²) | Effects |
|---|---|---|
| 65 | 2.2 | Body damage and bone ache |
| 125 | 4.2 | First level-critical burning |
| 200 | 6.7 | First level-Sevier burning |
| 250 | 8.3 | Level 2-burning |
| 325 | 1.8 | Level 3-burning |

**Table 2.** The rate of thermal radiation intensity and its effect on equipment and installations [13] [14].

| Radiation Intensity | Effects |
|---|---|
| 15 | Wood firing , when there is an ignition source |
| 25 | Automatic firing of wood |
| 4 | Ruins of glass |
| 12 | Plastic melting |
| 20 - 18 | Destruction of thermal isolation |
| 37.5 | Damage to the process equipment |
| 100 | Damage and breaking of the steel destruction |

**Table 3.** Characteristics of cases and scenarios.

| Parameter | Size/Type/Rate |
|---|---|
| Each tank diameter | 20 m |
| Tank height | 20 m |
| API Tanks interval based on standard | 15 m |
| Welding point | 27 centigrade |
| Relative humidity | 70 percent |
| Gasoline burning rate | 0.55% |
| Surface radiation power | 44 $Kw/m^2$ |
| Scenario description | Leakage from the upper level the tank |
| Capacity of each tank | 20,000 liters |
| Modeling target | Determining the proportional radiation intensity to distance from tank level<br>Determining the proportional radiation intensity to distance from ground level |

Here, we have $L$ for the flame length, $D$ is the flame diameter, $M$ is the rate of mass burning in $kg/m^3$ and $r_a$ is the air density in $kg/m^3$ and $g$ is the acceleration of gravity that equals to 9.8 $m/s^2$. It is worth noting that the calculation of eruptive fire characteristics can base on AP1521 [15].

The thermal radiation coefficient can be calculated according to the heat transfer from the source and on the following formula base:

$$t = 2.02 \times \left(P_w X\right)^{-0.09}$$

Here, $P_w$ is partial pressure of water vapor and $X$ is the distance [16].

Also, in this study, we use the following formula to determine maximum thermal radiation intensity in the horizontal distances from the effected source. Here, $I$ is equal to the thermal radiation in $KW/m^2$, $SEP$ is equal to the surface emission in $KW/m^2$, and $VF$ is equal to visibility factor and $t$ is equal to the thermal heat transfer power and the guidance wave in the water vapor [16].

$$I = SEP \times VF \times t$$

## 3. Result

On the basis of results, maximum flame length is 20 meters and maximum flame-to-flame length is 2.6 m. For determining the thermal radiation intensity in the ground level, the visibility coefficient item must be calculated. The method used to determine the visibility coefficient is based on two criteria that determine the reception amount on the ground level. The first standard is L + H, that here H is equal to the height of the tank and the second standard is equal to H. The difference between these two standards is equal to the visibility coefficient of flame. The nearest distance of the first tank is 15 meters (**Table 4**). **Table 5** shows that the maximum of thermal radiation intensity in this distance is equal to 13.7 $Kw/m^2$ and if the wind direction is in the path of the No. 2 storage tank, the thermal intensity will be much larger. According to the determined permissible limit for damage (37.5$kw/m^2$), it can be concluded that the determined distance on API standard [16] for the tanks is enough. If the fire caused in the first tank do not be extinguished quickly, or that the second tank is not cooled properly, the intensity of radiant heat of the first tank evaporates flammable material in the second tank. Then, the vapor flux in the second tank is increased, thereby; second tank causes the explosion in other tanks in a distance of 15 meters from each other.

## 4. Discussion

In order to manage the firing extents in every processing industry, proportional protection and control measures must be done, and continuously elimination of hazards situation be followed up. Therefore, quality control and inspection, risk assessment, service and maintenance of process pipes, isolator pumps, and safety valves are es-

**Table 4.** The thermal radiation intensity according to distance from ground.

| Distance from the Ground | Thermal Transfer Coefficient | Vision Coefficient | | | Thermal Radiation Intensity |
|---|---|---|---|---|---|
| | | L = L + H | L = H | Partial Vision Coefficient | |
| 15 | 0.783 | 0.404 | 0.392 | 0.212 | 0.4 |
| 20 | 0.763 | 0.294 | 0.268 | 0.230 | 0.9 |
| 30 | 0.763 | 0.184 | 0.141 | 0.043 | 1.4 |
| 40 | 0.717 | 0.126 | 0.084 | 0.043 | 1.3 |

**Table 5.** Thermal radiation intensity based on the distance from the tank.

| Tank Distance in Meter | 15 | 20 | 30 | 40 |
|---|---|---|---|---|
| The distance proportion radiation reception place to pool fire | 1.5 | 2 | 3 | 4 |
| Vision | 0.398 | 0.298 | 0.159 | 0.99 |
| Thermal transfer coefficient | 0783 | 0.763 | 0.763 | 0.717 |
| Radiation intensity (Kw/m$^2$) | 13.7 | 9.4 | 5.1 | 3.1 |

sential in order to prevent equipment corrosion.

This point is very important that all the safety and control systems, like the control valves, alarms, the alarming and extinguishing systems, must be tested and checked regularly and have high capability of performance in all the processing activities steps [17]. All the firing sources in the industries must be performed and documented; all of the procedures of firing risk assessment must be updated. Also, in order to control static electricity in the process units of the refineries, bonding and grounding are the simplest ways [18]. The phase pre-launched of systems and process industries is one of the most important phases because the nature of work has a very high risk rate of fire and explosion. So, for the risks control in this phase, the safety work policy and procedures, safe work permit systems, emergency response procedures and training programs for personnel must be considered before the implementation of this phase. Also, in order to control the spread of fire, it must be design a safe distance and the range for sources of flammable, explosive and toxic materials from other fire sources in the process industries. Determining the safe distance from the location of aid and rescue, fire-fighting units, fire extinguishing equipment and fire detection and alarm systems is very important. The water spreading system must be considered for protection against the thermal radiation to NFPA15 standard. Thus, the use of the water spreading systems for cooling of the tanks body containing flammable materials to prevent the occurrence of fire and explosion can be effective [19].

### 4.1. Fire Extinguishing

According to the determined scenarios, the strategies and equipments of fire extinguishing like hydrants, monitors, firefighting car (vehicles), hose reels, and the reaction equipment in the emergency conditions must be inspected and assessed. These designs must be programmed according to reaction design in the emergency conditions. All fire-fighting equipment must have high reliability. All workers in refineries and process industries must have been trained about risk of fire and explosion chemical material, toxic hazards and fire extinguishing.

### 4.2. The Final Results

A comprehensive planning to risks control is the fundamental approach to the risks management in the process industries. The aim of the is to process planning, assess and recognize fire risk, and assess the probability and severity of consequences, and so, develop guidelines and engineering controls to prioritize the control measures. In the study, current fire scenarios and methods of determining thermal radiation intensity arising from eruptive fire and pool fire and the limit of its damage relating pool fire in gasoline storage tanks in one of the South Pars area Petrochemicals was described. The process planning for managing process dangers in the refineries and petrochemicals includes identification of hazards and documenting the instructions and control of the engineering systems. Therefore, in order to manage and eliminate the risks of fire and explosion, implementation of safety audits, assessing regularly and periodically of process, simulation maneuvers of process and review of control procedures is essential.

## Acknowledgements

The authors would like to thank manager utility unit of South Pars area petrochemicals for their cooperation.

## Conflict of Interest Statement

The authors have no conflict of interests to declare.

## References

[1] SFPE (2006) Engineering Guide: Fire Risk Assessment. Society of Fire Protection Engineers.

[2] Robin, T. and Pitblado, R., Eds. (1996) Risk Assessment in the Process Industries. IChemE, New York.

[3] Gupta, K. (2010) Design of District Emergency Operations Centres, and the Case Study of Indian Oil Corporation Jaipur Depot Explosion. *International Journal of Emergency Management*, **7**, 221-232. http://dx.doi.org/10.1504/IJEM.2010.037007

[4] CCPS (2000) Guideline for Chemical Process Quantitative Risk Analysis. 2nd Edition, Center for Chemical Process Safety, New York.

[5] Casal, J. (2008) Evaluation of the Effects and Consequences of Major Accidents in Industrial Plants. Industrial Safety Series, Elsevier, Amsterdam.

[6] (1994) Draft British Standard Code of Practice for the Application of Fire Safety Engineering Principles to Fire Safety in Building, BSI.

[7] Advisory Committee on Major Hazards (1984) The Control of Major Hazards. Third Report. Her Majesty's Stationery Office, London, 1984.

[8] Levy, J.K. and Gopalakrishnan, C. (2010) Promoting Ecological Sustainability and Community Resilience in the US Gulf Coast after the 2010 Deepwater Horizon Oil Spill. *Journal of Natural Resources Policy Research*, **2**, 297-315. http://dx.doi.org/10.1080/19390459.2010.500462

[9] Greenstone, M. and Hanna, R. (2011) Environmental Regulations, Air and Water Pollution, and Infant Mortality in India (No. w17210). National Bureau of Economic Research, Cambridge. http://dx.doi.org/10.3386/w17210

[10] Lev, Y. (1991) Water Protection of Surfaces Exposed to Impinging LPG Jet Fires. *Journal of Loss Prevention in the Process Industries*, **4**, 252-259. http://dx.doi.org/10.1016/0950-4230(91)85009-I

[11] Mizner, G.A. and Eyre, J.A. (1982) Large-Scale LNG and LPG Pool Fires. EFCE Publication Series, European Federation of Chemical Engineering.

[12] Cowley, L.T. (1989) The Thermal Impact of LPG (Torch) Fires. *Proceedings of the LPGITA Seminar* on *Fire Protection of LPG Storage Vessels*, Solihull, October 1989, 123-131.

[13] Lees, F.P. (1980) Loss Prevention in the Process Industries. Butterworth-Heinemann, Oxford.

[14] Green Book (1989) Methods for the Determination of Possible Damage. Rep CPR 16E, Voorburg.

[15] Chamberlain, G.A. (1997) Development in Design Method for Predicting Thermal Radiation from Flares. *Chemical Engineering Research and Design*, **65**, 299-309.

[16] Yellow Book (1997) Methods for the Calculation of Physical Effects. 3rd Edition, Part 2, Rep CPR 14E, Voorburg.

[17] Hubert, P. and Pages, P. (1989) Risk Management for Hazardous Materials Transportation: A Local Study in Lyons. *Risk Analysis*, **9**, 445-451. http://dx.doi.org/10.1111/j.1539-6924.1989.tb01255.x

[18] Amir-Heidari, P., Ebrahemzadih, M., Farahani, H. and Khoubi, J. (2014) Quantitative Risk Assessment in Iran's Natural Gas Distribution Network. *Open Journal of Safety Science and Technology*, **4**, 59-72. http://dx.doi.org/10.4236/ojsst.2014.41008

[19] NFPA15: Standard for Water Spry Fixed System for Fire Protection, 1996.

# 5

# Relationships among Risk Assessment, Risk Perception and Acceptance Model of the Residents near Nuclear Power Plants in Japan

**Yuichi Otsuka**

Department of System safety, Nagaoka University of Technology, Nagaoka, Japan
Email: otsuka@vos.nagaokaut.ac.jp

## Abstract

**This study aims at revealing the relationships among risk assessment result, risk perception model and risk acceptance model of the residents near nuclear power plants in Japan. Risk acceptance model is based on the trust of resident to nuclear power plants (NPPs) companies. The risk perception of the residents is very fragile and is not always based on objective technical knowledge. In addition, there are two models of risk acceptance: traditional trust model and salient value similarity model. In order to discuss the risk communication framework between NPPs with residents, applicability of two modes on the perception of the residents should be investigated. The author collected questionnaire from local resident near Kashiwazaki-Kariwa NPP in Niigata, Japan. Severity of risk is significantly affected by the extents of risks, not by amount of knowledge. Radiation relating hazard is normally evaluated higher risk. Risk acceptance level is significantly related to the factors of social benefits and trust to the NPPs company, not by regional interest nor amount of knowledge. Furthermore, value similarity model was not significant in this study. The results demonstrated that risk acceptance model of local residents near NPPs could be arranged by traditional trust model. In order to establish mutual trustworthy relationships between local residents and NPPs engineer, expert knowledge in nuclear power, communication skills, and expertise in safety are necessary to the engineers.**

## Keywords

**Risk Assessment, Risk Perception, Risk Communication, Risk Acceptance Model**

## 1. Introduction

The Niigata Chuetsu-oki earthquake that occurred on July 16, 2007 caused severe damage to the Kashiwazaki-Kariwa region around the city of Kashiwazaki. Significantly, the Kashiwazaki-Kariwa nuclear power plant, whose capacity is the largest in the world, was affected. Images of black smoke rising from the power plant shocked local communities. The specific damage to the power plant concerned fire in the transformers, damage to buildings and equipment from the seismic activity, and traces of radiation leakage. Under these circumstances, it has not been possible to dispel the concerns and anxieties of the local population and resume the power plant's operation. The main reason for this is the difference between the perceptions of the technical expert and the citizens about the risks that nuclear power plants represent.

In the Special Measures Concerning Nuclear Emergencies, a nuclear disaster has been defined as "a situation in which radioactive material or radiations at an abnormal level has been released outside the nuclear site" [1]. Nuclear disasters are then synonymous with disasters due to radiation. However, this definition of nuclear disasters is not necessarily the same definition that local residents might have regarded "the risk of a nuclear power plant". The report by Niigata Prefecture about the damage caused by the Chuetsu-oki earthquake pointed out that the engineers at the nuclear power plant stated there was no need for an immediate response to the black smoke because the fire in the transformers was not a danger to the reactor. On the contrary the local residents were reportedly filled with anxiety because they did not understand the relationship of the smoke to the safety of the reactor. This difference in the perceptions of the risk influences the public's acceptance of nuclear power plants (NPPs). Local residents are skeptical about the extent to which crisis response measures have been made for the management of potentially fatal equipment. This attitude would obviously prejudice the public's acceptance of nuclear power plants. The difference in the perceptions of risk is not due to differences in technical knowledge, but the fact that risk assessment information is not shared between the technical experts and the public. For example, even if the mechanism of a hazardous event were not known, if the probability of an occurrence was foreseen and the countermeasures against it were common knowledge, the extent of anxiety by risk phenomena would likely be reduced. Understanding to share risk assessment information can build mutual confidence between engineers and the public.

The anxiety in the local residents caused by differences between the engineers' and the residents' perceptions of the risk posed by the nuclear power plant is an important issue in the Kashiwazaki-Kariwa region. Public anxiety casts a shadow over the ability to maintain the region's acceptance of nuclear power plants. Hence, it is necessary to clarify the ways that knowledge of the factors related to radiation and other factors as sources of risk affect public acceptance and trust of nuclear power plants. There are two models of risk acceptance: traditional trust model and salient value similarity model [2]-[4]. In order to discuss the risk communication framework between NPPs and residents, practical adaptively of the two models on local residents are indispensable. Therefore, this study aims at revealing the relationships among risk assessment result, risk perception model and risk acceptance model of the residents near nuclear power plants. The factors that determine acceptance examined in this study are risk perception, benefit recognition, and source of risk. Data were derived from responses to an Internet-based and mail questionnaire survey of the residents of the Kashiwazaki-Kariwa region. Based on the results of the survey, the study describes the role that the local residents want engineers to fill regarding risk assessment with respect to assessing the acceptability of nuclear power plants.

## 2. Investigation Method

### 2.1. Selection of Respondents for the Survey

A sample of 500 residents of Kashiwazaki City and Kariwa in Niigata Prefecture that were older than 20 years old was selected. The number of respondents from each city was first determined based on the population ratio and each municipality was classified according to gender and age. 476 subjects were then selected from Kashiwazaki (235 men and 241 women) as potential respondents for the survey and 24 were selected from Kariwa (12 men and 12 women). The electoral rolls of the municipalities were used to sample the names and addresses of the respondents. The respondents' names, genders, ages (to identify age groups) and addresses were extracted from the first print edition of the electoral roll by pre-determined random intervals.

### 2.2. Implementation of the Questionnaire

The questionnaire was sent by mail from February 22th to March 15th, 2010 with soliciting the cooperation of the respondents. The questionnaire could be completed either on the Internet or returned by mail, according to the choice of the respondent. Moreover, with respect to Internet responses, a function that recorded the input IP address was able to prevent the problems of duplicate posts. Furthermore, there was a function that prevented multiple submissions from the same personal computer, which meant that the Internet-based responses could be regarded as close to equivalent to the questionnaires received by mail.

## 2.3. Contents of the Questionnaire

Questions that were related to the acceptability, risk scenario by nuclear power plants, benefits to the local community, individual benefits, feelings of trust in power generation companies, and the extent of value similarity between the values of power producers and the respondents' values were included in the questionnaire. The details of the questions are shown as the follows.

1) Risk perception

Because hazards are associated with radiation, questions related to core meltdown, radiation leakage and the release of radioactive water were asked. Questions about failure of machinery and piping equipment and fire on the site were provided as examples of other hazards. For each hazard, the respondents were asked to evaluate three factors. The rating scale was excerpted from the international standard ISO14120 (JIS B9700) on machine safety [5].

a) The probability of the occurrence of risk (1 = *no possibility* through 5 = *occurring frequently*);

b) The extent of damage caused by risk (1 = *negligible damage* through 4 = *fatal damage*);

c) The severity of risk (1 = *acceptable* through 4 = *unacceptable*);

2) Benefits of power plants

Six things were listed as benefits to society and the respondents' region. For each question, the respondents were asked to select one of five options, where 1 = *I do not agree with* through 5 = *I agree with*

3) Trust in power plants

Seven questions were asked about the extent of trust in the company's intentions and abilities. For each question, the respondents were asked to select one of five options, where 1 = *I do not agree with* through 5 = *I agree with*. The above two types of questions were modifications of questions used by Kimura *et al.* (2003) [6].

4) Awareness of the values of the company

Five questions were asked about the benefits and the extent of commonality between the company and the respondents regarding values related to trust. For each question, one of five options was requested, 1 = *I do not agree with* through 5 = *I agree with*.

5) Requirements of engineers

There were 16 questions about the things that the respondents believed were necessary to enable engineers to understand the advantages and disadvantages of nuclear power plants. The questions consisted of proficiency in nuclear energy and safety, engineers' regional affinity, communication skills, and a sense of belonging to the organization. For each question, one of five options was requested, where 1 = *I do not agree with* through 5 = *I agree with*. These questions were selected by referring to Yagi *et al.* (2007) [7] and from the results of a preliminary survey.

6) Face sheet

Each respondent reported gender and age.

7) Other

A consent column was provided in which the respondents indicated that they understood the purpose of the survey and agreed to the results being used for academic and research purposes.

8) Amount of knowledge

To ascertain the objective knowledge of the respondents regarding nuclear power plants, there were eight questions to which the respondents answered by marking X or O. Examples of the questions are: 1.There are 7 units in Kashiwazaki Kariwa NPP (Answer: O); 2. Reactor type of Kashiwazaki Kariwa NPP is pressurized water reactor (PWR) (Answer: X)

# 3. Results and Discussion

## 3.1. Attributes of the Respondents

The total number of responses received by Internet and mail was 249 (response rate: 49.8%). Excluding the cases with item nonresponse, multiple answers to one question, and those that did not agree to the use of the results for academic and research purposes, the total number of respondents for use in the analysis was 208 (proportion of valid responses: 41.6%). Because the response rate of other similar studies is about 40% [6], an equivalent elicitation was accomplished by this survey.

The response rate by age distribution of the valid respondents is shown in **Table 1**. The response rate was relatively small among the respondents in their 20s or 30s in the age distribution. The numbers of persons in these two groups are smaller compared to the number of respondents in their 50s. Consequently, the response rates of those in their 20s and those older than age of 70 were lower than the average. The male to female ratio was approximately 1:1.

### 3.2. Response Situation with Respect to Risk Perception

**Figure 1** shows the respondents' perceptions of risk regarding the five types of hazards. The vertical axis shows the average extent of damage and the horizontal axis indicates the average probability of occurrence. Both factors were unacceptable to the respondents when the values were large.

Each point circle in **Figure 1** shows the extent of unacceptability of the risk. The correspondence between the point scale and the extents of unacceptability were: 2 = acceptable with countermeasures, 3 = undesirable, and 4 = not acceptable. First, there was a vast difference in the respondents' perceptions of the extent of damage depending on whether or not the hazard involved radiation leakage. Furthermore, the extent of the damage due to fire or failure of equipment or piping was perceived as relatively smaller than when there was radiation leakage. The fact that risk perceptions have a negative image indicates that the respondents determine the extents by combining the extent of damage with the probability of its occurrence. In short, risk perceptions of the resident are total based on rational basis.

### 3.3. Relationship between Amount of Knowledge and Risk Perception

The results of the multiple regression analysis performed on the relationship between risk perceptions and the amount of knowledge where the magnitude of the risk was the dependent variable and the frequency of occurrence of risk, the extent of damage, and the amount of knowledge were the independent variables are shown in **Table 2**. The symbols on the top right of the numbers indicate their statistical significance. First, the perceived extent of damage determines the perceived size of the risk. However, there may be factors that are not significant as far as their probabilities are concerned. In particular, if the probability of the hazard are low or if it is a frequently occurring hazard, then there is a possibility of a deterministic rather than probabilistic evaluation of the risk under consideration. On the other hand, none of the factors were statistically significant regarding the amount of knowledge. In addition, the adjusted coefficient of determination (the proportion of the variation in the dependent variable that can be explained by a given independent variable) is smaller for the effects of on-site fires than for the other factors.

### 3.4. Relationship of Risk Perception with Benefits and Trust

The data were summarized using factor analysis package R2.7.1. The number of factors in the factor analysis

**Table 1.** Age distribution of respondents (gender distribution: 101 = male, 102 = female, and 5 = unknown).

| Age group | Number of respondents | Response rate (%) |
|---|---|---|
| 20 | 18 | 29.5 |
| 30 | 29 | 40.3 |
| 40 | 36 | 46.8 |
| 50 | 42 | 43.7 |
| 60 | 39 | 51.3 |
| 71 and older | 43 | 36.4 |
| Unknown | 1 | - |
| Total | 208 | 41.6 |

**Table 2.** Relationship between the perception of the extent of the risk and the amount of knowledge.

| Type of hazard | Probability of occurrence | Extent of damage | Amount of knowledge | Adjusted coefficient of determination |
|---|---|---|---|---|
| Core meltdown | 0.033 | 0.540*** | 0.000 | 0.337 |
| Radiation leakage | 0.079+ | 0.502*** | −0.018 | 0.350 |
| Radioactive water release | 0.110** | 0.494*** | 0.004 | 0.356 |
| Failure of machinery and piping | 0.020 | 0.573*** | 0.019 | 0.344 |
| Fire at the site | 0.011 | 0.496*** | 0.002 | 0.212 |

* = $p < 0.05$, ** = $p < 0.01$, *** = $p < 0.001$.

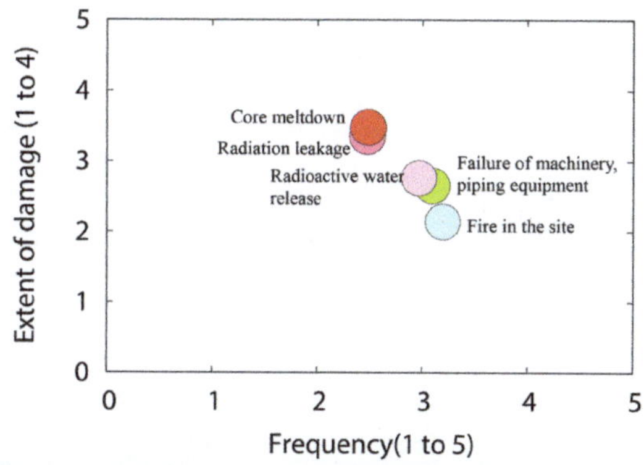


**Figure 1.** Respondents' risk perceptions (sizes of the circles indicate the extent of unacceptability) Size of data points show the extent of severy of hazard.

was specified as three to correspond with the factors found in previous studies. Applying a varimax rotation, the factor scores were calculated using maximum likelihood estimation. The results of the analysis are shown in **Table 3**. Items with loadings greater than 0.4 on one factor that did not show a loading greater than 0.4 on other factors were chosen. Because the third factor consists of items that indicate benefits to the nuclear energy industry, it was termed the "social benefit" factor. Because the second factor consists of items that result in benefits to the local community, it was termed the "regional interests" factor. Because the first factor consists of items that express the respondents' trust in the safe running of the nuclear power plant, it was termed the "trust in the company" factor. The value of the coefficient $\alpha$ indicating the internal consistency of the items that comprise afactor was close to 0.8 for all of the factors, which is considered a sufficient value. This composition of items is identical with that of Kimura (2003). Thus, since a factor for determining the acceptability of risk often comprises the risk as well as the benefits of and the confidence in the business entity, this study's composition factor for benefits and trust are appropriate. The arithmetic mean of the configured items was used as the factor score of the subordinate factors.

The combined impacts of the factors of "social benefits," "regional interests," and "trust in the company," and the amount of knowledge on the extent of unacceptability of the risk, were assessed with a multiple regression analysis. The results are shown in **Table 4**. The negative coefficients indicate that the extent of acceptability increases due to the influence of that factor. The measures of social benefits and trust in the company were statistically significant, whereas regional interests that might be expected to have direct advantages were not statisticily significant. Moreover, the amount of knowledge was not statistically significant for any of the items. Although an increase in knowledge tends to positively relate to trust in people with values that are similar to one's own, no such correlation was found in this study. A detailed study of the item of improvement of the amount of knowledge is necessary.

### 3.5. Awareness of the Values of the Company

Five questions were asked to learn about the benefits and the extents of commonality between the company and

**Table 3.** Factor analysis results of items related to benefits and trust.

| Question | Factor 1 ($\alpha = 0.908$) | Factor 2 ($\alpha = 0.842$) | Factor 3 ($\alpha = 0.791$) |
|---|---|---|---|
| A1: I think a nuclear power plant is necessary for stable power supply | 0.267 | 0.226 | 0.642 |
| A2: I think the use of nuclear power plants is an effective countermeasure against global warming | 0.171 | 0.158 | 0.864 |
| A3: I think a nuclear power will be able to provide relatively inexpensive electricity | 0.237 | 0.357 | 0.508 |
| A4: I think a nuclear power plant will increase employment opportunities in the region | 0.218 | 0.727 | 0.218 |
| A5: I think a nuclear power plant will enhance local facilities | 0.238 | 0.765 | 0.29 |
| A6: I think a nuclear power plant will make the population in the region increase | 0.163 | 0.748 | 0.116 |
| B2: I think the workers in nuclear power plants are well trained | 0.688 | 0.252 | 0.1 |
| B3: I think the problems that can occur in a nuclear power plant are well anticipated and suitable measures are being taken | 0.781 | 0.232 | 0.139 |
| B4: I think the emergency response system in the event of a disaster or accident in nuclear power plants has been well established | 0.805 | 0.208 | 0.216 |
| B5: I think the record of operations and inspection in nuclear power plants is well maintained | 0.784 | 0.107 | 0.184 |
| B6: I think that problems that occur within nuclear power plants are recorded properly and official announcements are made to the outside world | 0.79 | 0.19 | 0.273 |
| B7: I think the authorities in nuclear power plants learn from problems that occurred in other places and take preventive measures | 0.769 | 0.141 | 0.214 |
| Factor contribution extent | 3.851 | 2.102 | 1.792 |
| Factor contribution ratio | 0.321 | 0.175 | 0.149 |
| Cumulative contribution ratio | 0.321 | 0.496 | 0.645 |

**Table 4.** Relationship of the extent of risk with benefits and trust.

| Variable | Social benefits | Regional interests | Trust in the company | Amount of knowledge | Coefficient of determination |
|---|---|---|---|---|---|
| Magnitude of risk (unacceptability) | $-0.162^{***}$ | −0.004 | $-0.200^{***}$ | 0.014 | 0.238 |

$^{*} = p < 0.05$, $^{**} = p < 0.01$, $^{***} = p < 0.001$.

the respondents regarding values related to trust. For each question, respondents chose one of five options, where 1 = *I think so* through 5 = *I do not to think so*. The questions were: 1) Do you think the company must be trusted by the local community? 2) Do you think the company must not cause accidents? 3) Do you think the company must not cause radiation leakage? 4) Do you think the company must strive to provide a stable power supply? 5) Do you think the company must contribute to the community? The average of the responses to the five questions was greater than four points in the five-point scale and the standard deviation plus the mean exceeded the highest point, suggesting a ceiling effect. In other words, the respondents demonstrated a substantially uniform response to these questions. Although there was a lack of clarity regarding the respondents' opinions, such as, "I believe the company should be thinking in such and such a manner" or, "I am aware that the company should be doing such and such things," the positive responses obtained were unexpectedly high.

## 3.6. Awareness of Engineers' Requirements

There were 16 questions about factors that the respondents thought were necessary to enable engineers to understand the advantages and disadvantages of nuclear power plants. The questions consisted of four categories: 1) proficiency in nuclear energy and safety, 2) regional affinity of engineers, 3) communication skills, and 4) sense of belonging to the organization. Among these responses, a ceiling effect was observed with respect to proficiency in nuclear energy and safety and communication skills, just as was found in the case of awareness of the values of the company. This finding shows that the respondents had a nearly uniform response with respect to these items. In other words, because it is a matter of course that an engineer involved in risk communication is equipped with relevant knowledge about nuclear power and safety, it is expected that his or her communication skills are naturally high.

Regarding items such as regional affinity and commitment to the organization, there were no statistically significant correlations with the other items. This finding does not necessarily mean that these items have no relevance to public acceptance of nuclear power plants. Whereas the first set of questions concerned nuclear power plants and a specific company, this set of questions concerned general technical personnel. Thus, direct relevance to the respondents may not have been recognized. In any case, a more detailed investigation is necessary.

### 3.7. Regarding the Requirements of Engineers Involved in Risk Communication

It can clearly be seen from the risk perception map shown in **Figure 1** that the respondents had very rational perceptions of risk. In particular, the fact that they clearly distinguished between the extent of risk due to radiation and other kinds of risks is evidence of their anxiety about the damage caused by radiation, given their position as local residents. The probability (frequency) of an occurrence evidently did not have much of an impact on them. This result leads us to believe that if knowledge of the inherent risks in a power plant and the relevant measures that are being taken under the assumption of their occurrence (deterministically) were shared with local residents in a rational manner, then local residents' acceptance of the risk would increase.

On the other hand, although the risks due to factors other than radiation were not given much importance, we cannot conclude that measures against them are regarded as unnecessary. If the local public believes that measures for prevention and recurrence are not in place, their trust may be undermined, resulting in the risk becoming unacceptable to the public. In this sense, it is necessary to reveal to the public the fact that reasonable measures for prevention and recurrence have been taken, even regarding risks that are unrelated to radiation.

Because the extent of the risk was negatively correlated with the factors relating to social benefits of nuclear power and trust in the company, it is important to vigorously communicate the fact that trust and nuclear power plants have social benefits and that the nuclear power plant in Kashiwazaki, in particular, is useful for all of Japan. The greater the acknowledgement of social benefits, the more the acceptance of the risk of nuclear power plants may be expected to grow. As a prerequisite for this, it is essential that the activities of the power plant gain the trust of the local people. The responses in this survey indicate that it is necessary for the company to obtain the people's trust and, although efforts are being made and awareness is being created, continued efforts are needed. Regarding the answers to the questions on regional interests, Kimura *et al.* observed that a factor of risk tolerance is weakly correlated with the regional interests factor [6]. It is necessary to carefully examine whether this tendency is related to the respondents' characteristics.

The qualifications of engineers who are involved in risk communication need to be studied in detail. However, as was learned from the extent of awareness of the local residents, expert knowledge in nuclear power, communication skills, and expertise regarding safety are minimum desirable qualities. Furthermore, the relationship between engineers' regional affinity and their sense of belonging to their organization needs to be examined. Fukushima NPP disaster affected risk perception of not only in Japanese people, but also in Chinese or another international people [8]. Continuous observations on risk perception of residents are important to reveals the effectiveness of risk communications by engineers.

## 4. Conclusions

This study observed the relationship among risk assessment, risk perception model and risk perception model of local residents near NPP in Kashiwazaki-Kariwa region. The summary of the results is as follows.

1) The extent of the severity of risk is significantly affected by the perceived extent of damage due to a given hazard. In particular, radiation-related hazards are perceived as more severe than other hazards.

2) The extent of acceptance on the risks by NPPs is significantly related to the factors of social benefits by NPP and trust to NPP Company by the local residents. The benefits to the local region are not found to be statistically significant on risk acceptance model.

3) Respondents' amount of knowledge is not significantly correlated with any of the items. The results sug-

gest that only technical explanation to the resident is not sufficient to establish mutual reliable relationship between NPPs engineers and local residents.

4) Value similarity between NPPs with local residents is not statistically significant. Nuclear power plants company is perceived as making efforts to gain public trust in the region.

5) NPP engineers are expected to possess expert knowledge in nuclear power, communication skills and expertise in safety in order to conduct more smooth risk communication.

The result demonstrates that risk acceptance model of local residents to NPPs can be arranged by traditional trust model. The result by this study is beneficial for risk communicators to manage communication framework with local residents.

## Acknowledgements

This study was partly supported by a jointly funded research project of "regional contribution function" of the University by Niigata prefecture in 2009. The author greatly appreciates anonymous residents of Kashiwazaki, Niigata, and Kariwa for their cooperation in participating in this study.

## References

[1] IAEA (1996) Defense in Depth in Nuclear SSAFETY (INSAG-10). IAEA, Vienna.

[2] Slovic, P. (2000) The Perception of Risk. Earthscan Publishing Ltd., London.

[3] Morgan, M.G., Fischhoff, B., Bostrom, A. and Cynthia J. Atman, C.J. (2001) Risk communication. Cambridge University Press, Cambridge. http://dx.doi.org/10.1017/CBO9780511814679

[4] Nakayachi, K. and Cvetkovich, G. (2008) Trust of Risk Managers: An Integration of the SVS Model and the Traditional View of Trust. *The Japanese Journal of Social Psychology*, **23**, 259-268.

[5] ISO 14120:2002 (2002) Safety of Machinery—Guards—General Requirements for the Design and Construction of Fixed and Movable Guards.

[6] Kimura, H., Furuta, K. and Suzuki, A. (2003) Psychological Factors Affecting Public Acceptance of Nuclear Energy: Comparative Analysis Focusing on Regional Characteristics and Degree of Knowledge Prasad. *Transactions of the Atomic Energy Society of Japan*, **2**, 379-388.

[7] Yagi, E., Takahashi,S. and Kitamura, M. (2007) Clarification of Nuclear Risk Recognition Scheme through Dialogue Forum. *Transactions of the Atomic Energy Society of Japan*, **6**, 126-140.

[8] Huang, L., Zhou, Y., Han, Y.T., Hammitt, J.K., Bi, J. and Liu, Y. (2013) Effect of the Fukushima Nuclear Accident on the Risk Perception of Residents near a Nuclear Power Plant in China. *Proceedings of the National Academy of Sciences of the United States of America*, **110**, 19742-19747. http://dx.doi.org/10.1073/pnas.1313825110

# 6

# Lessons Learned from Recent Accidents in the Chemical Industry in Japan

**Akihiko Hidaka, Yuichiro Izato, Atsumi Miyake**

Graduate School of Environment and Information Sciences, Yokohama National University, Yokohama, Japan
Email: akihiko_hidaka@plamatels.co.jp

## Abstract

**Japanese chemical companies should extensively study the accidental fires and explosions that occurred in the industrial complexes of Japan in 2011 and 2012. We picked two major accidents and distributed questionnaires to large chemical companies in the Keihin Complex, Japan. Here, we report the results of our investigation of improvements in process safety.**

## Keywords

**Chemical Accident, Lessons Learned, Explosion, Fire**

## 1. Introduction

Chemical companies in Japan are currently facing major issues, as described below.

1) The lessons to be learned from the Tohoku Earthquake disaster, which occurred in March, 2011, are still being examined [1]-[3]. 2) Japanese chemical companies should study the accidental fires and explosions that occurred in industrial complexes in Japan in 2011 and 2012 [4] [5].

Here, we describe the safety improvements that should be implemented, based on the accidental fires and explosions in the two cases mentioned above.

The serious accidents that occurred in industrial complexes in Japan during 2011 and 2012 shocked the chemical industry. And also the Japanese chemical industry had a big subject that they should improve their level.

Accidents occurred at the Tosoh Corporation [4], Mitsui Chemicals [5], Nippon Shokubai [6], Cosmo Oil Chiba [7], Mitsubishi Chemical Kajima [8], and Showa Denko Oita. We should reflect on and reexamine these accidents, and improve our technological expertise to prevent such accidents. This should also apply to the development of new factories abroad from now on, and the results of investigations of these accidents and new security measures for disaster prevention should be applied globally. To investigate these accidents, a questionnaire was sent to large chemical corporations with high-pressure gas installations, located in the Kawasaki area in the Keihin complex, Japan. The questionnaire dealt with the accidents at Tosoh in the Nanyo district and Mitsui Chemicals in the Iwakuni-Ootake factory area. The theme we focused on in the investigation was "the emergency shutdown of production facilities, such as industrial complexes", because both these accidents continued for several hours, ~10 h, after from the emergency shutdown of the production facilities. We also tried to clarify what has already been done and what has not yet been done by Japanese chemical companies.

## 2. Tosoh Accident

This accident occurred on November 13, 2011. The process from which the accident arose is VCM production facilities (**Figure 1**). And the process which explosion and fire generated is HCl tower in it (**Figure 2**). The timeline is shown in **Table 1**. The displayed times show the progress from the time at which the emergency block valve was closed [4] (refer the **Table 1**).

### 2.1. Presumed Cause of the Explosion

HCl reacted with VCM to produce 1,1-EDC, catalyzed by $FeCl_3$, which was produced by HCl and iron rust inside the back-flow tank of the HCl tower.

**Table 1.** Timeline of Tosoh accident [4].

| Lapsed time | Phenomenon and action taken | Deficiency, problem |
|---|---|---|
| 0 (base) | The urgent discharge valve for disposal facilities of oxidation line A was broken. The valve suddenly opened during line A operation. | |
| 13 min after | Oxidation line A was stopped by the interlock function (oxidation line B was operational). | Was the partial stop appropriate? Was full examination of the partial stop made? |
| 14 min after | Stopped lines A and B of the EDC decomposition process to regulate the production of EDC, which was only produced by line B of the oxidation process. So, the operating line for the EDC decomposition process changed to line A only. As a result, the operating load changed from 100% to 45%. | Was manual operation of line B oxidation process + line C EDC decomposition performed? Was the operator trained? |
| 31 min after | Because the stoppage was partial, the temperature balance of the HCl tower changed and the middle tier temperature fell from 80°C to 57°C. In order to recover the temperature condition, the amount of heating steam was increased and the reflux volume was decreased. At that time, the operators did not realize that they should control the top and bottom temperatures of the HCl tower. | The operating standard of the chloride tower was insufficient. The operator training was inadequate. |
| 1 h 1 min after | Although the temperature at the top of the HCl tower is normally −24°C, it rose to 38°C. Then VCM was mixed in the upper part of the HCl tower and the back-flow tank of the HCl tower added HCl. The operators decided that the HCl tower condition had become stable because the temperature of the middle tower stage recovered to 80°C. | The alarm for detecting abnormal temperature was not set at the top tier of the HCl tower. It was not the point being monitored. The standard value of the temperature of each part of the HCl tower was not decided. |
| 2 h 18 min after | After VCM mixed in the back-flow tank of HCl, VCM mixed with the oxidation line B. The entire production facility was then stopped. The general manager, production manager, and section manager realized that the temperature at the top of the HCl tower stage was not normal and they assumed that VCM had mixed in the back-flow tank of the HCl tower. However, they did not know that 1,1-EDC was being generated, so they did perform any special operation. | There was no knowledge of the abnormal reaction and side reaction that are assumed to have occurred. Important information was not available. Important information was not included in the training manual. |
| 2 h 41 min after | The back-flow pump of the HCl tower was stopped. The level of the HCl back-flow tank was nearly 100%, and the tank was sealed. | There was no operating standard that ensured safety. |
| 5 h 1 min after | The HCl tower refrigerator was stopped. The back-flow tank of the HCl tower was disconnected from the HCl tower, so the back-flow tank of the HCl tower was sealed. | There was no operating standard that ensured safety. |
| 8 h 0 min after | The transfer of liquid from the back-flow tank of the HCl tower to the liquid HCl temporary receiving tank started. Although the pressure of liquid HCl in the temporary receiving tank was increasing, the operators were unaware of this phenomenon. | There was no operating standard that ensured that the safety pressure of the HCl receiving tank was monitored. |
| 11 h 20 min after | Operators realized that the pressure of the liquid HCl temporary receiving tank was rising, and they started pressure relief action. | The drop-in-pressure measure was insufficient. |
| 11 h 36 min after | The allophone sounded and white smoke was discharged from the top of the liquid HCl temporary tank during the pressure relief action. | |
| 11 h 44 min | The pressure of the back-flow tank of the HCl tower rose to 2.0 MPaG (the designed pressure of this tank is 1.9 MPaG). | |
| 11 h 45 min | The back-flow tank of the HCl tower ruptured, exploded, and burst into flames. | |

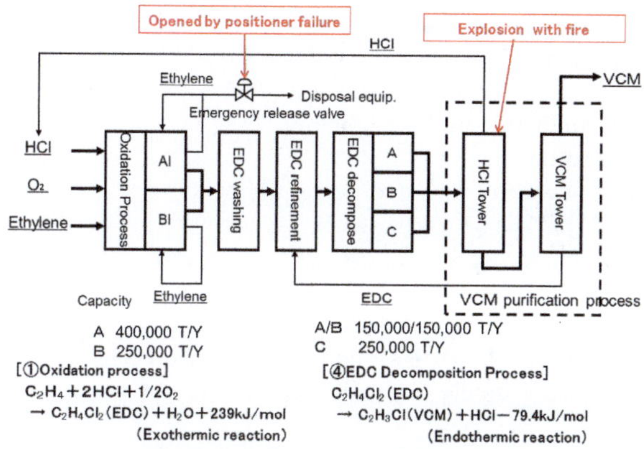


**Figure 1.** Flow chart of Tosoh VCM production facilities.

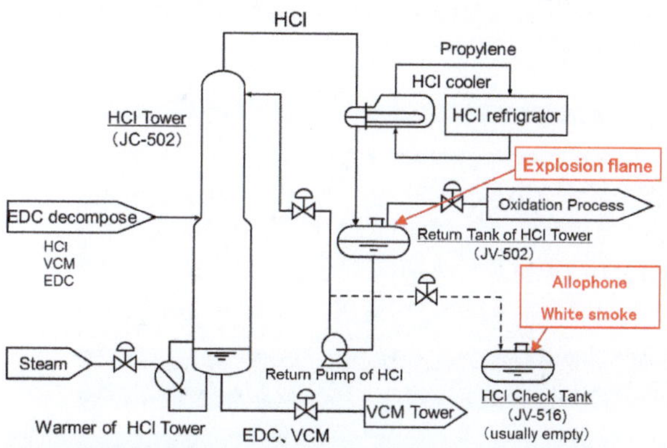


**Figure 2.** Flow chart around HCl tower of VCM process.

HCl + VCM → 1,1-EDC + 62 kJ/mol (exothermic reaction)

The temperature of the back-flow tank of the HCl tower increased, because this reaction is exothermic, and the reaction rate increased exponentially. After 6 h under these conditions, this reaction ruptured and destroyed the back-flow tank of the HCl tower.

## 2.2. Lessons to Be Learned from the Tosoh Accident

As mentioned above, we investigated the issues involved in this accident by sending questionnaires to large chemical corporations with high-pressure gas installations in the Kawasaki Keihin complex in Japan.

1) The main items in the questionnaire regarding the Tosoh accident were as follows.

a) Monitoring during normal operation

i) Does it clear the normal range of control for important equipment?

ii) Are the important alarms located separately from the DCS alarm?

iii) Can the slopes of important temperatures or pressures be easily identified?

b) Prior examination, and monitoring of emergency

i) Do you consider abnormal reactions or sub-reactions when operation is irregular?

ii) Do you consider the possibility of iron mixing, which could cause an abnormal reaction?

iii) Do you monitor the plant conditions after setting the appropriate control range for an abnormal reaction?

c) Preparation for emergency shutdown

i) Was the emergency shutdown manual prepared based on sufficient information?

ii) Is the person who is to initiate the emergency shutdown and evacuation identified?

iii) Is regular training regarding emergency shutdown provided?

d) Action after emergency shutdown

i) Are the plant safety conditions after an emergency shutdown clear?

ii) Are the procedures for reversing the emergency shutdown and returning to normal operation clear? Also, who is the person who approves this action?

2) The results from the main questionnaire are shown below.

a) Monitoring during normal operation

i) Does it cover the normal range of control for important equipment?

Yes: 11 facilities; No: 1 facility

**Figure 3** shows a typical monitoring during normal operation in Japan.

◆ Examples of reply

With regard to the operating conditions, the range is defined as a job standard or a process control standard value. Because abnormal values are defined by the DCS system, this can be detected by the alarm.

ii) Are the important alarms located separately from the DCS alarm?

Yes: 10 facilities; No: 2 facilities

**Figure 4** shows that the important alarm located separately from DCS in Japan.

iii) Can the slopes of important temperature or pressure be easily identified?

Yes: 12 facilities; No: 0 facilities

**Figure 5** shows typical trend records of conditions in Japan.

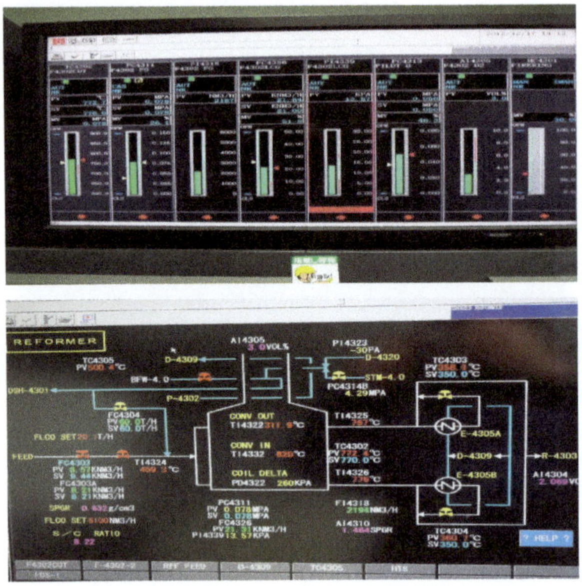


Monitoring during normal operation

**Figure 3.** Observation of operating condition by DCS [questionnaire a)-i)].

**Figure 4.** Important alarms located separately from DCS [questionnaire a)-ii)].

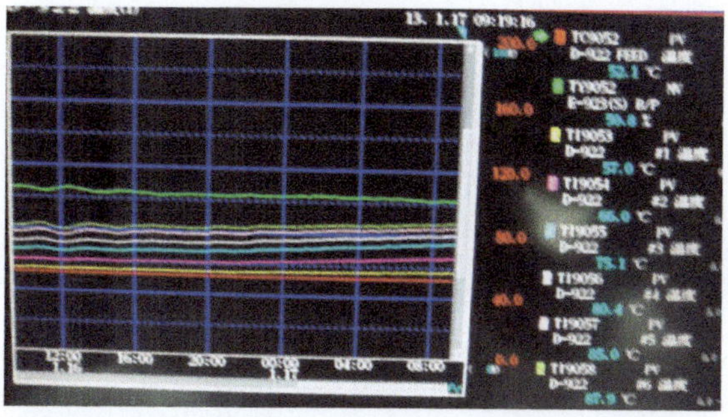

**Figure 5.** Trend records of conditions [questionnaire a)-iii)].

b) Prior examination, and monitoring of emergency

i) Do you consider abnormal reactions or sub-reactions when operation is abnormal?

Yes: 11 facilities; No: 1 facility

◆ Example of examination

The danger of material B polymerizing was examined by considering potential factors that could lead to temperature increases caused by heat of polymerization, leading to explosions. The factors that we considered are the material, operation, facility, and staff. Based on this analysis, we are now planning the necessary countermeasures.

ii) Do you consider the possibility of iron mixing, which could cause an abnormal reaction?

Yes: 11 facilities; No: 1 facility

◆ Examples of answer

Company A

We make it a rule in our standard operation to explore the influence of corrosion or damage to the installation and lower equipment. Previous experience and the process licenser's opinion suggest that there is almost no possibility of an abnormal reaction being caused by mixing of rust, so this is not examined.

Company B

In response to this accident, we examined the possibility that corrosion products (iron rust, etc.) could trigger an abnormal reaction.

iii) Are the control points for normal and abnormal reactions identified and monitored continuously?

Yes: 11 facilities; No: 1 facility

◆ Examples of answer

The controls are identified and are always supervised in DCS.

(Trend monitoring at control points, alarms set up at these points) A range deviation alarm is set up in four steps; interlock is set up in two steps.

c) Preparation for emergency shutdown

i) Was the emergency shutdown manual prepared based on sufficient information?

Yes: 12 facilities; No: 0 facilities

**Figure 6** shows the example of emergency procedure prepared card style.

◆ Examples of answer

We revised the emergency shutdown manual for the following situations:

-gas, water, and electricity supply stoppages

-abnormalities caused by a natural disaster

-plant abnormalities

ii) Is the person who makes the decision and the exemption-from-responsibility regulation for a scram defined?

Yes: 12 facilities; No: 0 facilities

**Figure 7** shows the example of exemption from responsibility regulation in Japan.

iii) Is regular training regarding emergency shutdown provided?

**Figure 8** shows the example of the training of emergency shutdown.

Yes: 12 facilities; No: 0 facilities

◆ Examples of answer

| | Emergency Shut Down Procedure (Ope1 –A1) |
|---|---|
| 1 | Reflect the emergency stop screen(CRT) |
| 2 | M-4140 stop |
| 3 | M-4160 stop |
| 4 | Confirm K-4030 A, B, C under the operation |
| 5 | Stop the O2 supply to T4 |
| 6 | Throw Nitrogen in to T4 for pressure balancing |
| 7 | Maintain the steam balance |
| 8 | Check the condition of T4 |

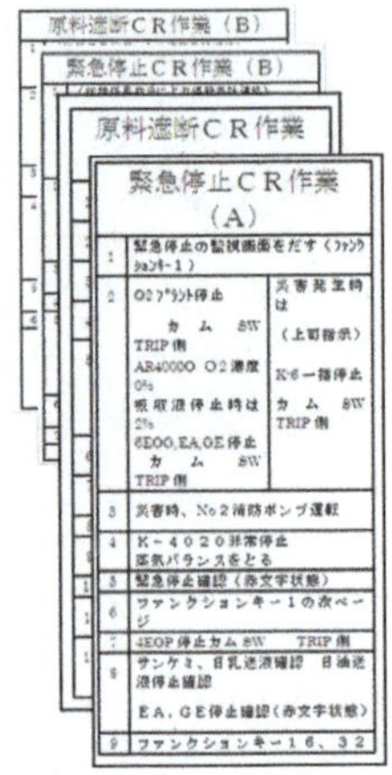


**Figure 6.** Example of emergency procedures prepared card style [questionnaire c)-i)].

Example of Exemption-from-responsibility regulation

| Number of rules K production− T − S − 3 0 1 | Revice Number Rev.13 | Manufacturing control standards | page 19/41 |
|---|---|---|---|

【The extract of a standard】

6. The measure in an emergency

Omission

6.3.1 **Judgment at the time of a scram**

Usually, the chief executive (section chief) of the plant operation team performs judgment and directions of urgent shutdown etc.

When requiring a holiday or emergency at night, after a group leader takes required measures, such as urgent shutdown, he shall report to a boss as soon as possible, and shall receive directions.

**In this case, the responsibility about judgment of the group leader of urgent shutdown shall notbe demanded.**

**Figure 7.** Example of exemption from responsibility regulation [questionnaire c)-ii)].

**Figure 8.** The training of emergency shutdown [questionnaire c)-iii)].

Training on troubleshooting is carried out regularly. However, because of generational changes, traditional technical information is currently required. The plant has now been operating continuously for four years, so staff have less experience of plant shutdown and startup. Moreover, operation is now stable, and because there is little experience of scram operation of equipment, the necessity for training is increasing. Troubleshooting training is therefore periodically conducted using training facilities etc.

d) Action after emergency shutdown

i) Is the safe state after an emergency shutdown clear?

Yes: 8 facilities; No: 4 facilities

**Figure 9** shows the example of the observation of plant conditions after emergency shutdown.

ii) Is the procedure for canceling an emergency shutdown clear? Is the decision maker clearly identified?

Have a procedure: 4 facilities;

Only decision maker specified: 3 facilities;

No procedure and no decision maker specified: 5 facilities.

**Figure 10** shows the example of the procedure of canceling for emergency shutdown condition.

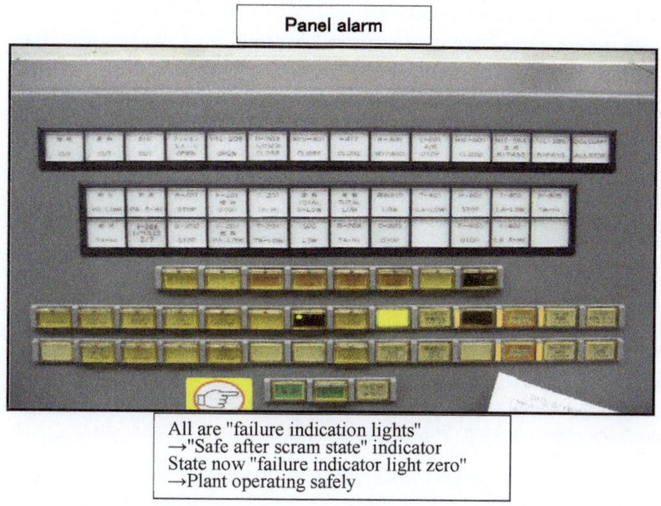


**Figure 9.** Observation of plant conditions after emergency shutdown [questionnaire d)-i)].

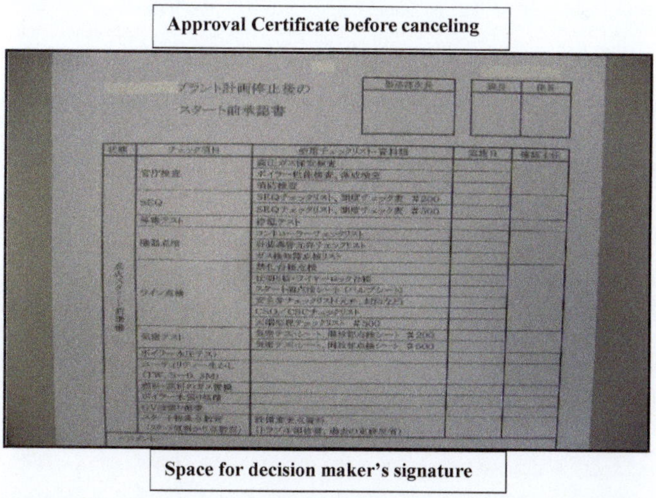


**Figure 10.** Procedure of canceling for emergency shutdown condition [questionnaire d)-ii)].

◆ Examples of answer

After discussion with the division manager, the plant manager and factory area manager decide conditions for canceling an emergency shutdown.

## 3. Mitsui Chemical Accident

The accident at the Mitsui Chemical Otake factory was examined as the second accident case study. The process from which the accident arose was RESORSINOL production facilities (**Figure 11**). And the equipment with which explosion and a fire broke out was an oxidational reactor of RS (**Figure 12**). The timeline in **Table 2** shows how the accident progressed. The displayed times show the progress at different times from the start time, at which the order was given to stop operation of the plant [5].

### 3.1. Presumed Cause of the Explosion (Direct Cause)

Although the emergency shutdown was initiated at the oxidation reactor, which generates an organic peroxide, the subsequent interlock released this oxidation reactor. Then, the nitrogen being supplied to the oxidation reactor stopped, and churning of the liquid phase stopped.

As a result, decomposition of the organic peroxide occurred in the liquid phase upper part, and this heat could not be removed, so the temperature of the upper part of the reactor rose.

**Table 2.** Timeline of Mitsui chemical accident [5].

| Lapsed time | Phenomenon and action taken | Deficiency, problem |
|---|---|---|
| 0 (base) | After 36 h of this batch reaction, which needs 40 h, operation of the steam-using plant was stopped because of problems with the steam. | |
| 12 min after | The Emergency Stop lever was operated; interlock change was effective, and the plant was shut down. The air supply to the oxidation reactor was stopped by the interlock operation, and nitrogen supply began. Cooling water changed from circulating water to emergency-cooling water. Nitrogen supply to the oxidation reactor continued and the temperature fell slowly. | |
| 1 h 20 min after | Interlock was canceled to change cooling water from emergency-cooling water to circulating water, because it was judged that the temperature of the oxidation reactor was not falling. By canceling the interlock, the nitrogen supply was also suspended and churning of the oxidation reactor also stopped. | The emergency shutdown manual was inadequate. Lack of understanding of the purpose of the interlock and of the theory of cooling and agitation. Lack of understanding of the importance of interlock. The interlock release system was inadequate. Training material was inadequate. |
| 1 h 24 min after | The temperature of the lower part of the liquid phase decreased, but at the upper part, the temperature began to increase gradually because the heat of decomposition of the organic peroxide could not be removed, because of lack of a cooling coil and agitation, as a result of stopping the nitrogen flow. | Monitoring of the upper part of the liquid phase was inadequate. The design of the cooling capacity and churning capability in the event of abnormalities was deficient. |
| 2 h 13 min after | Although the HI alarm in the upper part of the liquid phase sounded. Operators were convinced that the problem was the temperature of the gaseous part. The decomposition reaction proceeded and the temperature of the upper part of the liquid phase increased. | The operator was unaware that the liquid level was higher than its design base. The change control system was inadequate. |
| 2 h 51 min after | The decomposition reaction accelerated and the temperature and pressure increased abruptly. Although the pressure-regulating valve was fully opened manually, the pressure relief speed was insufficient. So, the pressure further increased abruptly. | |
| 2 h 55 min after | The oxidation reactor ruptured, exploded and caught fire. | |

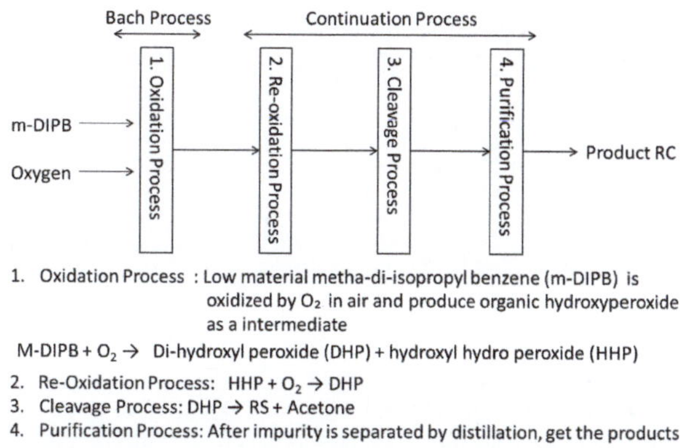


**Figure 11.** Block flow of resorsinol production facilities [5].

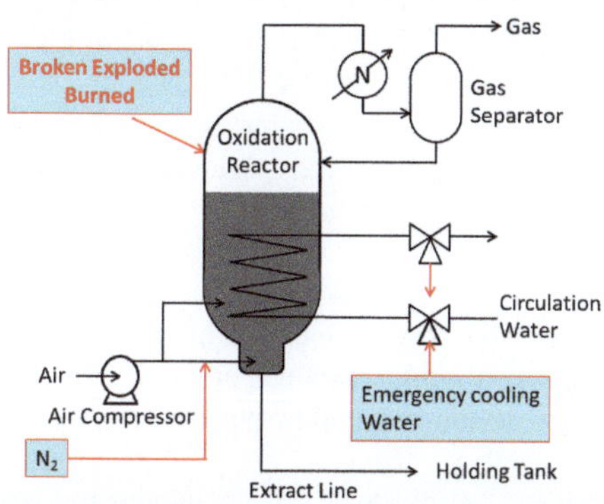


**Figure 12.** Flow sheet for oxidational reactor of RS [5].

The decomposition of the organic peroxide accelerated the heat increase. As a result, the pressure of the oxidation reactor increased, resulting in rupture, explosion, a fire explosion, and a fire.

### 3.2. Lessons to Be Learned from Mitsui Chemical Accident

1) The main questionnaire items concerning the Mitsui Chemical Otake factory accident are follows.

a) Cooling capacity

i) Is the cooling capacity sufficient to maintain safety and a stable reaction when cooling is required?

ii) Do the design and structure of the cooling system enable it to cool all the parts of the reaction vessel and a tank?

b) Two or more important hardware-monitoring installations

i) Are the operating conditions of important apparatus in the process always monitored and controlled at two or more places?

c) Interlock release

i) Even if the interlock is canceled, can important utilities, such as cooling water and nitrogen, be operated independently?

ii) Is the importance of interlock release fully understood by staff?

2) The questionnaire results are shown below.

a) Cooling capacity

i) Is the cooling capacity sufficient to maintain safety and a stable reaction when cooling is required?

Have sufficient capacity: 8 facilities

Do not need cooling or $N_2$ purge: 3 facilities

Insufficient: 1 facility

◆ Examples of answer

Required amount of water, resisting pressure performance, etc. are examined and designed.

In a scram, in order to stop the reaction but not cooling, the nitrogen or steam supply is set by interlock.

ii) Do the design and structure of the cooling system enable it to cool all the parts of the reaction vessel and a tank?

Yes: 7 facilities

Not applicable: 4 facilities (another safety method is used)

No: 1 facility

◆ Examples of answer

The exterior is cooled using a coil and a jacket. Moreover, reaction solution is circulated through the inner liquid with a stirrer, and attains equilibrium.

The facility design is inadequate, and all the portions generating reaction heat cannot be cooled.

b) Two or more important hardware-monitoring installations

i) Are the operating conditions of important apparatus in the process always monitored and controlled at two or more places?

Yes: 10 facilities; No: 2 facilities

◆ Example of answer

We monitor the top, middle, and bottom of the tower using several instruments, but we monitor a tank or heat exchanger using only one instrument.

c) Interlock release

i) Even if the interlock is canceled, can important utilities, such as cooling water and nitrogen, be operated independently?

Yes: 12 facilities; No: 0 facilities

◆ Examples of answer

Whether the interlock is ON or OFF, the utility can be operated manually.

We designed the important utility system as an independent system.

ii) Is the importance of interlock release fully understood by staff?

Yes: 12 facilities; No: 0 facilities

**Figure 13** shows the example of permit for the interlock setting change and by-pass operation.

◆ Example of answer

It is a security system and its importance is recognized. When it is necessary to cancel the interlock and continue operation, we perform change management and implement the required countermeasure.

## 4. The Issues Highlighted by This Examination Should Be Improved in the Future

In this study, two major accidental fires and explosions that occurred in industrial complexes in Japan were investigated. Each of these accidents developed during several hours after carrying out a scram of the manufacturing equipment [4] [5]. This poses the following questions. 1) Can we identify the deviation between stable operating conditions and abnormal conditions? Are there suitable sensing systems and methods for these condi-

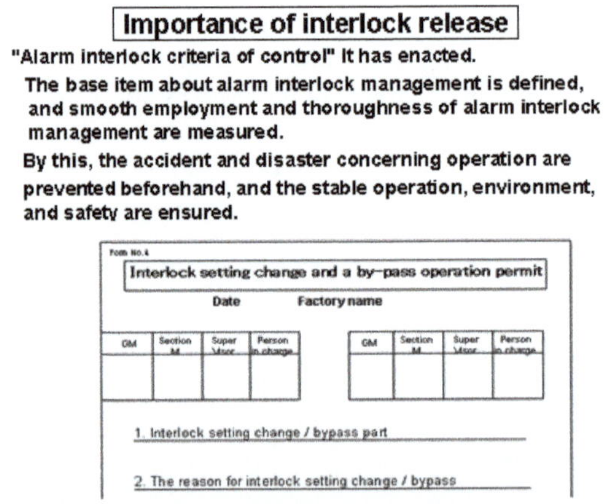


**Figure 13.** Permit for the interlock setting change and by-pass operation [questionnaire c)-ii)].

tions? 2) Is examination for abnormal plant conditions sufficiently made, or has the measure to be specified? 3) Is training in emergency shutdown performed daily? 4) Are the safe and stable states, even under emergency shutdown conditions, firmly understood? Moreover, is monitoring continued even after emergency shutdown, with a defined monitoring point? 5) Are the criteria and procedures for reversing an emergency shutdown clear? A subsequent reexamination was carried out after the accident at Mitsui Chemicals. 6) At a reaction vessel or a tank, are there devices or systems which can be used to identify or monitor reactions, even if they occur locally? From this examination, it is clear that the actual conditions at the company located in an industrial complex in the Kawasaki area in Japan are as follows.

1) Usually, an operation-monitoring method is established. Moreover, the difference between normal and abnormal conditions is clear. A detection method is also established. 2) In the case of inherent abnormal reactions in the plant, a general examination had already been made and inherited as a technical tradition from the past. However, because this cannot rule out new unexpected reactions in the future, the actual conditions must be reexamined using all resources. 3) Education and training for emergency shutdown are performed by companies that are experiencing a crisis because they have reached a change-of-generation stage, and because operators do not have experience in the shutdown or restarting of plants, because stable continuous operation has been achieved for two or four years. However, this effect has now been fully demonstrated, and a state is being been reached with which companies feel comfortable; this is now underway, and actual conditions are being improved.

However, there are two further viewpoints. 4) Is a safe state after an emergency shutdown ensured and is monitoring sufficient? 5) Were the criteria and procedure for restarting after a scram fully developed? Companies conclude that these have been inadequate. This is demonstrated by two questions and answers below.

◆ Measures after emergency shutdown

1) Is a safe state after an emergency shutdown ensured, and is monitoring adequate?

Yes: 8 facilities; No: 4 facilities

2) Were the criteria and procedures for restarting after an emergency shutdown fully developed?

Have a procedure: 4 facilities

Only acknowledger specified: 3 facilities

No procedure and no acknowledge specified: 5 facilities

The reason why this aspect is weak is because it is based on former beliefs. We believed that an emergency shutdown was a safe action. Because there was a shift to a safe state when carrying out the emergency shutdown, it was assumed that safety was achieved. These were blind spots. Furthermore, because operator skill has improved in Japan as a result of education and training in the operating sector, some problems can be solved by partial or transitory shutdowns, and there is a trend to avoid shutdowns of large plants. It is considered that part of the background to these accidents was the wish for partial continuation of operations and early resumption of operations, and confidence in doing this was too high. If abnormalities occur in part of a plant, after stopping all processes completely, the cause of the abnormality is analyzed; if it is removed, an accident will not happen.

It is also the case that the success of advanced technology was also insufficient, and the risks of performing continuous operation may have been added as a result of improvements in the skills of Japanese operators. I think that Japanese chemical companies need to choose one of the following two methods. The first method is to deal with plant accidents by partial shutdown or short-term plant idling, even if there is a risk of a major accident. The other method is to accept production or opportunity losses, in order to reduce the risk of a major accident.

Next, we consider the following questions. 6) At a reaction vessel or a tank, are there devices or systems which can be used to identify or monitor reactions, even if they occur locally?

It cannot be said that this has been fully achieved. Therefore, for containers that hold polymerization monomers, including liberating tanks, middle tanks, and even those that are not frequently used, the following equipment is required.

◆ Containers that hold polymerizable monomers (reservoirs, reaction vessels, separation towers, etc.)

1) The container should be designed to have thermometers and pressure gauges in the upper part, central part, and lower part of the container. The system should be continuously monitored for local heat increases. If a temperature distribution is detected by this system, an alarm should be triggered.

2) A churning function should cover the whole container and it should be designed so that it is not suspended in the event of a system scram.

3) A cooling device that covers the whole container should be designed so that it functions during a system

scram.

We have learned many things from these two accidents. I want to improve the level of safety of the chemical industry globally, by using the lessons learned.

## Acknowledgements

We acknowledge assistance from the High-Pressure Gas Safety Institute of Kanagawa prefecture. This examination was performed with a member of this institute.

## References

[1] Krausmann, E. and Cruz, A.M. (2013) Impact of the 11 March 2011, Great East Japan Earthquake and Tsunami on the Chemical Industry. *Natural Hazards*, **67**, 811-828. http://dx.doi.org/10.1007/s11069-013-0607-0

[2] Dobashi, R. (2014) Fire and Explosion Disasters Occurred Due to the Great East Japan Earthquake (March 11, 2011). *Journal of Loss Prevention in the Process Industries*, **31**, 121-126. http://dx.doi.org/10.1016/j.jlp.2014.03.001

[3] Li, X., Koseki, H. and Mannan, M.S. (2014) Case Study: Assessment on Large Scale LPG BLEVEs in the 2011 Tohoku Earthquakes. *Journal of Loss Prevention in the Process Industries*, in Press. http://dx.doi.org/10.1016/j.jlp.2014.07.015

[4] Tosoh Corporation (2012) Investigation Report of Explosion Accident at Nanyo Plant.

[5] Mitsui Chemicals Inc. (2013) Accident Investigation Committee Report on the Iwakuni-Ohtake Works Explosion and Fire.

[6] Nippon Shokubai Co. Ltd. (2013) Accident Investigation Committee Report on the Himeji Works Explosion and Fire.

[7] COSMO Oil Co. Ltd. (2013) Accident Investigation Committee Report on the Chiba Works Leakage.

[8] Mitsubishi Chemical Corporation (2008) Accident Investigation Committee Report on the Kajima Works Fire.

# 7

# The Research of Railway Line State Detection Signal Processing Method Based on EMD

**Wen-Fa Zhu[1], Hui-Zhen Ma[2], Xiao-Dong Chai[1], Shu-Bin Zhen[1]**

[1]College of Urban Railway Transportation, Shanghai University of Engineering Science, Shanghai, China
[2]Library, Shanghai University of Engineering Science, Shanghai, China
Email: zhuwenfa1986@163.com

## Abstract

**In this paper, an EMD de-noising algorithm is proposed based on the statistical feature of random noise, which can eliminate the noise impaction digital integrator generated by the collected railway line state detection signals using strap-down inertial technology. Firstly, the first IMF component of the noise-dominant modes treated by the process "random sort-sum-average-reconstruction", the signal-to-noise ratio is improved while the noise power is weakened in this process. Then the signal-to-noise cut-off can be determined according to the characters of noise autocorrelation function. Finally, the global threshold could be selected by the noise-dominant mode component, so as to realize the function of filtering. The simulation and validation based on the collected railway line acceleration data using the EMD de-noising algorithm show that the noise in railway line state acceleration detection signals can be effectively eliminated using this method.**

## Keywords

**Track Status, EMD, Autocorrelation Function, Statistical Properties, Filtering**

## 1. Introduction

In the use of Strap-down Inertial Navigation System [1] to detection track state, the collected acceleration signal should be have integrals two times to get the movement of the vector. The acceleration signal from the inertial measurement unit on the scene contains a certain amount of noise which is easy to cause the integrator saturation, so the acceleration signal effect of the reduction noise infects the accuracy of track long wave irregularity detection.

## 2. Hilbert-Huang Transform

Hilbert Huang transform [2] [3] is a signal processing method proposed by Dr. Hung of the America NASA. it has obvious advantages in dealing with nonlinear and non-stationary signal. HHT mainly contains the following two parts: Empirical Mode Decomposition (EMD) and Hilbert transform, the empirical mode decomposition (EMD) is core of the HHT. According to the fluctuation of real size step by step signal in the signal decomposition (for signal smoothing processing), EMD get a series of different characteristics of the scale of IMF (Intrinsic Mode Function, the intrinsic mode function). The IMF should meet the following two conditions:

1) the IMF, an number of extreme points and zero points should be equal or the maximum difference of 1;

2) signal locally symmetric on the time axis. the any point of signal, the mean of the envelope curve point determined by the signal local maxima and the envelope point determination and local minimum value is 0.

## 3. The EMD Denoising Algorithm Based on the Statistical Characteristic of Noise

### 3.1. Characteristics of the Random Noise Statistic

For discrete signal which length is $N$, the power is calculated as:

$$P_X = \frac{1}{N}\sum_{i=1}^{N} x^2(n_i) \tag{1}$$

If the original signal amplitude of each element was kept invariant, its position will be ordering up. According to the type 1, signal through random sort, its power remains unchanged. The first random noise of IMF component by EMD decomposition can still keep the random properties of approximation. However, power is obtained the first IMF component random sort the number which is after "random sort of cumulative average reconstruction" is inversely proportional, namely power decreases with the increase of random sort. as shown in **Figure 1**. Therefore, the noise energy can be reduced through the signal with noise is decomposed imfi "EMD of random scheduling and cumulative average".

### 3.2. The Statistical Properties of the Random Noise from the Autocorrelation Function

The random signal of the autocorrelation function is defined as:

$$R_X(t_1,t_2) = E\left[x(t_1)x(t_2)\right] \tag{1}$$

Here the use of normalized autocorrelation function to represent the the correlation degree of random signal in different time values. That is:

$$\rho_\tau = \frac{R_X(\tau)}{R_X(0)} \tag{3}$$

In the formula, $\tau = t_1 - t_2$

Although the normalized autocorrelation function of the random noise and signal get maximum value at zero, but the point of the time difference outside zero have following difference:

The random noise from the autocorrelation function have the maximum value at zero point. at the other point, the autocorrelation function decays rapidly decrease small on the contrary. but the general signal autocorrelation function does not have such a characteristic. According to the difference of random noise and general signal autocorrelation function can accurately judge the signal-to-noise cutoff point between the dominant mode of K.

### 3.3. Denoising Algorithm of Constructing EMD

According to the two points of the statistical characteristics to random noise the above, the construction of a new denoising method-EMD denoising method based on the noise statistical characteristics [4]-[6], the specific steps are as follows in **Figure 1**.

1) After the signal with noise is decomposed by EMD, the N of intrinsic mode components in IMF were got. the decomposed the last trend component is recorded as the first N IMF.

2) Record $n(t) = imf_1(t), x(t) = \sum_{i=2}^{N} imf_i(t), n_{cumulate}(t) = n(t)$

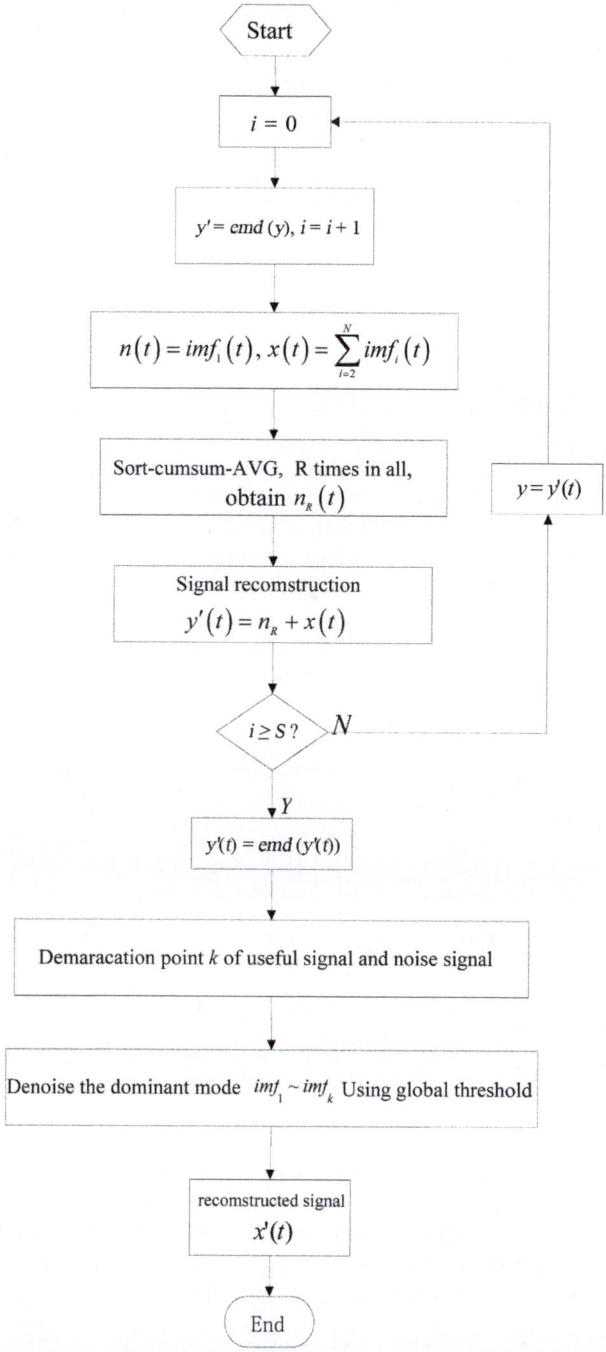


**Figure 1.** Algorithm flow chart.

3) A new component is obtained after the random ordering, and the cumulative,

$$recordn_{cumulate}(t) = n_{cumulate}(t) + n_1(t)$$

4) Repeat step 3 R times, a new power weakened noise dominant mode was got after calculated accumulation average. record $n_R(t) = \dfrac{n_{cumulate}(t)}{R+1}$

5) Get a new SNR signal with noise is improved from the reconstruction. Record $y'(t)=n_R(t)+x(t)$

6) Received signal after inhibition will be regarded as the noise repeated S times steps after 1 - 5.

7) The EMD decomposition, calculate the autocorrelation function of N imf component, according to the characteristics of the autocorrelation function obtained signal-to-noise leading the dividing point between modes K

(8) Global threshold processing method selection to the dominant modal components of noise record

$$imf_i'^{(t)} = \begin{cases} \operatorname{sgn}(imf_i(j)) \cdot (|imf_i(j)| - T_i), |imf_i(j)| > T_i \\ 0, |imf_i(j)| \le T_i \end{cases}$$

In the formula, $T_i = \sigma_i \sqrt{2 \ln L}$ as the components of the threshold, $L$ is the length of the signal, $\sigma_i$ is the standard deviation.

9) Obtain the signal after denoising, get the obtained and reconstructed.

### 3.4. Simulation Analysis

Signal denoising by using the statistical characteristic of noise in EMD based de-noising method respectively, which is obtained by the superposition of signal after Gauss white noise after the EMD decomposition, take the first IMF component is "random scheduling and cumulative average" R times, then with the other IMF component reconstruction, continue to repeat S times (), the noise power is weakened, improved SNR signal were got, as shown in **Figure 2**.

On the EMD decomposition of IMF, calculation of the various components of the autocorrelation function x corr (imf), in **Figure 3**. According to the random noise of the autocorrelation function to the statistical proper-

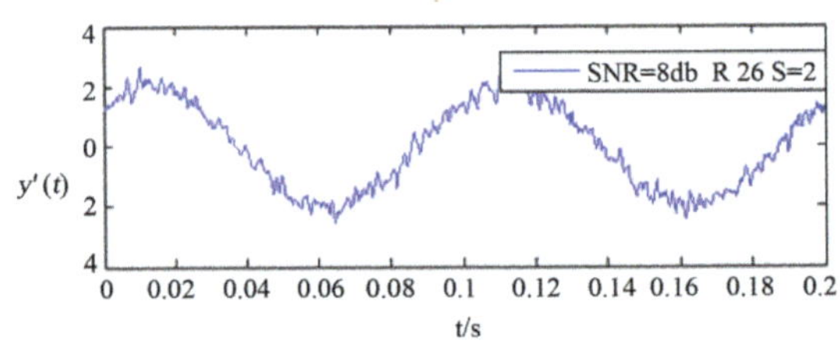


**Figure 2.** Signal to noise ratio improvement after SNR.

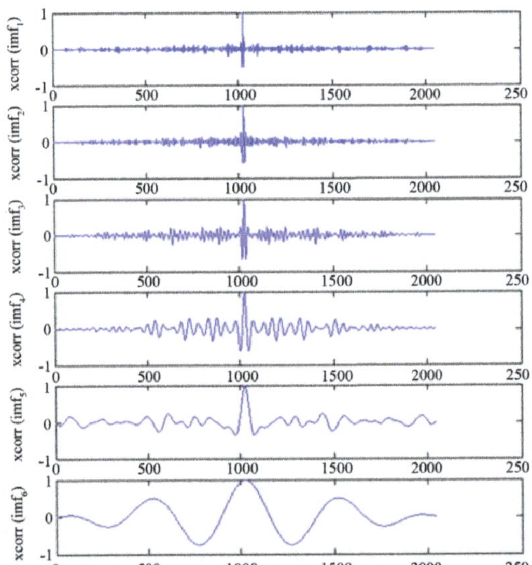


**Figure 3.** Each component of the normalized autocorrelation function.

ties, the signal-to-noise cutoff point can be determined. Therefore the global threshold selection method for denoising is obtained, after reconstruction, the signal after denoising was got, as shown in **Figure 4**.

## 4. Experimental Verification

In the experiment to test the car loaded with inertial measurement unit through a simulation of line section, collecting sensitive acceleration signal to trolley moving along X, Y, Z axis. First of all, using the method of mean filter DC quantity contained in the micro accelerometer signal. Using EMD de-noising method based on the statistical characteristic of noise, the acceleration signal is on denoising processing after eliminating direct flow. And the waveform and frequency spectrum of signals before and after respectively were got as shown in **Figures 5-7**. Above, denoising method based on noise statistical characteristics can remove random noise effectively from the acceleration signal.

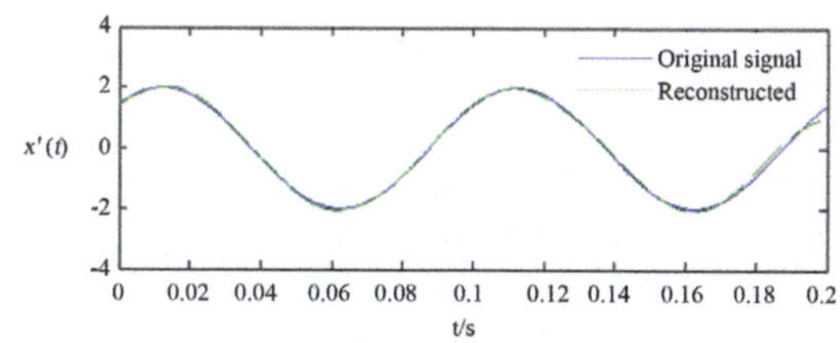


**Figure 4**. After denoising.

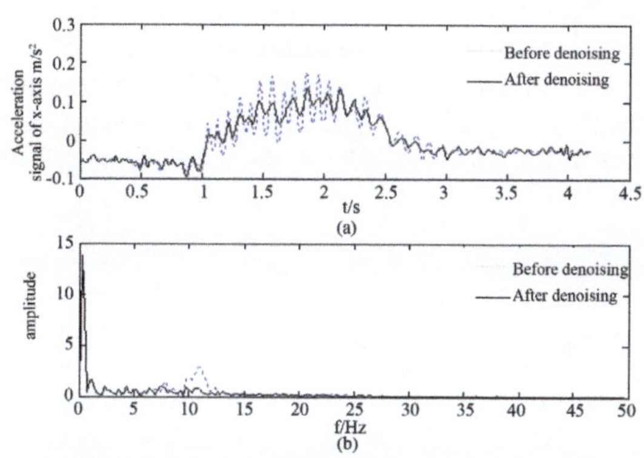


**Figure 5**. Waveform and frequency spectrum of X axis acceleration signal before and after denoising.

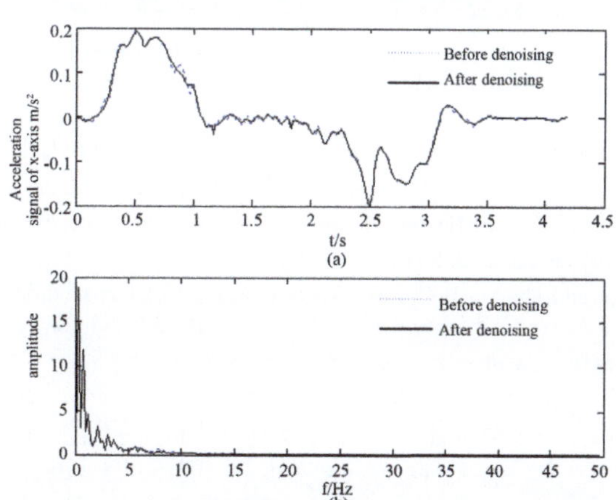


**Figure 6**. Waveform and frequency spectrum of Y axis acceleration signal before and after denoising.

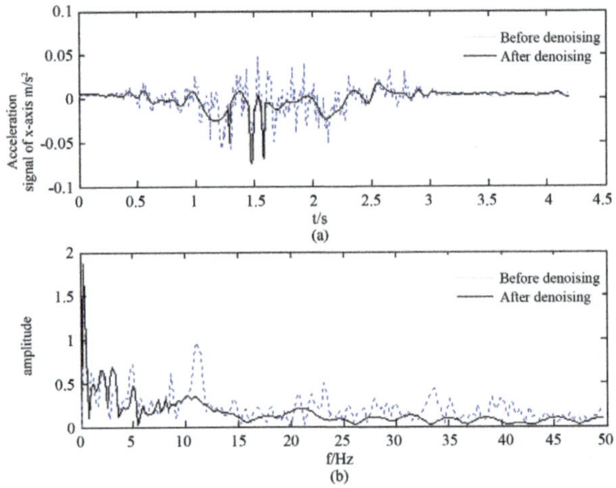


**Figure 7.** Waveform and frequency spectrum of Z axis acceleration signal before and after denoising.

## 5. Conclusion

This article is on the correlation function of studying the statistical properties of random noise. EMD filtering method is proposed based on the statistical characteristics. Using this method the acceleration signal exported an inertial measurement unit to deal with denoising. The acceleration signal denoised is integral operated. Through the attitude matrix the motion of the information is converted from the carrier coordinates system to the geographic coordinate system. The car trajectory can be got [7] [8]. Experiments show: EMD filtering method based on the statistical characteristics of the noise can remove the noise signal effectively. The requirements of detection accuracy can be met better.

## Acknowledgements

The project is jointly supported by the Shanghai Tertiary Education Specialized Fund for Planning to Support Young Teacher's Trainings (ZZGJD13072), National Natural Science Foundation of China (51478258), and the Shanghai Graduate Education Innovation Project in Layout and Construction Project (13sc002).

## References

[1] Zheng, S.B., Lin, J.H. and Lin, G.B. (2007) Implementation of Detecting Maglev Track Long Wave Irregularity Based on Inertial Measurement Principle. *Journal of Electronic Measurement and Instrument*, **21**, 61-65.

[2] Huang, N.E., Shen, Z., Long, S.R., Wu, M.C., Shih, H.H., Zheng, Q., *et al.* (1998) The Empirical Mode Decomposition and the Hilbert Spectrum for Nonlinear and Non-Stationary Time Series Analysis. *Proceedings of the Royal Society of London A*, **454**, 56-78.

[3] Rilling, G. and Flandfin, P. (2008) One or Two Frequencies? The Empirical Mode Decomposition Answers. *IEEE Transactions on Signal Processing*, **56**, 85-96. http://dx.doi.org/10.1109/TSP.2007.906771

[4] Boudraa, A.-O. and Cexus, J.-C. (2007) EMD-Based Signal Filtering. *IEEE Transactions on Instrumentation and Measurement*, **56**, 2196-2202. http://dx.doi.org/10.1109/TIM.2007.907967

[5] Sun, J.-B., Zhu, Y.-L., Liu, L.-Q. and Huang, J.-C. (2010) Study on De-Noising Method Based on EMD Used in Insulator Leakage Current. *Journal of North China Electric Power University*, **37**, 1-5, 22.

[6] Wang, Z. (2010) Research on EMD Algorithm and Its Application in Signal Denoising. *Harbin Engineering University*, 34-36.

[7] Zhu, W.F., Chai, X.D., Zheng, S.B., Li, L.M. and Luo, Y.J. (2012) Developing IMU Signal Acquisition and Processing System Based on LabVIEW. *Computer Measurement & Control*, **20**, 1697-1698, 1706.

[8] Liu, X.-C., Chai, X.D., Zheng, S.-B., Zhu, W.-F. and Jiang, Q.-Z. (2013) De-noising Method for Acceleration Signal Measured by Inertial Measurement Unit. *Instrument Technique and Sensor*, **10**, 97-99.

# 8

# Characteristics of Seismic Hazard in a V-Shaped Valley and Hazard Mitigation Methods

**Yan-Ling Qiu[1], Ling-Kan Yao[1,2*], Quan Yuan[3]**

[1]School of Civil Engineering, Southwest Jiaotong University, Chengdu, China
[2]Road and Railway Engineering Research Institute, Sichuan Key Laboratory of Aseismic Engineering and Technology, Chengdu, China
[3]Xingshu Company, Sichuan Provincial Transport Department, Chengdu, China
Email: *yaolk@home.swjtu.edu.cn

## Abstract

**The distribution of hazards triggered by strong earthquakes showed great differences with traditional experience that gained from normal condition or moderate earthquake. This paper researches the differences and discusses the effect on road in V-shaped valleys. The distribution map of slope failures triggered by Lushan earthquake along roads is worked out by detailed investigation. Abnormal areas of seismic hazards are found and marked in the map. Then, co-seismic hazards and post-seismic hazards are researched in detail along Yingxiu-Wolong section of S303 road. V-shaped valleys are abnormal areas in strong earthquakes, where the geological hazards can be extreme large. Rock avalanche is the major type of co-seismic hazard, and debris flow is prone to occur and is the most harmful post-seismic hazard. For new built roads, the line elevation is as high as it can be if the economy and technique are acceptable. In strong earthquake-stricken area, makeshift roads are adopted in damage trunk roads to ensure the traffic. Reconstruction of formal roads should not be started until the end of the active period. The line elevation with enough vertical clearance over the river should be adopted to prevent the post-seismic debris flow hazard.**

## Keywords

**Co-Seismic Hazards, Post-Seismic Hazards, V-Shaped Valley, Route Scheme**

## 1. Introduction

In rugged mountainous areas, rivers regions are the most significant topographic features, along which most

*Corresponding author.

human settlements are located. In order to ensure that these settlements have convenient and fast road transportation, and moreover, take full advantage of the natural gradient of riverbed to overcome elevations, roads in mountainous areas extend along riversides as much as possible. This is the basic design principle and common practice for route selection of transportation lines in mountainous areas [1]. However, river valleys are places where a variety of adverse geological phenomena develop. In order to minimize the impact of geological hazards, geological route selection techniques were developed [1]-[3]. Sites with rock masses of higher strength and less discontinuity are less disaster-prone from an engineering point of view. Specifically, intrusive igneous rocks formed under high temperature and pressure such as granite and diorite have very high mechanical strength, little discontinuity and quite good weathering resistance [4]. Roads built in these rock masses experience much less geological hazards, except high *in situ* stress conditions. In the past few years, several strong earthquakes have happened at the eastern margin of Tibetan Plateau, where stratigraphy and lithology vary greatly. Geological hazards triggered by strong earthquakes have shown great differences in various lithologies, which doesn't conform to traditional experience that gained from normal condition or moderate earthquake. This paper discusses this issue based on the actual situation of seismic hazards.

## 2. Abnormal Areas with Seismic Hazards

Valleys are formed by a process of erosion in the long geological history. Flowing water and its entrained sands cut down into the riverbed to carve out a valley. The forms of valleys can be divided as U-shaped or V-shaped in cross section. A V-shaped valley is a deep, narrow valley with very steep sides. In comparison with a V-shaped valley, the U-shaped valley floor is wider with developed floodplain or fluvial terrace. Most valleys belong to one of these two main types or their mixture [5]. The exact shape depends on the rate of uplift of rock, the type of rock, the type of climate, and the stage of the river. Ultimately, the characteristics of erosive activity determine the type of the valley shape. Valleys with V-shaped cross section are in the youth stage of river development process [6] [7], where erosion cuts downwards more than it does sideways. A valley with U-shaped cross section evolves from V-shaped valley, where erosion cuts sideways more than it does downwards. Along a valley, the cross section usually alternates between V-shaped and U-shaped, as the geological structure varies. Hard rock layers have strong anti-sideways erosion capacity, and they always are V-shaped valleys. Soft rock layers have weak resistance to erosion, so rock layers are worn away fast and U-shaped valleys are evolved into. Because of the lithology, river through hard rock areas such as magmatic, carbonate, breccia and metamorphic rocks, usually form V-shaped valleys.

Earthquakes can trigger all kinds of slope movements. Slope failures such as rock falls, landslides, flows and rock avalanches can occur during seismic shaking. The general distribution situation of these geological hazards triggered by earthquake attenuates with the increase of distance from an epicenter [8]. The scale of geological hazards triggered by earthquake has a close relationship with earthquake intensity. According to empirical data, no slope failure of any significance has been experienced for earthquake intensities <VIII [9]. Specifically, there is hardly any slope failure in intensity of VI, and there are no massive slope failures in intensity of VII. However, geomorphic unit of V-shaped valleys do not meet the above rules. A large number of geological hazards are triggered and serious damages are caused in the abnormal areas.

An earthquake with a magnitude of 7.0 occurred in Lushan, Sichuan, China on April 12, 2013. By these researchers, detailed investigation was carried out on slope failures triggered by the earthquake along national roads, provincial roads and main town roads, which amounted to 300 km in the earthquake-stricken area. The investigated roads and the distribution of slope failures with a volume of more than 100 $m^3$ are showed in **Figure 1**. Stratigraphy of outcrops along these roads is marked only strong hard rocks, which mainly distribute in three zones. One is the NE strike Upper Cretaceous-Paleocene Dasi rock group, the lithology is breccia. Three roads include provincial road S210, town roads of Lushan-Shuangshi and Longmen-Taiping cross through this zone. Another is Proterozoic intrusive igneous rocks, Permian and Devonian carbonate rocks, which are extensive outcrops in Lingguan-Fengtongzhai section of road S210. The last one is Proterozoic intrusive igneous rocks distributed in national road G318 and Labahe scenic road.

Two obvious abnormal areas of slope failures along roads appeared in Lushan earthquake. One is Labahe Nature Reserve. It is far from the epicenter, with an earthquake intensity of VI. 12 slope failures are triggered along 14 km Labahe scenic road, where the lithology is mainly granite and amphibolite metamorphosed from basalt and gabbro. The other one is in accordance with the second zone marked hard rocks, with an earthquake

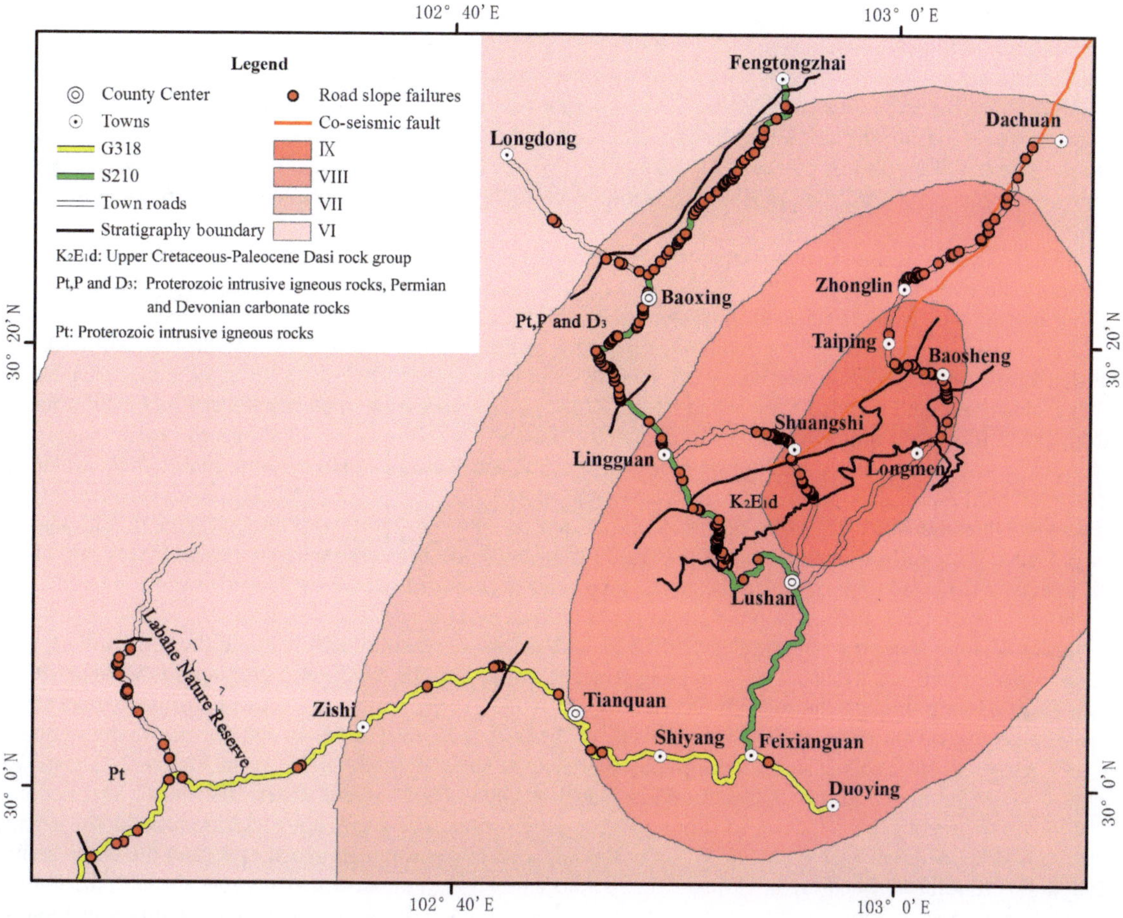


**Figure 1.** The distribution map of slope failures triggered by Ms7.0 Lushan earthquake along roads.

intensity of VII. Numerous slope failures densely distribute along S210. Slope failures in this area are in greater number and larger than that in the epicenter area with earthquake intensities of IX or VIII. Both of the two abnormal areas are tributary river valleys with V-shaped cross section. The two areas have similar characteristics in lithology and topography, which are strong hard rock lithology and V-shaped valleys topography. The trunk river valley has a longer history of evolution, and therefore has less slope failures.

In addition to the above, V-shaped valley sections of the three roads across Dasi rock group also have large scale seismic slope failures. A large number of seismic slope failures distribute along the line of Dachuan-Shuangshi, because they are zonal distribution along the co-seismic fault.

## 3. Co-Seismic Hazards and Related Road Damages in a V-shaped Valley

Where the earth crust uplifts significantly, the regional erosion and planation are active. As a result, the original rocks buried deeply are gradually approaching and exposing to the surface. In the regional vertical unloading process, once high *in suit* stress of the original rock is constantly decreasing. A series of near-horizontal unloading fissures are formed in the shallow part of the crust, by unloading rebound. These fissures make the intact rock mass disintegrate, and change the mechanical behavior similarly to layered rock mass [10]. The evolution of the valley is an advanced unloading process of rock mass in regional vertical unloading environment. During the process of valley incising and unloading, the balanced state of rock mass is broke and the stress field is adjusted and redistributed. Because of the stress relief, the slope rock mass takes place unloading rebound towards free-direction of valley and forms fissures [11].

The strong hard rock grade such as igneous, carbonate, breccia and metamorphic rocks has high strength. The mechanical response to forces is in a brittle, elastic way [12] [13]. If ultimate strength is reached, the compressive strength is drastically reduced and may even fall to values approaching zero. The brittle behavior implies an almost instantaneous loss of strength with little or no plastic deformation. Because the rock mass can subject to large loads, they can store high strain energy in geomorphic evolution [14]-[16]. During the seismic shaking, the increasing stress adds on the unloading fissures and the total stress reaches to or exceeds the strength limitation. Then the brittle failure occurs. Because of the momentary loading of seismic stress, unloading fissures storing certain strain energy in hard rock mass make brittle fracture to release energy rapidly, resulting in seismic hazards.

A V-shaped valley is in the youth stage of valley evolution process, which is a most active unloading process of geomorphic evolution [6] [7]. Due to earthquake, stress and strain accumulated in unloading fissures in the shallow part of the crust are released in a short time. It ahead and together releases the stress and strain that would release in a period of valley evolution time, and causes serious damages to the surface. Soft rock and semi-hard rock grades have ductile behavior or brittle-ductile behavior. They have lower ultimate strength and higher residual strength than hard rock grade does. With a small load, these rock masses can make deformation and breakage to release unloading stress in evolution process. Therefore, geological hazards easily happen on these rock masses in normal condition or moderate earthquake. But under strong earthquake, rock avalanche triggered in hard rock valley become the most serious hazards.

The Ms 8.0 Wenchuan earthquake occurred at 2:28 pm on May 12, 2008, with the epicenter in Yingxiu, Sichuan, China. Yingxiu-Wolong section of provincial road S303 locates in the epicenter area. It was one of the most seriously damaged trunk roads in the earthquake. Various kinds of geological hazards were triggered along the whole valleys, and road damage phenomena could be seen everywhere.

Yingxiu-Wolong section of provincial road S303 with a total of 45km length is along the left bank of Yuzi Stream. Yuzi Stream is a tributary of Mingjiang River, flowing into the trunk river at Yingxiu. Maoxian-Wenchuan Fault passing through the road in the vicinity of Gengda is the boundary between two geological structures. The geological structure on the northwest side of the fault is Xiaojin fold belt, and the one on the southeast side is Cathaysian structural system. Stratigraphy of outcrops along the road is shown in **Figure 2**. The downstream of Yuzi Stream is on the southeast of Maoxian-Wenchuan Fault, where widely distributes intermediate-mafic intrusive igneous rocks formed in Jinning-Chengjiang period. These igneous rocks are hard biotite granite and granodiorite. Silurian Maoxian rock group in the midstream is soft or semi-hard rock, and the other groups belonging to Xiaojin fold belt are semi-hard rock. Although valley in the downstream goes through the longest time of erosion for Yuzi Stream, the cross sections of the valley are deeper and narrower than that in Silurian

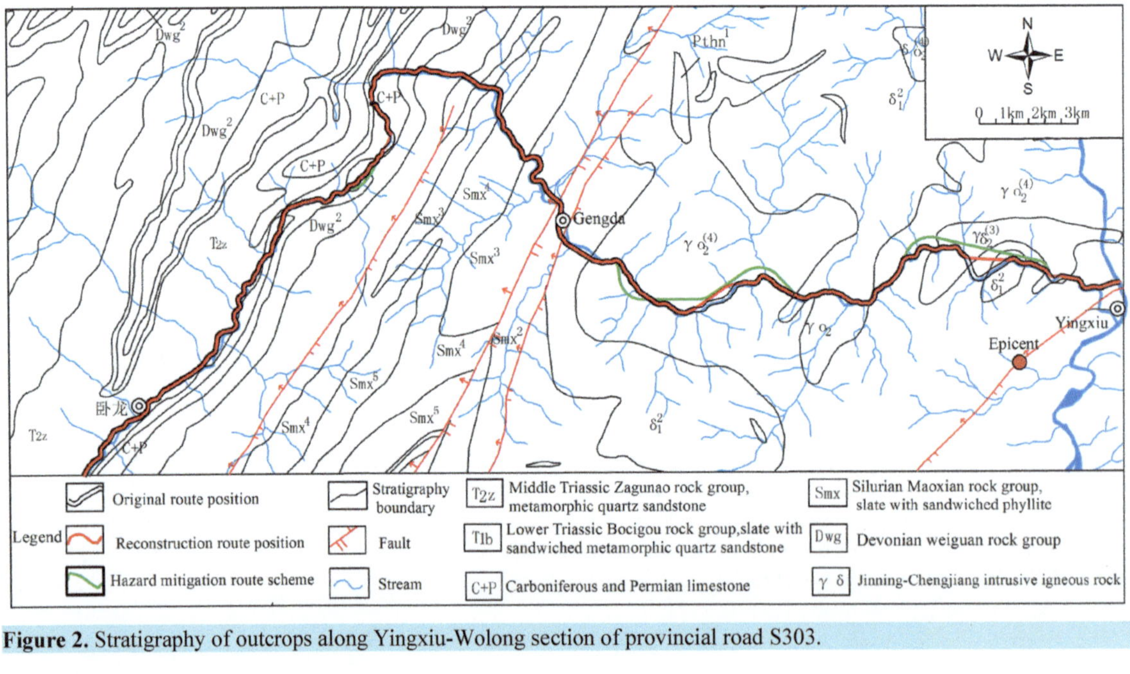


**Figure 2.** Stratigraphy of outcrops along Yingxiu-Wolong section of provincial road S303.

Group because of lithology. The valley in the downstream is still typical V-shaped with steep slopes and narrow floor, while the valley in Silurian Group has eroded into U-shaped with a wide floor.

Co-seismic geological hazards along Yuzi Stream were serious. Road S303 along the valley was seriously damaged. Especially in the downstream, where once vegetation-covered slopes widely exposed inner rocks, almost all roads were buried or severely damaged except Panlongshan tunnel and Gengda tunnel. The two sides of Maoxian-Wenchuan Fault provided a startling contrast to co-seismic hazards. On the southeast of the fault, co-seismic hazards were triggered on both sides of the valley continuously. While on the northwest, co-seismic hazards were triggered only on partial limited slopes separately. Co-seismic geological hazards damaged the road on the two sides are shown in **Table 1**. On the southeast of Maoxian-Wenchuan Fault, the largest scale and the most harmful co-seismic geological hazard was rock avalanche. It rapidly falls masses of rocks and debris that detach themselves from steep slopes. These rock blocks with different sizes deposited in the valley floor and formed quake lakes (**Figure 3**). Road damage by these rock avalanches was extremely huge. Embankments were buried and bridges were destroyed (**Figures 4-6**). 46 rock avalanche deposits buried over 10 km of a total 18 km length road. Moreover, throughout the road in the downstream region, countless rock falls of various sizes dashed down to the roadbed and made serious damages. Landslide is hardly occurred in igneous rocks zone. There was only one landslide on the upper slope. On the northwest, co-seismic geological hazards decreased substantially. The road damage was negligible. 26 deposits formed by debris avalanche buried about 3 km of a total 25 km length road. The three landslides were all old ones reactivated by seismic forces.

**Table 1.** Co-seismic geological hazards damaged Yingxiu-Wolong section of provincial road S303 on the two sides of Maoxian-Wenchuan Fault.

| | On the southeast of fault | On the northwest of fault |
|---|---|---|
| Length of the road (except tunnels) | 18 km | 25 km |
| Number of avalanche | 46 rock avalanches | 26 debris avalanches |
| Number of landslide | 1 rock landslide | 3 old landslides |
| Number of quake lake | 7 | 1 |
| Number of road washout | 4 | 2 |
| Number of debris flow | 9 gully-type and 2 slope-type | 2 gully-type and 4 slope-type |

**Figure 3.** A quake lake formed by rock avalanche deposits submerged a hydropower station.

**Figure 4.** Rock avalanche buried the embankments.

**Figure 5.** Rock avalanche destoryed the bridge.

**Figure 6.** Rock avalanche fell from the steep slopes continuously, destoryed the bridge and buried a large section of road.

## 4. Post-Seismic Hazards and Relating Road Damages in a V-shaped Valley

After an earthquake in mountainous areas, the commonest geological hazard is rock fall. It usually occurs in places where rock avalanches had once happened. This kind of post-seismic geological hazard always threatens pedestrians and vehicles in a long period. But the most harmful post-seismic hazard for road in V-shaped valley is debris flow triggered by rainstorm.

During the seismic shaking, the pre-existing discontinuity planes such as tectonic, bedding surface, tension cracks extend and rock masses disintegrate to massive blocks on the steep slope surface. These rock blocks provide a wealth of loose materials for the source of debris flows. The rainfall threshold for debris flow is lower than the original level prior to the earthquake [17]. These co-seismic hurt slopes can easily form massive debris flows in the rainstorms. The carrying capacity of a V-shaped valley is negligible because of small volume of narrow valley floor. The rock blocks can hardly be carried away and accumulate in the river, which form dammed lake, rise water lever, change the hydrodynamic conditions. All these hazards significantly threaten the road in the V-shaped valley.

The reconstruction of Yingxiu-Wolong section of road S303 began in May 2009. It was expected to be completed by the end of 2010, with a budget of 811.9 million yuan. The reconstruction project was mainly along the route before Wenchuan earthquake, but newly built 3 tunnels and 6 bridges. The original and reconstruction route are described in **Figure 2**. Until August 13, 2010, the main structures of reconstruction were completed except one tunnel.

A rainstorm every 130 years hit the area on August 13, 2010, with a rainfall of 160 mm in eight hours. The extreme rainstorm induced massive debris flows in most tributaries of Yuzi Stream. Along Yinxiu-Wulong section of road S303, 60 debris flows including 32 gully-type and 28 slope-type debris flows damaged the road. Though the scales of debris flows on the two sides of Maoxian-Wenchuan Fault were similar, the road sections on the southeast of fault were more seriously damaged: 24 gully-type and 14 slope-type debris flows. These debris flows deposited on the V-shaped valley floor and formed dammed lakes, which silting river and rising water level. The 3 gully-type debris flows-Xiaojiagou, Xiezigou, and Wasigou-completely clogged the river. The reconstruction project in the length of 6.5 km was directly buried or submerged. Moreover, large sections indirectly affected had to be abandon. The reconstruction project was totally destroyed.

The Xiaojiagou gully is very close to the portal of newly-built Nanhua Tunnel. An excessive large-scale debris flow of 0.5 million·m³ outbroke in the gully, accumulating together with the slope debris flow near tunnel portal. The deposits buried the entire tunnel portal and stuffed the whole tunnel. The river silted up and the riverbed uplifted 30 m high. The water level rose above the route elevation more than 20 m and reached the roof of the tunnel. More than 1km length of roadbed was buried and submerged. It was one of the most affected sections by debris flow hazard. **Figure 7** and **Figure 8** show the portal before and after the event. The Xiaojiagou

**Figure 7.** The portal of Nanhua tunnel before August 13, 2010.

**Figure 8.** The portal of Nanhua tunnel was buried by debris flow deposits after August 13, 2010.

gully is an old debris flow with a watershed area of 0.24 km², and there has a debris flow drainage canal before the earthquake. According to a survey, loose or non-consolidated materials for originating were estimated to be up to 3.57 million·m³ after Wenchuan earthquake. The huge material threatened the road continuously.

The Xiezigou was another excessive large-scale debris flow, with a deposit of 0.42 million·m³ at this time. The riverbed uplifted more than 10m. The dammed lake formed by the deposit submerged not only the road but also the headquarters of the reconstruction. The Xiezigou debris flow had happened once before this time. There is a potential gully debris flow named Xiangjiagou just 250 m away from Xiezigou. The two close gullies contain loose material with a total volume of 4.01 million·m³.

There were more debris flows on the one valley side such as Wasigou caused serious damage to the road. Moreover, some debris flows occurred on the other side flowed into the river and narrowed the water path. When water was through the narrow river cross-section, it reached to the road side and its velocity increased. As the result of the hydraulics changing, there were washouts along road embankment. Taking into consideration of the existing and potential hazards, route spaces had to shift up from the valley to mountain and raise line elevation substantially (show in **Figure 2**).

## 5. Suggestion for Hazard Mitigation Methods

Route selection in high earthquake intensity mountainous areas is a multi-objective decision-making process. It co-ordinates the contradiction between saving investment and reducing seismic hazard risk in the future. The concept of risk-control must be maintained consistently and systematically in the whole process of route selection. In V-shaped valleys, the location of road is influenced in the first instance by topography. Generally, the location makes use of the narrow zonal topography that is fluvial terrace or accumulation area on foot slope. As a result, the road line elevation is little higher than the water level. It is the project with the least cost, labor work, as well as the influence on geological environment. But such project usually has less hazard-resistant capacity. In order to increase the hazard-resistant capacity, the location of highway and railway avoids hazard-prone foot slope by location adjustment in the plane and elevation, by taking advantage of bridges and tunnels. Otherwise, corrective or stabilization measures are adopted to prevent hazards and mitigate damages. For new road, the scheme with higher line elevation is better than the other schemes, when the differences of economy and technique are in the acceptable range.

In strong earthquake-stricken area, it is not suitable for engineering construction in the active period of post-seismic hazards. The conclusion is learned from a profound lesson of the reconstruction project of S303. Transportation is the basic of the recovery after an earthquake. Makeshift roads are adopted in damage trunk roads to ensure the traffic. Reconstruction of formal roads shall be delayed until the end of the active period. Roads in

V-shaped valley need pay attention to post-seismic debris flow hazard, which may form dammed lake and rise the water level substantially. The design of the basic line elevation needs a series of study on the hazard: A survey on the amount of material source, a calculation of the deposit in recovery period, an estimate of the rising water level. Earthquake shaking generates a lot of fresh discontinuities and expands the pre-existing discontinuities in inner hard rock. These discontinuities expand in the weathering process, and finally disintegrate into blocks and fall down. Rock fall is the most common and the longest duration hazard after an earthquake. Structures stabilization or protection against rock fall shall be considered in design to mitigate the damage, and tunnels are proposed to avoid areas of rock fall.

Recovery after an earthquake is a long-term process. A comprehensive recovery includes physical recovery, economic recovery and life recovery. Among of them, physical recovery is the base of the other two [18] [19]. Natural environment deteriorates after the earthquake, especially in mountainous valleys [20] [21]. The initial years after the strong earthquake become a very active period of post-seismic hazards that large-scale geological hazards occur constantly, because of the massive loose materials. Research has shown that the active period may last 5 years in earthquake intensity of X and XI [22]. Vegetation effectively helps to prevent soil erosion, so vegetation coverage reflects this active period to some extent. Data of some events indicates that the vegetation recovery process needs several years or more [23]-[25]. Over time, the number and the activity of unstable slopes are decreasing. But it also needs enough time to recover to the original level, which is decade for landslides, and is more for debris flows in earthquake intensity of X and XI [22].

## 6. Summary

Geomorphic unit of V-shaped valleys are abnormal areas in strong earthquakes, where the geological hazards can be extreme larger than other places. Because of the storage of high strain energy and brittle behavior of the rock mass, co-seismic hazard of rock avalanche are triggered massively in V-shaped valleys. As a wealth of loose materials, rainstorms trigger massive debris flows. Because of the negligible carrying capacity, debris flows become the most harmful post-seismic hazard for road in V-shaped valleys. For new built roads, the line elevation is as high as it can be if the economy and technique are acceptable. For strong earthquake-stricken area, makeshift roads are adopted in damage trunk roads to ensure the traffic. Reconstruction of formal roads should not be started until the end of the active period. The line elevation with enough vertical clearance over the river should be adopted to prevent the post-seismic debris flow hazard.

## Acknowledgements

The study is supported by the Research and Development of Science and Technology Project of China Railway Corporation (Grant No. 2013G014-A and No. 2014G004-A).

## References

[1] He, Z.N. (2010) On Geological Location and Main Technical Principles for Railway. In: Zhu, Y., Ed., *Proceedings of Railway Location and Overall Design in Complex Mountainous Area*, China Railway Press, Beijing, 97-102.

[2] Konietzky, H., te Kamp, L., Hammer, H. and Niedermeyer, S. (2001) Numerical Modelling of *in Situ* Stress Conditions as an Aid in Route Selection for Rail Tunnels in Complex Geological Formations in South Germany. *Computers and Geotechnics*, **28**, 495-516. http://dx.doi.org/10.1016/S0266-352X(01)00009-X

[3] Podverbniy, V. and Filatov, E. (2012) Design of Protective Constructions on East Siberian Railway. *The 3rd International Symposium on Innovation and Sustainability of Modern Railway*, Nanchang, 20-21 September 2012, 206-215.

[4] Bell, F.G. (2007) Engineering Geology. 2nd Edition, Butterworth-Heinemann, Oxford.

[5] Pavlopoulos, K., Evelpidou, N. and Vassilopoulos, A. (2009) Mapping Geomorphological Environments. Springer-Verlag, Berlin. http://dx.doi.org/10.1007/978-3-642-01950-0

[6] Strahler, A.N. (1952) Hypsometric (Area-Altitude) Analysis of Erosional Topography. *Geological Society of America Bulletin*, **63**, 1117-1142. http://dx.doi.org/10.1130/0016-7606(1952)63[1117:HAAOET]2.0.CO;2

[7] Schumm, S.A. (1973) Geomorphic Thresholds and Complex Response of Drainage Systems. In: Morisawa, M., Ed., *Fluvial Geomorphology*, Publications of Geomorphology, State University of New York, Binghamton, 299-310.

[8] Cui, P., Chen, X.Q., Zhu, Y.Y., Su, F.H., Wei, F.Q., Han, Y.S., Liu, H.J. and Zhuang, J.Q. (2011) The Wenchuan Earthquake (May 12, 2008), Sichuan Province, China, and Resulting Geohazards. *Natural Hazards*, **56**, 19-36. http://dx.doi.org/10.1007/s11069-009-9392-1

[9] Vallejo, L.G. and Ferrer, M. (2011) Geological Engineering. CRC Press, London.

[10] Tao, M., Li, X.B. and Wu, C.Q. (2012) Characteristics of the Unloading Process of Rocks under High Initial Stress, *Computers and Geotechnics*, **45**, 83-92. http://dx.doi.org/10.1016/j.compgeo.2012.05.002

[11] Xia, M. and Ren, G.M. (2013) Characteristics and Mechanism of Concentrated Unloading in Bank Slope of Yangqu Hydropower Station. *Journal of the Geological Society of India*, **82**, 421-429. http://dx.doi.org/10.1007/s12594-013-0169-5

[12] Wawersik, W.R. and Fairhurst, C. (1970) A Study of Brittle Rock Fracture in Laboratory Compression Experiments. *International Journal of Rock Mechanics and Mining Sciences*, **7**, 561-575. http://dx.doi.org/10.1016/0148-9062(70)90007-0

[13] Miao, J.L., Jia, X.N. and Cheng, C. (2011) The Failure Characteristics of Granite under True Triaxial Unloading Condition. *Procedia Engineering*, **26**, 1620-1625. http://dx.doi.org/10.1016/j.proeng.2011.11.2346

[14] Yin, Z.Q., Li, X.B., Jin, J.F., He, X.Q. and Du, K. (2012) Failure Characteristics of High Stress Rock Induced by Impact Disturbance under Confining Pressure Unloading. *Transactions of Nonferrous Metals Society of China*, **22**, 175-184. http://dx.doi.org/10.1016/S1003-6326(11)61158-8

[15] Hua, A.Z. and You, M.Q. (2001) Rock Failure due to Energy Release during Unloading and Application to Underground Rock Burst Control. *Tunnelling and Underground Space Technology*, **16**, 241-246. http://dx.doi.org/10.1016/S0886-7798(01)00046-3

[16] Tao, M., Li, X.B. and Li, D.Y. (2013) Rock Failure Induced by Dynamic Unloading under 3D Stress State. *Theoretical and Applied Fracture Mechanics*, **65**, 47-54. http://dx.doi.org/10.1016/j.tafmec.2013.05.007

[17] Shieh, C.L., Chen, Y.S., Tsai, Y.J. and Wu, J.H. (2009) Variability in Rainfall Threshold for Debris Flow after the Chi-Chi Earthquake in Central Taiwan, China. *International Journal of Sediment Research*, **24**, 177-188. http://dx.doi.org/10.1016/S1001-6279(09)60025-1

[18] Hayashi, H. (2009) Long-Term Disaster Recovery Processes: Lessons Learned from the 1995 Kobe Earthquake. *International Conference on Earthquake Engineering, The First Anniversary of Wenchuan Earthquake*, Chengdu, 9-12 May 2009, 670-676.

[19] Frazier, A.E., Renschler, C.S. and Miles, S.B. (2013) Evaluating Post-Disaster Ecosystem Resilience Using MODIS GPP Data. *International Journal of Applied Earth Observation and Geoinformation*, **21**, 43-52. http://dx.doi.org/10.1016/j.jag.2012.07.019

[20] Wang, Z.Y., Cui, P. and Wang, R.Y. (2009) Mass Movements Triggered by the Wenchuan Earthquake and Management Strategies of Quake Lakes. *International Journal of River Basin Management*, **7**, 391-402. http://dx.doi.org/10.1080/15715124.2009.9635397

[21] Di, B.F., Zeng, H.G., Zhang, M.H., Ustin, S.L., Tang, Y., Wang, Z.Y., Chen, N.S. and Zhang, B. (2010) Quantifying the Spatial Distribution of Soil Mass Wasting Processes after the 2008 Earthquake in Wenchuan, China: A Case Study of the Longmenshan Area. *Remote Sensing of Environment*, **114**, 761-771. http://dx.doi.org/10.1016/j.rse.2009.11.011

[22] Cui, P., He, S.M., Yao, L.K., Wang, Z.Y. and Chen, X.Q. (2011) Formation Mechanism and Risk Control of Mountain Hazards Induced by Wenchuan Earthquake. Science Press, Beijing.

[23] Lin, W.T., Lin, C.Y. and Chou, W.C. (2006) Assessment of Vegetation Recovery and Soil Erosion at Landslides Caused by a Catastrophic Earthquake: A Case Study in Central Taiwan. *Ecological Engineering*, **28**, 79-89. http://dx.doi.org/10.1016/j.ecoleng.2006.04.005

[24] Lin, C.Y., Lo, H.M., Chou, W.C. and Lin, W.T. (2004) Vegetation Recovery Assessment at the Jou-Jou Mountain Landslide Area Caused by the 921 Earthquake in Central Taiwan. *Ecological Modelling*, **176**, 75-81. http://dx.doi.org/10.1016/j.ecolmodel.2003.12.037

[25] Lin, W.T., Chou, W.C. and Lin, C.Y. (2005) Vegetation Recovery Monitoring and Assessment at Landslides Caused by Earthquake in Central Taiwan. *Forest Ecology and Management*, **210**, 55-66. http://dx.doi.org/10.1016/j.foreco.2005.02.026

# 9

# Safety Culture, Policies and Practices in Nigerian Maritime Industry: The Exxon-Mobil Experience

**John Nsikan Efiok[1*], Opeoluwa Oluseye[2], Tommy Uduak[3], Raimi Olalekan[4]**

[1]Faculty of Business Administration, University of Uyo, Uyo, Nigeria
[2]Warsash Maritime Academy, Southampton Solent University, Southampton, UK
[3]Faculty of Management Sciences, University of Calabar, Calabar, Nigeria
[4]Department of Environmental Health Management, University of Uyo, Uyo, Nigeria
Email: *nsikiboy5000@yahoo.com

## Abstract

**Over the years, issues on safety culture and policies in maritime industry have received widespread research attention. Some maritime firms have even gone further to anchor their safety policies on learned corporate culture. However, whether these policies are fully embraced and complied with by seamen is a different kettle of fish altogether. This study therefore, examines safety culture, policies and practices in Nigerian maritime industry with a view to ascertaining: the extent of freight men' awareness and compliance to safety policies, identifying hindrances to conformity with safety policies and practices, and determining intervention options that management of Exxon-Mobil could use to foster safety compliance. The study adopts a survey approach, with structured questionnaire, and gathering primary data from seamen onboard. The findings reveal that workers are aware of the various safety policies and practices put in place by management, but adherence is impeded by several factors-inadequate staff strength, crew fatigue, and poor wages. Others include contract labour, inadequate safety training, complacency, and high risk tolerance. Surprisingly, improper handover and supervisory interference were not potent enough to hinder compliance with safety regulations at sea.**

## Keywords

**Safety Culture, Marine Vessel, Safety Policy and Strategy, Maritime Industry**

*Corresponding author.

## 1. Introduction

In the global maritime industry, ensuring safety of vessels and crew entails crafting and judiciously implementing sound safety policies and practices. However, the policies and practices necessary to achieve incident-free marine operation must be anchored on the culture of work safety prevailing in the firm at any given time. Culture is the acceptable social norms, values, and behaviour pattern of the people in a particular setting or work group, it is imperative for learning and act as a source of solving groups' problem. According to [1], culture is a series of social assumptions considered valuable to be taught to new members as a correct way to perceive, think and feel, in relation to group's problems. Thus, safety culture can be visualized as series of values, beliefs, customs, and norms about a phenomenon affecting the safety of work group. It is associated with the manner of behaviour, and appropriate attitudes in marine environment. Workers' perception of safety in the work place can be positive, negative or neutral, and when learned and appropriately internalized, safety culture creates a climate that influences how well people plan, communicate and make decisions concerning their health and safety on board.

Policy is a definite course or strategy selected by individuals, government agencies or private organizations from among alternatives to guide and usually determine present and future decisions [2]. Thus, policies exist to serve as a guide to appropriate action that may be needed to achieve expected results. In practical terms, safety policy consists of courses of actions and measures deliberately taken by shipping firms or its agents to direct the affairs of its people (all categories of seafarers) towards the minimization of vessel accidents and crew injuries [3]. Thus, once a safety policy is conceived to meet an end, its cycle has been kick-started. However, the extent by which marine safety policy can be adjudged to have failed or succeeded is among other factors dependent on the extent of awareness of sea workers whose support and compliance is sought in the first place, hindrances to compliance, and the safety culture interventions/initiatives of management particularly regarding compliant enforcement.

It may be argued that a company that upholds higher level safety cultures has the tendency to be committed to workers' and client safety as a strategic priority. The commitments so displayed also translate to actions and attitudes which are evident throughout the organization [4]. Apart from industry efforts to contend vessel and crew accidents, safety culture in maritime business have received empirical consideration in so many other ways. But there are clearly some issues which still need special attention. For instance, in a bid to prevent damage to vessels, cargo and crew injuries, management intervene by putting in place safety policies and regulations. However, crew awareness of applicable safety regulations, crew compliance behavior, and the factors that may make or mar compliance among others, have not been given thorough empirical examination and documentation. Though these issues are often expressed marine safety literature, they are rarely actually studied in depth.

When accident occurs, the extent of damage may range from less damage to vessel and cargo to entire damage to cargo, massive crew injuries and even death [5]. Accordingly, vessel safety policies and regulations as well as their enforcement seek to prevent and reduce the severity of vessel accidents. To that extent, vessel safety rules and regulations are enacted and enforced by signatory nation states. Policy and regulatory enactment have an acknowledged impact on safety performance, but excellence can only be achieved by factoring purposeful interventions and awareness into the organizational culture. Therefore, it is necessary to connect methodologies of safety management to the prevailing organizational culture.

In general, this study therefore examines firms' safety policies with emphasis on prevailing safety culture in maritime industry. Consequently, the study sought to achieve the following specific objectives. To:

1) Examine the extent of awareness of and compliance with safety culture policies and practices in Exxon-Mobil Company.

2) Identify the hindrances to conformity with safety policies and practices in Exxon-Mobil Company.

3) Assess the Initiatives/interventions by management of Exxon-Mobil to foster safety culture and policy compliance.

The following research questions are applicable:

1) To what extent are staff aware and comply with safety policies and practices in Exxon-Mobil Company?

2) What are the hindrances to conformity with safety policies and practices in Exxon-Mobil Company?

3) What intervention options exist for management of Exxon-Mobil to foster safety culture and policy compliance?

Apart from the introductory aspect, the remaining part of this report is structured as follows: Section two discusses theoretical issues in the safety literature, section three describes the methodological blueprint through which the study was carried out and documented, and in section four, we present results of data analysis and discuss the findings, drawing applicable implications where necessary. The study is brought to conclusion in section five

## 2. Theoretical Framework

### 2.1. Corporate Culture

According to [6], culture is "a complex framework of national, organizational, and professional attitudes and values within which groups and individuals function." Corporate culture is often referred to as the glue that holds an organization together by socializing workers in a way that increases commitment to the goals of the entity [7]. It embodies the philosophy of top management, which is translated into, and affects the behaviors of employees. Although some schools of thought focus on the role of executives, others note that middle management likely plays a substantial role as well, conveying the culture to front-line workers in any organization [8].

Culture is critical to understanding any society or group [9]. It can be viewed as the sum total of the attainments and activities of any specific period, race or people. Through the process of socialization, individuals align into groups while groups have the capacity to shape, influence, and determine members' perception, inputs, attitudes and indeed behaviours [10]. [11] asserts that organisational culture is the collection of values and norms that are shared by people and groups in an organisation and what control the way they interact with each other and with stakeholders outside the organisation. This therefore implies that organisational culture is the fundamental assumption that guide and determine how organization solves their problem. [12] are of the view that organisational culture is one of the factors responsible for corporate effectiveness. [13] identified organizational culture as the basis for the differences among Organisations. Culture provides stability to an organization and it gives employees a clear understanding of 'the way things are done around the organization [14]. Hence, it is very important that employees learn the existing culture of an organisation in order to achieve success.

### 2.2. Safety Culture

While an exact definition of a safety culture does not exist, a recurring theme in the literature is that organizations with effective safety cultures share a constant commitment to safety as a top-level priority, which permeates the entire organization. Accordingly, the following dimensions of safety culture are noticed in the literature 1) acknowledgment of the high risk, error-prone nature of an organization's activities, 2) blame-free environment where individuals are able to report errors or close calls without punishment, 3) expectation of collaboration across ranks to seek solutions to vulnerabilities, and 4) willingness on the part of the organization to direct resources to address safety concerns [4]. Based on extensive field work in multiple organizations, [13] have observed several common, cultural values in safety enhancing organizations: "interpersonal responsibility; person centeredness; [co-workers] helpful and supportive of one another; friendly, open sensitive personal relations; creativity; achieving goals, strong feelings of credibility; strong feelings of interpersonal trust; and resiliency[7]."

Researchers have not also reached a consensus on the dimensions that constitute a safety culture. Studies have however identified safety culture dimensions ranging from management to risk awareness [15] and attitudes and perceptions of the safety climate. A dissertation on predictors of work-related injuries in the maritime industry cited dimensions of a positive safety culture to include: "commitment by management and workforce, leadership style and communication, individual responsibility, management responsibility, risk awareness and risk-taking" [16].

According to [6], the aspect of organizational safety culture that may be visible or measurable is sometimes referred to as the safety "climate," which includes management systems, safety systems, and individual attitudes and perceptions. Maritime organizations are now adapting safety culture and climate surveys from other industries to benchmark and identify potential deficiencies in their unique safety culture. Such safety culture surveys take into consideration several input variables such as production pressures, risk tolerance, communication lapses, and safety consequences on morale and performance. In this direction, [7] noted that previous work in assessing organizational culture effects on total quality management, and organizational culture in high reliability organizations may also be pertinent to efforts to measure culture and its consequences for safety at sea.

## 3. Materials and Procedures

The study employed a survey design. This type of design allows generalization to be made from large population when representative samples are drawn. Survey design was used in this study for its economy, rapid data collection and ability to understand the characteristics of the population under study [17]. The fact that this study is quantitative in nature, and involves an attempt to describe the safety culture policies and practices in Exxon Mobil Nigeria made the use of descriptive survey design appropriate for this study. Exxon-Mobil is one of the largest oil producers in Nigeria. It operates a joint venture agreement with the Federal Government of Nigeria, through the Nigerian National Petroleum Corporation (NNPC).

The survey approach involved using structure but self-design questionnaire as the research instrument to gather data from the relatively large population of Exxon-Mobil questionnaire. Data collection was mainly from primary source using the questionnaire augmented with pieces of information from periodicals, journals and other internal materials from the firm. Items in the questionnaire were drawn from the extensive maritime safety literature examined throughout the study thus, ensuring high level of instrument reliability and scale validity.

The targeted population consisted of all seagoing employees including engineers working at Exxon –Mobil oil prospecting floating vessel in Nigeria, and simple random sampling technique was used to draw sixty two (62) respondents as sample size for this study. The respondents comprised forty two (42) junior level engineers, marine deck personnel and seamen, 6 shift supervisors, 5 top level managers, and 7 auxiliary contract employees. Among the 62 respondents, 12 either could not complete their questionnaire on schedule or were not accurate in answering the questions as required. Thus, the remaining 50 copies of the questionnaire were found useful for data analysis. Data was analyzed using frequency count and percentages supported with charts to enhance lucid analysis and interpretation. Data collected were processed through the use of statistical package for social science (SPSS) version 20.

## 4. Results and Discussion

### 4.1. Socio-Demographic Characteristics

In terms of socio-demographic characteristics of respondents, **Table 1** shows that 47 male and 3 female respondents participated in the study adding up to 50 respondents. From the 47 male respondents, 7 (14%) were between 20 - 30 years, 28 (56%) were between the ages of 31 - 40, 10 (20%) respondent was between age 41 - 50, and 2 (4%) were 51 - 60 years old. Similarly, Out of the 3 female participants in this study, 1 (2%) was between the age of 20 - 30 years, and were 2 (4%) were between the ages of 31 - 40. Other age groupings had no female respondent. This implies that the study was made up of more male than female respondents, and majority of respondents (both male and female) were between 30 - 40 years. The low percentage of female maritime workers in the company raise crucial concern as workplace safety issues has been shown to be amenable to gender sensitivity. Female workers are generally regarded to be temperate in matters involving safety at workplace [18].

**Table 2** shows cross tabulation of the respondents' educational qualification and length of work experience. It is clear from the Table that graduates with HND, B. Sc or B.A degree constituted the highest number (29) of respondents in the study while OND/NCE certificate holders were 14 and Master degree holders were seven participants. This is an indication of the fact that the study was carried out among highly educated groups of people who were expected to make meaningful contributions to solving the problem at hand. Similarly, 19 (38%) respondents had between 1 - 5 years working experience, 20 (40%) had between 6 - 10 years of work experience with the organization, and 11 (22%) respondents has work with the organization between 11 - 15 years. Thus, majority of respondents has a good number of years of working experience necessary to understand the subject

**Table 1.** Age of the respondents classified by their sex.

| | | | Age of respondent | | | | Total |
|---|---|---|---|---|---|---|---|
| | | | 20 - 30 yrs | 31 - 40 yrs | 41 - 50 yrs | 51 - 60 yrs | |
| Respondent's Sex | **Male** | Frequency | 7 | 28 | 10 | 2 | 47 |
| | | Percent | 14% | 56% | 20% | 4% | 100.0% |
| | **Female** | Frequency | 1 | 2 | 0 | 0 | 3 |
| | | Percent | 2% | 4% | 0.0% | 0.0% | 100.0% |
| **Total** | | Frequency | 8 | 30 | 10 | 6 | 50 |
| | | Percent | 16.0% | 60.0% | 20.0% | 12% | 100.0% |

Source: Author's compilation from field survey, 2015.

matter of the study and to contribute meaningfully.

**Table 3** shows distribution of respondents according to work position. Results indicate that 6% of respondents were ship managers, 12% were supervisors, 8% were crew engineers, and 6% were specialist/advisors. Similarly, 36% of participants in the survey were onboard operators/technician, 4% are mechanic, 6% were deck foreman on shift, 16% were crane operators and 10% were marine deck personnel working in the Exxon-Mobil floating vessel. Thus, many work position were considered while carrying out the study. As the study cut across different job specification and position, the credibility and generalization of findings is enhanced [17].

## 4.2. Extent of Awareness and Compliance with Safety Measures

Awareness and compliance to safety policies and practices is essential to marine safety performance. Consequently, we sought to find out the extent of workers awareness to safety policies and practices overtime in Exxon Mobil. Results are presented on **Table 4**.

It can be observed that most respondents were aware of the numerous safety policies and practices commonly in operation in the surveyed organization. For instance 32% majority of respondents were aware of the use of safety induction video to monitor safety related operations carried out by employees. In like manner, 16% were familiar hazard identification, and 12% were familiar with loss prevention. It is noted that the least familiar principles were those related to use of vessel safety manual, Vessel Trim and Stability Booklet, and use of personal Protective Equipment with 4.0% of respondents each claiming knowledge and familiarity. This implies that the extent to which marine employees adhere to safety regulations on board may depend on the level with which they are aware of existing safety culture policies and practices. Accordingly, management is encouraged to ensure increased awareness on some of the safety policies shown not to be popular with marine workers.

**Table 2.** Educational qualification of respondent classified by their length of marine work experience.

| | | | Respondents' years of experience | | | Total |
|---|---|---|---|---|---|---|
| | | | 1 - 5 yrs | 6 - 10 yrs | 11 - 15 yrs | |
| Respondent's Educational Qualification | OND/NCE | Frequency | 6 | 5 | 3 | 14 |
| | | Percent (%) | 43% | 36% | 21% | 100.0% |
| | HND/B.Sc/B.A | Frequency | 12 | 10 | 7 | 29 |
| | | Percent (%) | 41% | 35% | 24% | 100.0% |
| | M.Sc/MBA | Frequency | 1 | 5 | 1 | 7 |
| | | Percent (%) | 14% | 71% | 14% | 100.0% |
| **Total** | | Frequency | 19 | 20 | 11 | 50 |
| | | Percent (%) | 38% | 40% | 22% | 100.0% |

Source: Author's compilation from field survey, 2015.

**Table 3.** Work position of respondents

| **Job position** | **Frequency** | **Percentage** |
|---|---|---|
| Manager | 3 | 6 |
| Supervisor | 6 | 12 |
| Engineer | 4 | 8 |
| Specialist/Advisor | 3 | 6 |
| Operator/Technician | 18 | 36 |
| Mechanic | 2 | 4 |
| Deck foremen | 3 | 6 |
| Crane Operator | 8 | 16 |
| Marine Deck Personnel | 5 | 10 |
| Total | 50 | 100 |

Source: Author's compilation from field survey, 2015.

**Table 4.** Seafarers' awareness of safety culture policies and practices in exxon mobil.

| Dimensions | Frequency | Percent |
|---|---|---|
| Safety Induction Video | 16 | 32 |
| Hazard Identification (U-SEE, U-ACT) | 8 | 16 |
| Loss Prevention Observation (LPO) | 6 | 12 |
| Last Minute Risk Assessment | 5 | 10 |
| Job Intervention Programme | 3 | 6 |
| Step Back 5x5 | 3 | 6 |
| Job Safety Analysis | 3 | 6 |
| Vessel Safety Manual | 2 | 4 |
| Vessel Trim and Stability Booklet | 2 | 4 |
| Personal Protective Equipment | 2 | 4 |
| **Total** | **50** | **100** |

Source: Author's compilation from field survey, 2015.

### 4.3. Hindrances to Conformity with Safety Policies and Practices

Certain factors may conspire to hinder strict compliance to marine safety policies and regulations at sea. In an attempt to identify these ills with the view to suggesting means of addressing them, **Table 5** shows respondents opinion on the hindrances to conformity with safety policies and practices in Exxon-Mobil Company.

On **Table 5**, respondents give their opinions on the factors that have impeded strict adherence to safety regulations in Exxon Mobil. Among the factors identified, inadequate staff strength (30.0%) top respondents' suggestions. This is followed by crew fatigue (20%), and issues of poor wages (16%). Issues and problem relating to contract labour and inadequate safety training for marine workers occupied the fourth position with 8% each from respondents. It is also observed that the two factors; improper handover 8%, and under supervisory interference 8% occupied the bottom of the list, showing that these factor did not receive much consideration from respondents.

In addition, the findings may have suggested that improper handover and supervisory interference were not considered potent enough factors to significantly hinder strict adherence and compliance to safety regulations at sea. One important implication drawn from the findings is the strength of impediment of the top three factors (inadequate staff strength, crew fatigue, and poor wages). Inadequate staffing of vessel has been reported to cause fatigue, sleeplessness, and lack of concentration [10] [13], and has been associated with some of the major accidents in the maritime and other sectors [19]. Work fatigue is dangerous to sea men, the main cause being work overload and has been fingered in many shipwrecks causing injury to crews, damage to property and loss of vessels. Similarly, poor wages is tantamount to service exploitation; it dampens workers' morale and declines commitment. It is one of the causes of unsafe and dysfunctional behavior in the workplace and causes job dissatisfaction. A dissatisfied marine worker is a safety risk to himself, his coworker, and the organization as a whole.

### 4.4. Initiatives/Interventions to Foster Safety Culture and Policy Compliance

Results on **Table 6** show various interventions by the company under investigation to enforce awareness and compliance to safety standard. By the results, it can be stated that openness to communication (32.0%) was given highest priority by respondents. This espouses the need to adequately keep open various lines of formal and informal communication between workers and the employers in the organization. From the results, openness and communication is followed by feedback on safety errors (20%), support for safety implementation by management (12%), and support for the injured (12%). The importance of top management support in enhancing safety compliance can never be overemphasized. Management support is needed in terms of training on safety, safety culture learning, compensation for the injured, ensuring team building and intervention in unsafe acts and behavior. Management support and dedicated participation enhances employee's safety consciousness and lead to internalizing safety policies and programs.

**Table 5.** Hindrances to conformity with safety policies and practices in exxon mobil.

| Dimensions | Frequency | Percent |
|---|---|---|
| Inadequate staff strength | 15 | 30 |
| Crew fatigue | 10 | 20 |
| Poor wages | 8 | 16 |
| Contract staffing issues | 4 | 8 |
| Inadequate training | 4 | 8 |
| Non availability of spares and work materials | 3 | 6 |
| Complacency | 2 | 4 |
| High risk tolerance | 2 | 4 |
| Improper handover | 1 | 2 |
| Un due interference from superior | 1 | 2 |
| **Total** | **50** | **100** |

Source: Author's compilation from Field Survey, 2015.

**Table 6.** Initiatives/interventions to foster safety culture and policy compliance.

| Dimensions | Frequency | Percent |
|---|---|---|
| Openness and communication | 16 | 32 |
| Feedback and communication about safety errors. | 10 | 20 |
| Management support for safety implementation | 6 | 12 |
| Management support for the injured | 6 | 12 |
| Culture of learning/improvement | 4 | 8 |
| Crew training and staffing | 4 | 8 |
| Teamwork within units | 2 | 4 |
| Intervention in unsafe acts | 2 | 4 |
| **Total** | **50** | **100** |

Source: Author's compilation from field survey, 2015.

## 5. Conclusions

The main thrust of this report was to examine safety culture policies and practices in maritime industry in Nigeria. The focus was Exxon-Mobil incorporated, an international offshore oil prospecting firm based in Nigerian coastal region. To give direction to the study, three specific objectives were considered: Examine the extent of awareness of and compliance with safety culture policies and practices in Exxon-Mobil Company; identify the hindrances to conformity with safety policies and practices in Exxon-Mobil Company; and access the Initiatives/interventions by management of Exxon-Mobil to foster safety culture and policy compliance.

A lot of interesting findings arises from the analysis of data to aid the conclusion of this study. To start with, the study has established that employee awareness is necessary for successful implementation of marine safety regulations. Our study has also found that workers are aware of the various safety culture policies and practices put in place by maritime practitioners in the company. But adherence to these policies is impeded by several factors key among which are inadequate staff strength, crew fatigue, and poor wages. Other factors include contract labour and inadequate safety training for marine workers, complacency, and high risk tolerance. On the other hand, improper handover and supervisory interference were not potent enough to hinder compliance with safety regulations at sea.

This study made significant contributions to marine safety management literature by identifying various intervention and initiative managers can bring on board to ensure safety compliance and robust safety practice. For instance, it has been found that adequate lines of formal and informal communication between workers and the employers in the organization is vital to enhancing safety, reducing marine accidents and minimizing injuries and dead at sea.

In addition, from the study we conclude that maintaining a system of feedback on safety errors and support for safety implementation by management leads to drastic reduction of marine incident. Furthermore, importance of management support for the injured has been shown to enhance safety compliance, boost morale, and increase satisfaction with the job. Thus, to ensure strict compliance and reduce safety risk in global maritime industry, management support is needed in terms of training on safety, safety culture learning, compensation for the injured, ensuring team building and intervention in unsafe acts and behavior.

## References

[1] Schein, E.H. (2004) Organizational Culture and Leadership. 3rd Edition, Jossey-Bass, San Francisco.

[2] Dess, M., Lumpkin, K. and Eisner, B. (2000) Business Strategy and Performance in Malaysian SMEs: A Recent Survey. *Malaysian Management Review*, **24**, 1-10.

[3] Horck, J. (2004) An Analysis of Decision-Making Processes in Multicultural Maritime Scenarios. *Maritime Policy Management*, **31**, 15-29. http://dx.doi.org/10.1080/0308883031000164202l

[4] Frank, W.L. (2005) Essential Elements of a Sound Safety Culture. Process Plant Safety Symposium, Atlanta.

[5] Bloor, M. and Sampson, H. (2009) Regulatory Enforcement of Labor Standards in an Outsourcing Globalized Industry: The Case of the Shipping Industry. *Work Employment and Society*, **23**, 711-726. http://dx.doi.org/10.1177/0950017009344915

[6] Kanev, D. (2005) Seaborne Trade Effects of International Terrorism and Effectiveness of the Anti-Terrorist Policy. *Accident Analysis & Prevention*, **40**, 594-601.

[7] Håvold, J. (2007) National Cultures and Safety Orientation: A Study of Seafarers Working for Norwegian Shipping Companies. *Work & Stress*, **21**, 173-195. http://dx.doi.org/10.1080/02678370701424594

[8] Kohn, J.P. and Mark, A.F. (1996) Fundamental of Occupational Safety and Health. Rockville.

[9] Mansell, J. (2009) Choice of Flag State. Flag State Responsibility. *Journal of Safety Research*, **37**, 401-411. http://dx.doi.org/10.1007/978-3-540-92933-8_7

[10] Barsan, E., Surugiu, F. and Dragomir, C. (2012) Factors of Human Resources Competitiveness in Maritime Transport. *International Journal on Marine Navigation and Safety of Sea Transportation*, **6**, 126-130.

[11] Hänninen, M. (2008) Analysis of Human and Organizational Factors in Marine Traffic Risk Modeling (Literature Review). Helsinki University of Technology, Helsinki.

[12] Magramo, M. and Gellada, L. (2009) A Noble Profession Called Seafaring: The Making of an Officer. *International Journal on Marine Navigation and Safety of Sea Transportation*, **23**, 124-154. http://dx.doi.org/10.1201/9780203869345.ch113

[13] Bloor, M., Datta, R. Gilinsky, Y. and Horlick-Jones, T. (2006) Unicorn among the Cedars: On the Possibility of Effective Smart Regulation of the Globalised Shipping Industry. *Social and Legal Studies*, **15**, 537-554. http://dx.doi.org/10.1177/0964663906069546

[14] Baylon, A. and Santos, V. (2011) The Challenges in Philippine Maritime Education and Training. *International Journal of Innovative Interdisciplinary Research*, **1**, 349-356.

[15] Lane, T. (1997) Globalization, Deregulation and Crew Competence in World Shipping in McConville, James (Edit). Transport Regulation Matters, London.

[16] Ding, J. and Liang, G. (2005) The Choices of Employing Seafarers for the National Ship-Owners in Taiwan: An Empirical Study. *Maritime Policy & Management*, **32**, 123-137. http://dx.doi.org/10.1080/09523980500062643

[17] Sekaran, U. (2001) Research Methods for Business: A Skills Building Approach. 2nd Edition, John Wiley and Sons, Inc., New York.

[18] Galam, R. (2011) Navigating Lives: The Spatio temporality of the Gender Identity, Agency and Subjectivity of Filipino Seamen's Wives. Ph.D. Thesis, Cardiff University, Cardiff.

[19] Benton, G. (2005) Multicultural Crews and the Culture of Globalization. *The International Association of Maritime Universities (IAMU) 6th Annual General Assembly and Conference*, Malmö.

# 10

# Selection of Suitable Safety Activities and Experience of Safety Condition Establishment

**Akihiko Hidaka, Atsumi Miyake**

Graduate School of Environment and Information Sciences, Yokohama National University, Yokohama, Japan
Email: akihiko_hidaka@plamatels.co.jp

## Abstract

**In recent years, Japanese manufacturing firms have been building factories in the developing countries of Southeast Asia and employing local personnel. Using local employees who were basically people of the agricultural countries implies difficulties in both technical and safety terms. This study examines how quickly and effectively the companies going into the developing countries establish a standard of safety. The authors focus on Vietnam and Thailand as representative countries. To their surprise, the authors find that all 27 companies that answered their questionnaire had introduced "5S activities" at startup and were continuing to use them. In addition, nearly 80% of the companies began using "KY activities", "Safety proposal activities", and "Examinations of accidents" soon after startup. One of the authors Akihiko Hidaka, when serving as the president of the Thai subsidiary of a Japanese chemical company, developed safety activities to reduce work-related accidents. And he knew that the kinds of safety activities he developed were same as above 4 activities. So he evaluated these activities and judged them to be effective. The authors conclude that "5S activities", "KY activities", "Safety proposal activities", and "Examinations of accidents" were the activities best able to improve safety effectively and quickly in developing countries. The authors describe the safety activities carried out at the Thai company and then describe the most important features of the methods used to establish safety conditions at the firm.**

## Keywords

**Safety Activity, Developing Countries, East Asian Country, 5S Activity, KY Activity, Safety Proposal Activity, Examinations of Accidents**

## 1. Introduction

For 30 years, Akihiko Hidaka worked at a major Japanese chemical company, focusing mainly on the manufacturing and research and development aspects of synthetic resin. Many changes in technology and business practices occurred during that period. In the latter years, many Japanese companies relocated outside the country, for two main reasons. First, both labor and construction costs were lower elsewhere in East Asia; second, relocation increased the size of the market and potential consumers at a time when the demand for synthetic resin in Japan was diminishing. Hidaka's company participated in this shift in manufacturing, and the author spent the second half of his professional career as the president of its Thai subsidiary. It had 150 employees and focused on the coloring and compounding of resin as well as marketing it to consumers in Southeast Asia. As the firm's only Japanese technical personnel, the author's duties ranged from overseeing production, technical operations, and quality control of products, facility maintenance, security management, hygiene supervision, and accident prevention.

Despite relocating to developing country, all manufacturing firms [1] [2] should maintain safety. This paper examines how to establish a safety system or standard by which safety conditions can be ensured quickly and effectively.

This paper seeks to select the most effective and suitable safety activity by investigating the activities Japanese-affiliated companies operating in developing countries adopted when first established, using Thailand and Vietnam as representative nations. A questionnaire survey on safety was conducted on the managers of Japanese subsidiary companies located in these two developing countries. Using the survey results and the author's Thai experience, the authors selected the safety activities required in the factories of developing countries as well as the most effective ways of introducing them and some other key points about them.

## 2. Questionnaire for Japanese Companies in Thailand and Vietnam

### 2.1. Contents of the Questionnaire

1) What is the safety level of your parent company?

5 4 3 2 1 (5: Superior 3: Industry average 1: Too low)

2) What is the safety level of your company located in Southeast Asia?

5 4 3 2 1 (5: Superior 3: Industry average 1: Too low)

3) What safety activity has your company adopted? If your company is using several activities, please check all of them.

| | | |
|---|---|---|
| a) 5S activity (including 3S and 6S) | **※1)** Adopted | Not adopted |
| b) KY (including KYT) | **※2)** Adopted | Not adopted |
| c) Shisa Kosyou (point and call) | Adopted | Not adopted |
| d) HHK (Hiyari-Hatto Teian) | **※3)** Adopted | Not adopted |
| e) Safety proposal activity | Adopted | Not adopted |
| f) Examination of accidents | Adopted | Not adopted |
| g) OHSMS | Adopted | Not adopted |
| h) An activity other than the above a) - g) | Adopted | Not adopted |

If you have chosen h), please fill out the following columns:

※1) "Arrangement" "Put into order" "Cleaning" "Sanitary" "Discipline"

※2) Hazard prediction

※3) A proposal activity: fill in what you felt a shiver or anxious about

4) Please indicate when the activities identified above began, in chronological order.

(Please indicate using the item numbers above, starting with the earliest)

( )→( )→( )→( )→( )→( )→( )→( )

5) Please indicate if there are ongoing activities to improve safety at your factory.

### 2.2. Results of the Questionnaire

1) The evaluation results for the safety levels of the each parent company of subsidiaries in Thailand or Vietnam (**Table 1**).

2) The evaluation results for the safety levels of the subsidiary companies in Thailand or Vietnam (**Table 2**).

The evaluation results for the safety levels of the each parent company of subsidiaries in Thailand or Vietnam appear in **Table 1**. For both Thailand and Vietnam, the average value is nearly 4. In the result for his own subsidiary company, seen in **Table 2**, the average value is nearly 3. Thus, the management of a Japanese company

**Table 1.** The safety levels of the each parent company (the number of the company).

| Level of Safety | Rank 5 | Rank 4 | Rank 3 | Rank 2 | Rank 1 |
|---|---|---|---|---|---|
| Thailand | 2 | 8 | 4 | 0 | 0 |
| Vietnam | 4 | 4 | 5 | 0 | 0 |

Rank 5: superior, rank 3: industry average, rank 1: too low.

**Table 2.** The safety level of each subsidiary company (the number of the company).

| Level of Safety | Rank 5 | Rank 4 | Rank 3 | Rank 2 | Rank 1 |
|---|---|---|---|---|---|
| Thailand | 0 | 3 | 10 | 1 | 0 |
| Vietnam | 3 | 3 | 3 | 4 | 0 |

Rank 5: superior, rank 3: industry average, rank 1: too low.

located in developing countries striving to improve safety judged that the safety level of his own company was one degree lower than that of his parent. However, the new subsidiary had also managed to improve safety to within only one degree of the parent, which had a longer history and consisted entirely of Japanese employees. The author was proud of this achievement.

3) Safety activities adopted by companies in developing countries.

**Table 3** summarizes the responses given by 14 Thai and 13 Vietnamese companies.

This surprising result shows that all of the 27 Thai and Vietnamese companies that answered the questionnaire adopted "5S activities" [3] [4]. The fact that the Japanese companies adopted "5S activities" was very significant, as are the directionalities. Many companies also adopted "Safety proposal activity" and "Examinations of accidents". The former was adopted by all 14 Thai companies and nine Vietnamese companies, a total of 85% of all firms. "Examinations of accidents" was adopted by all 14 Thai companies and nine Vietnamese firms, again a total of 85%. Next was "KY activities" [5] [6], performed by 11 Thai companies and nine Vietnamese companies, for a total of 74%.

4) The chronological order of safety activity adoption.

Next, the authors examined the safety activities' order of adoption. Native workers were hired when the subsidiaries were established. They were trained first in plant operations and then in safety awareness. Safety activities were adopted in a meaningful sequence. The top managers of newly established companies conduct safety work according to an order that maximizes the profits allowable within their limited time span. They must select the activities that are essential for the native workers to know and that will improve their safety awareness. It is thus important to know the order in which the activities were adopted. The results allow us to plot the order in which safety standards were established.

a) The chronological order of safety activity adoption in the Thai companies.

The chronological order of safety activity adoption in the Thai companies is shown in **Table 4**. The numbers corresponding to the safety activities are as follows:

1. 5S activities
2. KY activities
3. Point and call
4. HHK
5. Safety proposal activity
6. Examinations of accidents
7. OHSMS
8. Another activity

As **Table 4** shows, one company had no records, and two began all their activities at once. Eleven companies provided the order in which their safety activities began.

Surprisingly, nine companies began with "5S activities". Activities 2, 6, and 5 stand out within third of each company.

b) The chronological order of safety activity adoption in the Vietnamese companies.

Eight Vietnamese companies responded about the order of their activities, shown in **Table 5**. All responded that "5S activities" came first, as with the Thai companies. The importance of "5S activity" thus needs to be understood.

**Table 3.** Safety activity adopted by Japanese companies in Thailand and Vietnam.

| Safety activity | Thailand | Vietnam | Total | Adoption % |
|---|---|---|---|---|
| 1. 5S activity | 14 | 13 | 27 | 100% **※1)** |
| 2. KY activity | 11 | 9 | 20 | 74% |
| 3. Point and call | 6 | 3 | 9 | 33% |
| 4. HHK | 6 | 7 | 13 | 48% |
| 5. Safety proposal activity | 14 | 9 | 23 | 85% |
| 6. Examinations of accidents | 13 | 10 | 23 | 85% |
| 7. OHSMS | 8 | 1 | 9 | 33% |
| 8. Another activity | 5 | 6 | 11 | 41% |

**※1)** The adoption percentage is based on the total of the two countries.

**Table 4.** The chronological order of safety activity adoption in the Thai companies.

| Company name | Order of adoption (early ← → late) |
|---|---|
| A | No record; can't answer (now ongoing: 1, 2, 3, 4, 5, 6, 7) |
| B | All of them started at once 18 years ago (now ongoing: 1, 2, 5, 6, 7) |
| C | 7→4→6→1→5→2→3→8 |
| D | 1→5→6→8 |
| E | All of them started at once 20 years ago (now ongoing: 1, 2, 5, 6, 7) |
| F | 1→2→3→5→4→6 |
| G | 1→6→6 |
| H | 1→6→2→7→5 |
| I | 1→6→5→7 |
| J | 1→5→6→2 |
| K | 1→2→6→4→5→3 |
| L | 1→2→6→5 |
| M | 2→3→4→5→6→1→7→8 |
| N | 1→2→5→6→8 |

**Table 5.** The chronological order of safety activity adoption in the Vietnamese companies.

| Company name | The order of adoption (early ← → late) |
|---|---|
| O | 1→5→6 |
| P | 1→8→6→5→2→4 |
| Q | 1→2→4→8→6 |
| R | All of them started at once one year ago (now ongoing: 1, 2, 4, 6, 8) |
| S | 1→5→4→2→6 |
| T | All of them started at once one year ago (now ongoing: 1, 2, 3, 4, 5, 6) |
| U | All of them started at once one year ago (now ongoing: 1, 2, 3, 5) |
| V | 1→2→3 |
| W | All of them started at once one year ago (now ongoing: 1, 6) |
| X | 1→2→5→4→6 |
| Y | 1, 5→6→2, 4, 8→3→8 |
| Z | 1→5→2 |

1) One company didn't describe the order of their activities.

## 3. Selection of Suitable Safety Activities for New Companies in Developing Countries

The questionnaire results indicate that most of the Thai and Vietnamese companies adopted "5S activities" when first established and then introduced "Examinations of accidents", "Safety proposal activity", and "KY activities" one after the other. This result reflects Hidaka's management work in Thailand, during which he improved his firm's safety level. The activities he used and their order are nearly as they appear in the survey results.

The questionnaire asked "Please indicate if there are ongoing activities to improve safety at your factory".

Two responses are described below.

1) A Thai company reported that almost no discipline or morale is provided by the parent, which follows something close to a noninterference policy.

2) A Vietnamese company reported that they lack safety awareness. They were trained but did not understand the training. They do their work their own way, which is very dangerous by Japanese standards.

These examples show that it is very difficult for native workers to operate in safe conditions when using the newest equipment and machines at newly established companies. As native employees are from agricultural countries, they are engaging in industrial work for the first time. It is thus important for newly established factories in developing countries to create a system for establishing safe conditions quickly and effectively through suitable safety activities. This study intends to select the most suitable activity.

The most important priority is "5S activities". The new employees of newly established factories have little safety awareness, as they have had no need of one before being hired. They are also undisciplined and do not obey the rules, the most basic requirement of safety, in the authors' view. The basis of the safety culture [7] [8] is this "Obey the rules", which requires discipline. The top managers [9] [10] of developing-nation firms who responded to the questionnaire recognized this lack of discipline and considered it fatal to the establishment of new companies. First, therefore, it was important to begin "5S activities" early in order to educate and discipline and then to build up the organization until its employees were ready to "Obey the rules". "5S Activities" have five behavior standards: "*Seiri*", "*Seiton*", "*Seiketsu*", "*Seisou*", and "*Shitsuke*". As they all begin with "S" we call them the "5S activities". The words mean "Arrangement", "Put into order", "Cleaning" "Sanitary", and "Discipline", respectively. These standards ask all employees to keep the factory and office clean after examining what needs to be done—for example, around a machine—to achieve that goal and understanding its importance. Through this behavior, the worker acquires discipline. Although this activity doesn't constitute direct action to improve or maintain safety, it teaches the workers about a basic feature of safety awareness; it is a shortcut to safety awareness rather than a practical path to safe behavior.

It is then important to proceed to "Examinations of accidents", "Safety proposal activity", and "KY activities". The safety lessons begun through the "5S activities" can instill a certain degree of discipline, leading to "obeying the rule" and an awareness of safety. Therefore, there is a need for safety education. Employees in the manufacturing industry need to know that they can be injured using machines or through equipment failure. Learning about the reasons for industrial accidents and injuries will help factory workers gain the safety knowledge they need to prevent them. "Examinations of accidents" corresponds to this process [11]. "KY activities (*i.e.* hazard prediction)" also provide safety education. In this process, employees learn that there is a risk at any location and with any process and then study how to predict hazards and thus prevent accidents. They come to understand that they can eliminate risks through KY activities; they are safety improvement measures based on knowledge. "Safety proposal activity" also improves safety by identifying safety failures, which employees can do after having obtained safety awareness and education. "Safety proposal activity" thus enhances employees' safety environment.

Chemical firms in developing countries must thus foster, first, safety awareness, then safety education, and third, a safety environment in order to improve its safety level rapidly and effectively. "5S activities", "Examinations of accidents", "KY activities", and "Safety proposal activity" must thus all be implemented.

## 4. Safety Process Establishment Experiences in East Asian Countries

The experience at a Thai chemical company will now be described. As mentioned, the survey results reflect Hidaka's management experience at the company, which managed to improve its safety level. The firm adopted the activities described in the survey, in nearly the same order. First, Hidaka's Thai firm will be described.

## 4.1. Company Profile and Safety Risk Concerning Factory Work

1) Company profile

a) The company performed plastic coloring and compounding.

b) The factory mainly produced engineering and multipurpose plastics.

c) The working framework consisted of three groups in two shifts on a rotation basis.

d) The basic raw materials used in the factory were petrochemical resins and additives. The resins were flammable, and the workers handled objects that were as dangerous as the base materials. Thus, there was a high risk of fire. There was also a risk of fine powder explosion because the factory treated the resin powders. The factory used heavy machines (such as the extruder and the tumbler), rotating equipment, and high temperature facilities. Workers risked accidents using this equipment (such as being pinched, trapped, burned, cut, or knocked to the ground).

2) Level of the employees

a) The management, or upper staff, level consisted of university graduates.

b) The factory workers were mostly junior or senior high school graduates.

c) The factory had no maintenance specialists with expertise to repair the equipment or facilities. Thus, all the employees required extensive training.

3) Safety target and method at the Thai factory

a) Target

The target was to bring the employees' safety awareness up to par with that of the employees at the Japanese parent company.

b) Method

The method of improving safety awareness was an orderly, step-by-step process of improving the most important aspect of safety first and then proceeding in order of importance.

## 4.2. Safety Activities

**1) The first activity is 5S**

All of the employees were trained to understand and use the 5S ideas and to thoroughly apply them to all factory processes, from buying basic materials and production of the plastics to delivery of the product to customers. Implementing the 5S activities changed the factory, making it bright and easy to move around in, where problems were easily found and accidents, such as stumbling into someone or dropping something, easily avoided by keeping the factory clean, although polymer powder and coloring agents were still problems. The 5S process also led employees to obey the company's policies and pay attention to safety matters.

a) The first step was translating the concepts of the 5S, explaining them, and educating the employees about them. To this end, employees were given specific examples of usage and activities that demonstrated the concepts [*Seiri*] (meaning "arrangement") and [*Seiton*] (meaning "to put into order").

b) Next, 5S committee member from each section and from among the employees were selected. The 5S committee was tasked with choosing the target and plan of action. It was important that 5S activities be promoted not only by top-down policies but also by the bottom-up activities of the 5S committee. Committee members were selected from all sections, including the staff section of the office. The 5S committee promoted many safety activities, such as writing the 5S patrol plan for monitoring 5S conditions, evaluating the plan, and commending those groups that show good results. These tasks were performed using a rally process called the "5S rally" in which groups competed against each other to win evaluation point for their activities.

c) Third, management patrolled the factory area to verify the results of the 5S implementation and the 5S committee's work. Those in charge in each section coached their members and evaluated the 5S level of their section and were responsible for the success of their group.

d) As the firm president, Hidaka patrolled the entire factory area. During one patrol, he photographed a 5S failure, printed the photos, and distributed copies around the factory to let the employees know what to avoid and to educate them in order to help them achieve the safety level the author demanded.

e) Despite the daily 5S activities, dirt and dust accumulated in the factory; for instance, there was dust or cobwebs at the high points that could not be reached by hand. To remedy this problem, the firm held major semiannual cleaning events in which factory operations were halted while the members cleaned the factory premises. The factory was divided into 20 areas, the office into 10, and the warehouse and waste facility into 30 areas. Each group cleaned its area all day. At the end of the day, the group leaders patrolled and checked all the areas. In this significant 5S event, called the "Big Cleaning Day", everyone attains a sense of achievement by cooperating. When employees assessed the status of each area, they were reassured about their 5S standards. Thus, Big Cleaning Day was fruitful in many ways.

**2) KY: an important activity**

The KY process, in which K stands for *kiken* ("danger") and Y for *yochi* ("provision"), is a well-known practice in Japan. The firm believed that it was very important that all employees understood its meaning, kept it in mind, and applied it.

a) First, we taught KY

First, the firm taught KY to the leader class member. Next, it taught group KY to the working groups. This group KY, called "KYT", is a group training method used to teach KY concepts and train employees in its methods.

b) At the beginning of the KY activity, a KY training sheet was used in every operational group as a way to practice the identification of a danger, determine its importance, and consider possible countermeasures using the KYT method. During this training, the firm recorded the names of all participating employees or used a KY CARD in which the group members described the results of KY activities in order to confirm that all workers were participated in the training.

**3) Examinations of accidents**

Early on, the firm implemented examinations of accidents at the Japanese parent company and how they had been handled to prevent similar accidents at the Thai factory. Focusing on rotating material and compounding factory accidents, employees studied data on their causes and considered prevention methods. This activity required translating the accident reports from Japanese into English and then into Thai.

## 4.3. Evaluation of the Results of These Safety Activities

1) The safety policy

The firm's safety policy should be posted at the office, in the meeting room, and in the control room. There should be occasional training sessions in which the policy is explained and the employees educated on it. Participating in 5S and KY training sessions will help employees understand the implementation and spirit of the policy.

2) 5S activity

The Thai employees understood the purposes of the 5S activity and what it required of them, and every representative of every section had worked with the 5S committee. The Safety Officer managed the committee, and the committee (comprising 15 representatives of each section) formulated the annual plan. Once the plan was in place, the monthly safety targets were determined at the monthly meetings.

a) At the monthly meetings, the responsible person reported on any countermeasures for the problems identified by the 5S patrol the previous month. After reviewing these reports, they confirmed the important target for the current month and then patrolled in two or three groups. b) The patrol member marked with red paper the location or piece of equipment that the 5S patrol had indicated was problematic. c) After finishing the patrol, members discussed these trouble points and chose countermeasures with members of the responsible section. d) Twice a year, the group and employee with the best 5S records were selected for an award. The monthly meetings, patrols, group and individual evaluations, and semiannual selection of group and individual awards were faithfully repeated. The 5S committee members were rotated each year to reduce the top–down pressure exerted by taking direct charge of the 5S work. As mentioned, the compounding factory was dirty because of the powdery nature of the coloring agent, the plastic snips that accumulate, and the scattering of plastic pellets. It is challenging to maintain a clean and safe work environment. However, Hidaka strongly believes that the effective management of any factory begins with cleanliness, which is also necessary for increasing safety levels. Cleanliness was very effective for another reason, as indicated by the term [*shitsuke*] ("well-bred"). Moreover, the 5S activities improved the quality of the firm's plastic pellets, which, in turn, helped improve the customers' confidence in the quality of all the firm's products. In the compounding process, it was difficult to prevent contaminating the powder with another type of resin, burnt resin, or waste. Success was achieved by applying 5S in the same direction as the activity used to prevent the contamination of the pellets. The result was that maintaining good 5S conditions contributed to resuming ties with Japanese automobile parts makers that expected the firm's products to be extremely high quality.

3) KY activity

All the Thai employees were able to understand the concepts of KY through explanations and education. However, as the firm used readymade training sheets, the training was quite different from the actual processes. Moreover, because the KYT method is a group method, the firm was unable to assess employees' knowledge at the individual level. Therefore, this method is suitable for studying the effectiveness only of KYT because the employees consider only one theme as a group. However, it is inadequate for evaluating the effectiveness of individuals. Thus, individual KY effectiveness cannot be assessed nor the accident prevention levels evaluated. This is a problem even in Japan and is quite difficult to solve. However, it must be addressed, even at the Thai factory.

4) Examinations of other accidents

The Thai compounding factory performed [examinations of in-house accidents]. After eliminating the accidents that were too dissimilar to be helpful, a series of worker accidents was chosen as teaching material. First, Hidaka translated the Japanese accident reports into English, and then the Safety Officer translated them into Thai. Then, the Thai reports were distributed to each group. The place and timing of the study sessions were different for each group, but they all examined the reports at their safety meetings as a safety topic. They examined another case during the meeting of shift workers, also as a safety topic. These study activities started with the translation of an accident report. Hidaka immediately addressed cases of accidents that could have occurred at the Thai factory (e.g., cases of pinching or being trapped) and requested an intensive examination of the case. This educational material indicated to the Thai employees the types of safety problems and accidents that occurred in the Japanese parent company or in the worldwide group company, the types of problems that needed to be eliminated, and the measures required to eliminate the risks. The materials also indicated the level of safety required of the workers by the parent group. As a result, this activity was also beneficial for the Thai firm.

## 5. Conclusions

1) A questionnaire survey on Japanese-affiliated companies operating in Thailand and Vietnam indicated that all the respondent companies have been engaging in "5S activities" since being founded. The authors conclude that the firms' top management wishes to instill discipline in their employees and make them obey the firms' rules, thereby improving their safety awareness. In addition, "Examinations of accidents", "KY activities", and "Safety proposal activity" were adopted by nearly 80% of the companies; these activities were also effective. These activities should all be adopted by new factories operating in developing countries.

2) The results of the actions to improve safety at Hidaka's Thai factory are as follows.

Initially, the target was to improve the safety at the Thai subsidiary by raising the level of the employees' safety awareness to parity with that of the employees at the Japanese parent company. It was important not to overwhelm them with activities while doing this. Instead, the Thai subsidiary introduced one safety activity at a time according to the order of importance and moved forward only after determining the extent of their comprehension and achievement.

The safety activities introduced first were 5S and KY, with particular emphasis on 5S. The employees created 5S conditions through [arrangement], [putting into order], [cleaning], [sanitation], and [being well-bred]. Hidaka believed that these 5S activities were the foundation of all production work as well as safety. The concepts and methods of KYT were also taught. The group method of KYT was used to generate group outcomes. This activity taught employees that risks may be present anyplace and in any situation and that they can reduce this risk through KY. The firm also developed prevention plans for accidents similar to those that had occurred at the Japanese parent company.

As a result of the above activities, the safety level of the Thai workers reached between 70% and 80% of that of the Japanese workers. However, differences between the Japanese and Thai workers' fixation ratios suggest that safety education must be differently approached in Thailand, where workers would leave the firm after having been educated at a considerable cost in time and money; new workers constantly needed to be educated. This problem is less likely to exist in Japan. The solution may be to educate a Thai "key" person and train him or her to educate new employees. However, only one loss-time accident and only two or three no-loss-time accidents would occur at the Thai factory each year, indicating that the workers' safety awareness had grown and the number of accidents had decreased a great deal since the firm had begun. Therefore, if the Thai subsidiary accomplishes one more step forward, it will be close to its first target: raising the employees' safety awareness to the same level as that of the Japanese parent company.

## Acknowledgements

The authors wish to express their gratitude to several significant contributors to this paper. First, we would like to thank all of the members of the Thai subsidiary who tried to improve their safety practices. Second, we would like to thank the members of our parent company, who supported the management of the Thai subsidiary. In particular, we must say many thanks to Mr. Nobukata Ochiai, the first president of the company. Finally, we sincerely thank the top managements of the Thai and Vietnamese companies who answered our questionnaire.

## References

[1] Duijm, N.J., Fievez, C., Gerbec, M., Hauptmanns, U. and Konstandinidou, M. (2008) Management of Health, Safety and Environment in Process Industry. *Safety Science*, **46**, 908-920. http://dx.doi.org/10.1016/j.ssci.2007.11.003

[2] Cox, S. and Cox, T. (1991) The Structure of Employee Attitudes to Safety: A European Example. *Work & Stress*, **5**, 93-106. http://dx.doi.org/10.1080/02678379108257007

[3] Matsuzaki, H. and Yamana, Y. (2005) 5S, Kaizen and English Expression. Sansyusya, Tokyo.

[4] Rahman, M.N.A., Khamis, N.K., Zain, R.M., Deros, B.M. and Mahmood, W.H.W. (2010) Implementation of 5S Practice in the Manufacturing Companies: A Case Study. *American Journal of Applied Sciences*, **7**, 1182-1189. http://dx.doi.org/10.3844/ajassp.2010.1182.1189

[5] Tanabe, H. (2004) Beginner's Book of KY Prediction. Japan Industrial Safety and Health Association, Tokyo.

[6] Tanabe, H. (2004) Training of KY Prediction. Japan Industrial Safety and Health Association, Tokyo.

[7] Cheyne, A., Cox, S., Oliver, A. and Tomas, J.M. (1998) Modeling Safety Climate in the Prediction of Level of Safety Activity. *Work and Stress*, **12**, 255-271. http://dx.doi.org/10.1080/02678379808256865

[8] Guldenmund, F.W. (2000) The Nature of Safety Culture: A Review of Theory and Research. *Safety Science*, **34**, 215-257. http://dx.doi.org/10.1016/S0925-7535(00)00014-X

[9] Dea, A.O. and Fin, R. (2001) Site Manager and Safety Leadership in the Offshore Oil and Gas Industry. *Safety Science*, **37**, 39-57. http://dx.doi.org/10.1016/S0925-7535(00)00049-7

[10] Pousette, A., Larsson, S. and Tomer, M. (2008) Safety Climate Cross-Validation, Strength and Prediction of Safety Behavior. *Safety Science*, **46**, 398-404. http://dx.doi.org/10.1016/j.ssci.2007.06.016

[11] Hidaka, A., Izato, Y. and Miyake, A. (2014) Lessons Learned from Recent Accidents in the Chemical Industry in Japan. *Open Journal of Safety Science and Technology*, **4**, 145-156. http://dx.doi.org/10.4236/ojsst.2014.43016

# 11

# Finite Size Scaling of Flame Spread along Randomly Distributed Combustible Cubes

**Yukinobu Yoshida, Akihiko Ito, Hiroyuki Torikai**

Graduate School of Science and Technology, Hirosaki University, Hirosakicity, Japan
Email: aito@cc.hirosaki-u.ac.jp

## Abstract

**Flame spread in urban fire depends on quantity of combustible materials and their placement. The purpose of this study is to predict the flame-spread route by the rate of combustible area. In this experiment, we used combustible cube made from filter paper as a test sample, and combustible cubes and pores were randomly distributed in experimental area. Additionally, we examined non-uniform flame spread along combustible cubes in various porosities and experimental areas. Experimental results show that the flame-spread probability falls with increasing porosity, and despite uniform porosity, the flame-spread probability differs with the experimental area. When expanding the experimental area, less porosity is required to obtain zero flame-spread probability. To investigate based on finite size scaling of percolation theory for the change in porosity, we concluded that different thresholds exist for areas of all sizes. Because of the flame-spread progress stops in the thresholds, these porosities may be useful to predict fire hazard.**

## Keywords

**Percolation Theory, Finite Size Scaling, Flame-Spread Probability, Experimental Area, Porosity**

## 1. Introduction

After a huge earthquake, multiple fires proliferate simultaneously in urban areas, while secondary damage is exacerbated by the spread of fire. In fact, several fires broke out after the Great Kanto Earthquake, Great Hanshin Earthquake, Great East Japan Earthquake and others. Urban regions include combustible areas such as wooden structures and plants, as well as the non-combustible, e.g. concrete buildings, roads, parks, and open areas, which mean that flames spread non-uniformly. Moreover, it is difficult to predict the flame-spread route because it may also be affected by the wind direction and terrain and the density of combustible materials. **Figure 1** shows a photograph of an example—urban area in Japan and a binary image, where the urban area is divided into combustible and non-combustible areas. As shown in **Figure 1**, combustible and non-combustible areas of various sizes are randomly distributed in an urban area. When a fire starts somewhere, the phenomenon is divided into two cases. One involves the flame spreading and combustible materials burning out, and the other involves the flame self-extinguishing on the way. The threshold for burning out or self-extinguishing may be determined by the quantity and placement of combustible materials. Establishing safety by predicting the flame-spread route is important to reduce secondary damage caused by urban fires.

Many studies have been conducted focusing on flame spread along a uniform solid fuel load as a fundamental component of fire research [1]-[3], but few have examined non-uniform flame spread in mixed combustible and non-combustible materials [4]. Some studies have focused on the application of percolation theory to non-uniform flame spread [5]-[7], while a few numerical simulations have also been conducted using a square lattice model [8]. However, discussing from an experimental perspective is rare. Accordingly, in our previous studies [9]-[11], we conducted experiments concerning flame spread along a thin combustible solid with randomly distributed pores of different shapes and sizes. We also discussed the relationship between porosity and flame-spread probability based on percolation theory which treats connection of random media and refers to the "Introduction to Percolation Theory" written by Stauffer D. *et al.* about percolation theory [12] [13]. The latter is a probability theory dealing with properties of the connection of random media. A square lattice is shown in **Figure 2(a)**, and the dots are written in some squares in **Figure 2(b)**. This dot is called a site, and the group of neighboring squares occupied by these dots is circled as shown in **Figure 2(c)**. Many of the sites become connected into a single large cluster, which spans the entire area vertically or horizontally due to an increase in site occupancy. This is called critical probability $P_c$ and known as the percolation threshold. The value of $P_c$ differs depending on the lattice shape and is about 59.3% in a square lattice of infinite size.

In this paper, we studied the flame-spread probability along combustible cubes with randomly distributed pores of various porosities. In addition, performing flame-spread experiments by changing the experimental area, we studied the effect of the size of the experimental area on flame spread by investigating the relationship between porosity and flame-spread probability. Furthermore, we introduce the porosity of flame-spread probability set to zero, and discuss the hazards of urban fire.

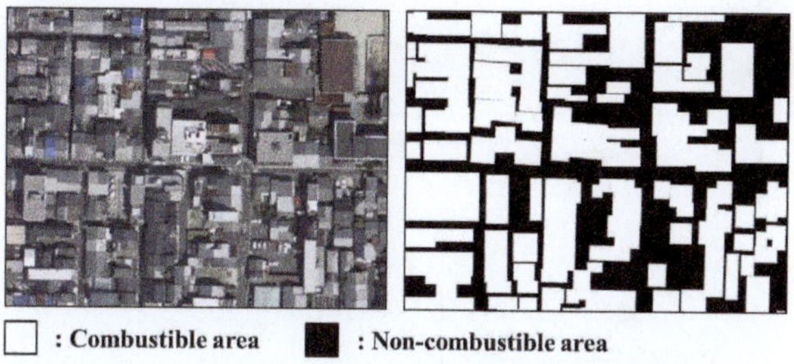


**Figure 1**. A photo of an urban area and a binary image of the urban area which is represented by combustible and non-combustible regions.

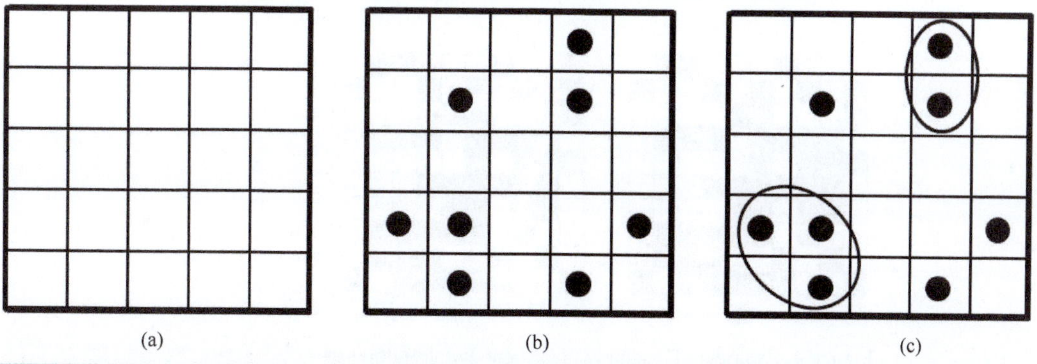


**Figure 2.** Definition of percolation and its site and cluster.

## 2. Experimental Methods

### 2.1. Experimental Apparatus and Procedure

A schematic of the experimental apparatus and combustible cube are shown in **Figure 3**. In this study, we used combustible cubes made from filter paper (ADVANTEC, No. 131, 250 μm thick) as a test sample. The experimental area was divided into a square grid, combustible cubes and pores were randomly arranged in each grid and the sample which arranged corner of the experimental area was ignited by a pilot flame. Experiments were conducted in atmosphere and a video camera (SONY, HDR-11, 30 fps) was set up above the experimental area to record the top view image of the flame spread. The experiment apparatus was also covered in vinyl sheeting to eliminate ambient disturbance of the flame spread. The test sample was a hollow cube of sides 20 mm, while the experimental area was square lattice of equivalent size, like the sample width. The number of samples to be placed to obtain arbitrary porosity was specified and their position was determined, using a random function to produce an experimental area of non-uniformly distributed pores and samples.

### 2.2. Experimental Condition

In this study, we selected ten different experimental areas, while the number of grids was 10 × 10, 20 × 20, 30 × 30, 40 × 40, 50 × 50, 60 × 60, 70 × 70, 80 × 80, 90 × 90, and 100 × 100. The flame-spread test was initiated by igniting a combustible cube placed in a corner of the experimental area. **Figure 4** shows an example sample placement, where we defined the flame spread as the point when the flame had progressed from the top left (ignited sample) to the right or bottom sides. The flame-spread probability is defined as the value obtained by dividing the number of flame spreads by the number of flame-spread tests, as revealed by Equation (1):

$$\text{Flame-spread probability}\left(F_s\right) = \frac{\text{Number of flame spreads}\left(N_s\right)}{\text{Number of flame-spread tests}} \quad (1)$$

Number of flame-spread tests is the number of sample arrangement pattern which is 50 in this study. Porosity is defined as the percentage total pore area in the experimental area, as given by Equation (2):

$$\text{Porosity}\left(P\right) = \frac{\text{Total pore area}}{\text{Experimental area}} \times 100[\%] \quad (2)$$

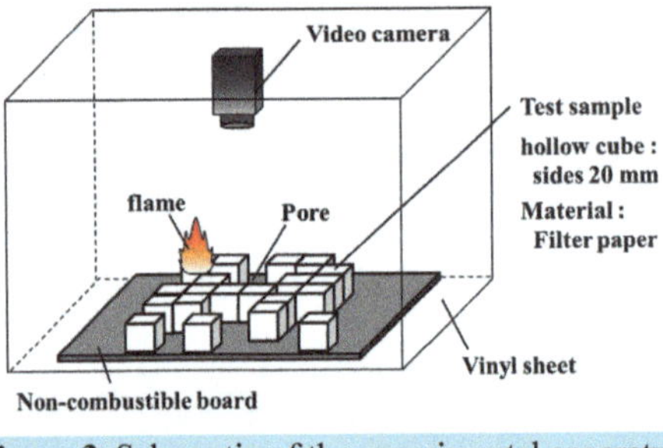


**Figure 3.** Schematic of the experimental apparatus.

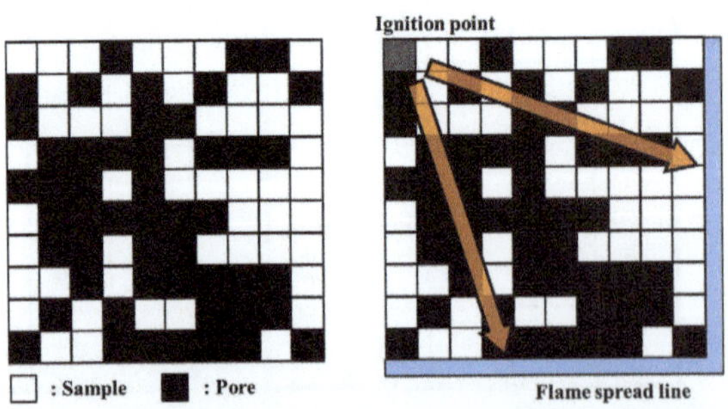


**Figure 4.** Example of a sample placement and definition of the flame spread.

In this study, porosity was varied from 40% to 70% to investigate the relationship between flame-spread probability and changes in porosity.

#### 2.2.1. Jump of Flame

The previous study [9]-[11] confirmed the phenomenon whereby the flame spread of a thin combustible solid jumped over a non-combustible area. In this study using combustible cubes, flame spread also featured jumping over non-combustible areas. A flame-jumping test was conducted using two combustible cubes with a non-combustible area as a preliminary test shown in **Figure 5** and ignited one sample. It was defined as jumping success if the flame of ignited sample was spread to the unburned sample, and the flame-jumping probability is defined as Equation (3):

$$\text{Flame-jumping probability}\left(F_j\right) = \frac{\text{Number of jumping successes}\left(N_j\right)}{\text{Number of flame-jumpingtests}} \quad (3)$$

The relationship between flame-jumping probability and non-combustible distance was plotted in **Figure 5**. When the non-combustible distance is less than 5 mm, the flame can always propagate, but when the non-combustible distance exceeds 8 mm, the flame cannot propagate. Flame-jumping is related to the pre-heat length, $L_h$, ahead of the flame leading edge, which means a flame can traverse a non-combustible area in $d < L_h$, and not $d > L_h$ as shown in **Figure 6**. We introduced the scale ratio $S \equiv d/L_h$ ($d$: non-combustible distance, $L_h$: pre-heat length) to classify whether the behavior was flame jumping or not, namely whether the flame could jump the non-combustible area for $S < 1$, and not for $S > 1$. From these results, in this study with the square lattice of 20 mm width, no flame spread to traverse a pore.

#### 2.2.2. Assumption of $S < 1$

The condition to satisfy $S < 1$, namely requiring $d < L_h$, was not met in this study. To study the effect of scale ratio on flame-spread probability, we assumed that the flame spread to the position shown in **Figure 7**, providing the condition of $S < 1$. **Figure 7** is shown as the flame spread area of $S > 1$ and the flame-jumping area of $S < 1$. Where $S < 1$, in addition to the eight samples in contact with the combusting sample, we assumed the flame would spread to the four samples shown in orange color even if there were a pore between the samples.

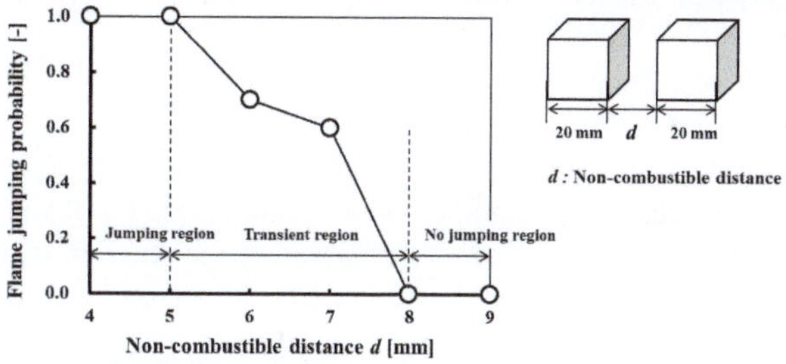


**Figure 5.** Relationship between flame jumping probability and non-combustible distance.

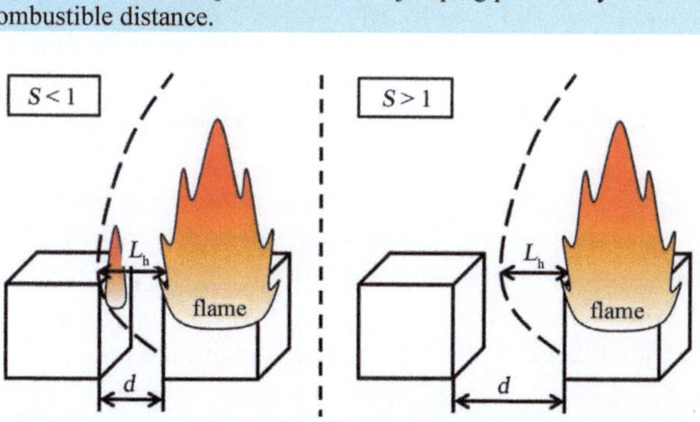


**Figure 6.** Schematic of flame jumping over a non-combustible distance.

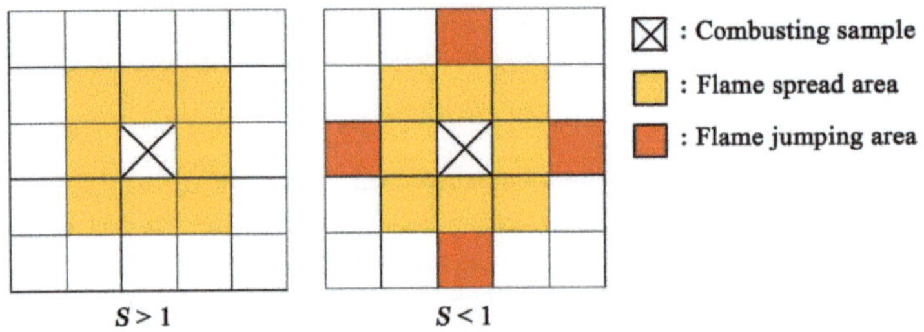


**Figure 7.** Flame spread and flame jumping areas.

## 3. Results and Discussion

### 3.1. Flame-Spread Trajectory

The cluster distribution for three different porosities and the flame-spread trajectory are shown in **Figure 8**. The image processing techniques of use for the flame-spread trajectory include normal speed photography (30 frames per second). In this figure, the black line is the flame-spread trajectory, which is depicted at intervals of ten seconds after ignition, while clusters mean area so finter connected pores, which prevent the flame-spread progressing. Pores form large clusters with increasing porosity until the flame is eventually extinguished when the cluster forms around the ignition point. For 40% porosity, the combustible area accounted for a large proportion, with little pore connection. For 50% porosity, the percentage combustible and non-combustible areas are half and half, with a large pore connection formed compared to the 40% porosity. Finally, in the case of 60% porosity, with a large pore proportion, the flame is swiftly extinguished. **Figure 9** shows an example sample placement and the flame-spread trajectory at the time. Here, the L-shaped slit represents the cluster surrounding the flame source, where self-extinguishing occurred. In **Figure 9**, the red dashed line represents the L-shaped slit, while the samples shown in orange color are estimated to reach the flame spread. Accordingly, we conceived a method of calculating the extent of flame spread for the presence or absence of the L-shaped slit. In fact, there was no difference in the flame-spread areas to be calculated from the sample placement and observed from the flame-spread test respectively, therefore the following results were obtained from this method.

### 3.2. Flame-Spread Probability

**Figure 10(a)** plots the flame-spread probability as a function of different porosities in the number of grids, 20 × 20, 40 × 40, 60 × 60, 80 × 80 and 100 × 100 when $S > 1$. The flame-spread probability declines with increasing porosity in any number of grids. However, as the number of grids increases, the flame-spread probability becomes zero shows small porosity. To study the effect of experimental area in terms of its impact on flame-spread probability, we investigated porosity when the flame-spread probability become zero at any number of grids. **Figure 10(b)** shows the relationship of the number of grids and porosity when the flame-spread probability becomes zero. It can be seen that porosity when the flame-spread probability becomes zero declines with an increasing number of grids, and asymptotically approaches approximately 60%. It can be seen that in **Figure 11(a)**, also $S < 1$, the flame-spread probability declines with increasing porosity as in the case of $S > 1$. However, the expanding flame-spread range means more porosity is needed to self-extinguish. In addition, the porosity when the flame-spread probability becomes zero declines with increasing number of grids, as in the case of $S > 1$ shown in **Figure 11(b)**.

### 3.3. Scaling

We studied based on the percolation theory for the relationship shown in **Figure 10(b)** and **Figure 11(b)**. According to the percolation theory, finite size scaling exists, which takes into account the finiteness of the area, and the following relationship is known between the threshold of the infinite region $P_c$ and the effective threshold of the finite region $P(L)$:

$$P(L) - P_c \propto L^{-\frac{1}{\nu}} \tag{4}$$

40% porosity 50% porosity 60% porosity

**Figure 8.** Flame-spread trajectory at 40%, 50%, and 60%, porosity of 10 × 10 grids.

**Figure 9.** Comparison of flame-spread test and flame-spread areas to be calculated from sample placement.

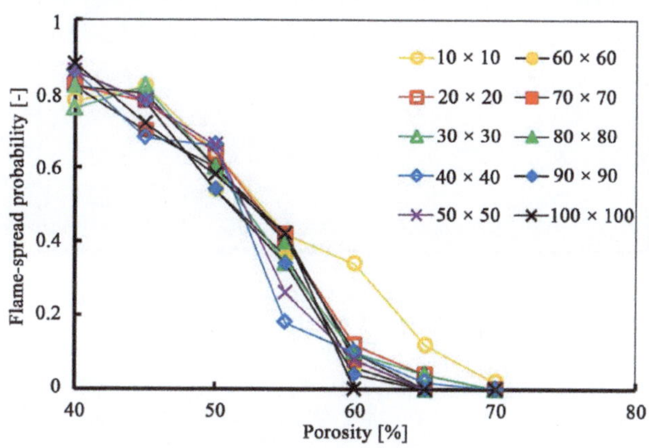


(a)

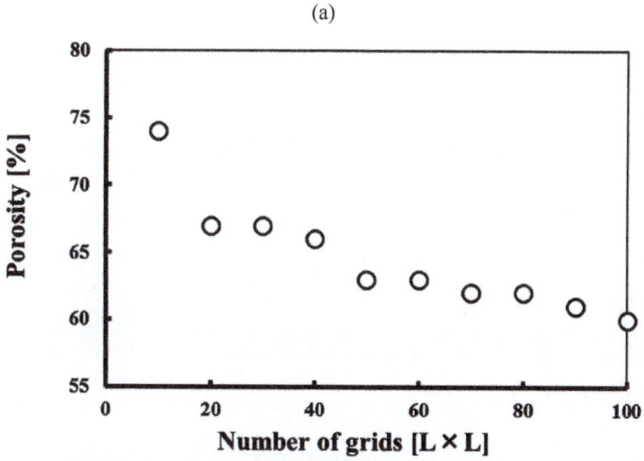


(b)

**Figure 10.** (a) Relations of flame-spread probability and porosities at $S > 1$; (b) porosity with zero flame-spread probability and number of grids at $S > 1$.

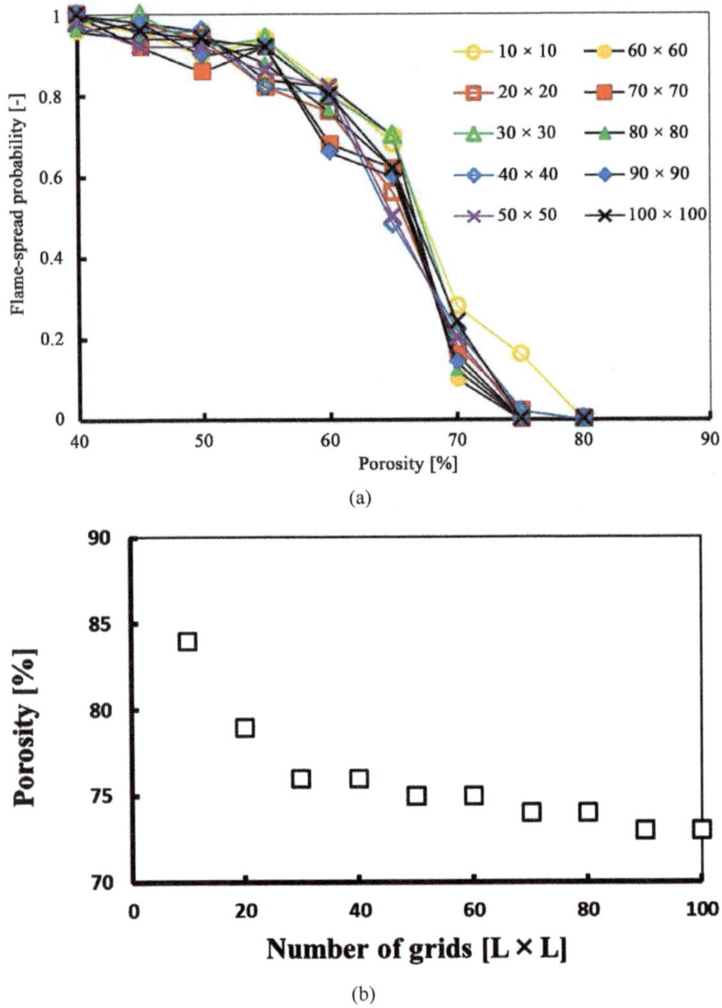


**Figure 11.** (a) Relations of flame-spread probability and porosities at $S < 1$; (b) porosity with zero flame-spread probability and number of grids at $S < 1$.

where $P_c$ is the percolation threshold, $L$ is the size of the area, and $v$ is a critical exponent (4/3 in two dimensions). Using a proportional constant $C$, Equation (4) can be rewritten as follows:

$$P(L) = P_c + CL^{-\frac{1}{v}} \tag{5}$$

In this study, pores need to lead to the entire experimental area to stop flame spread, so we used the relation by correcting the coefficient ($C$) and regarded $P(L)$ as porosity for which the flame-spread probability is zero. When approximating the experimental results by selecting a suitable value for $C$, $P(L)$ can be expressed by Equation (6) for $S > 1$:

$$P(L)_{S>1} = 59.3 + 0.73L^{-\frac{1}{v}} \tag{6}$$

If the condition is $S < 1$, we found that the effective threshold at this time was represented by Equation (7):

$$P(L)_{S<1} = 71.0 + 0.73L^{-\frac{1}{v}} \tag{7}$$

Here, the threshold ($P_c$) has clearly changed compared to $S > 1$. The threshold of $S < 1$ exceeds that of $S > 1$ by about 10%, namely the risk of expanding flame spread increases. We considered that if the placement of the

$S > 1$ and $S < 1$ samples were mixed, the threshold would be between both equations. These relationships are shown in **Figure 12**, where circular and square plots show zero porosity of flame-spread probability at $S > 1$ and $S < 1$, respectively. Solid and dotted lines represent the approximate curve of each of the plots.

## 4. Conclusions

In this study, we conducted flame-spread experiments using cubes of filter paper and mixed areas of pores. The following conclusions can be summarized here:

1) In the $S > 1$, L-shaped slit, the formed probability rises with increasing porosity, whereupon the flame-spread probability declines. However, where the number of grids differs, a difference also emerges in porosity when the flame-spread probability is zero. With extinguishing over 100 × 100 for 60% porosity and 20 × 20 for 70% porosity, just as described, it emerged that the porosity with zero flame-spread probability declined with increasing grids.

2) Similarly for $S < 1$, the flame-spread probability declined with increasing porosity, and the porosity when the flame-spread probability became zero decreased along with the increase of the number of grids. But, when compared with the same porosity, the flame-spread probability at the $S < 1$ is high value than the $S > 1$. And the porosity when the flame-spread probability becomes zero at the $S < 1$ is high percentage than the $S > 1$.

3) We determined the threshold of the porosity requirement for self-extinguishment within the finite region based on the percolation theory. The effective threshold can be expressed as follows:

$$P(L)_{S>1} = 59.3 + 0.73L^{-\frac{1}{\nu}}\,(S > 1) \tag{8}$$

$$P(L)_{S<1} = 71.0 + 0.73L^{-\frac{1}{\nu}}\,(S < 1) \tag{9}$$

We considered the fact that if $S > 1$ and $S < 1$ are mixed, the curve of the effective threshold is assumed to be between the above two equations.

4) We also considered the fact that the placement of combustible materials in the urban area is irregular, and the positions of $S > 1$ and $S < 1$ are mixed. Therefore, by providing a non-combustible area within the range of 60% to 70%, self-extinguishment result due to a non-combustible area, even if a fire occurs somewhere, will not generate significant fire-spread damage.

It is the future work of this study to make combustible areas more close to actual urban area by changing height of the combustible material and tilting the experimental area.

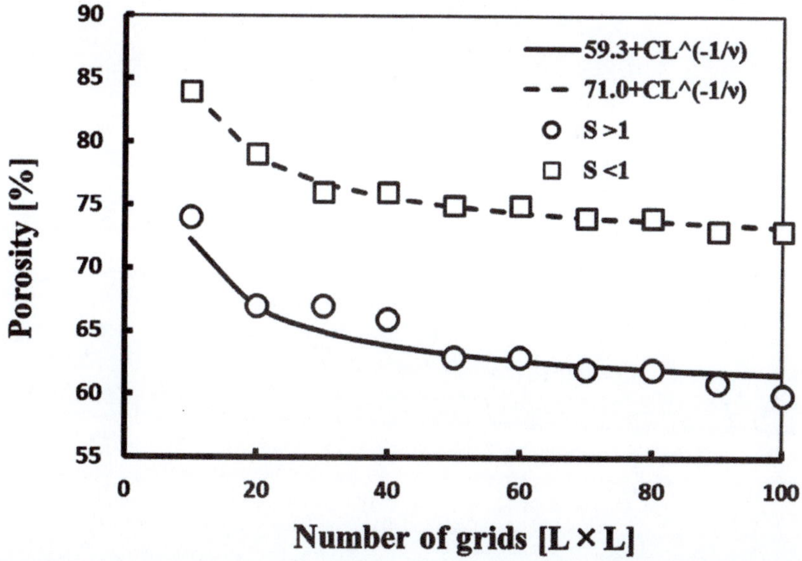


**Figure 12.** Relationship of porosity which flame-spread probability and number of grids, and the approximated curve.

## Acknowledgements

We would like to thank S. Abe for his help with experiments and in preparing the manuscript.

## References

[1] Fernandez-Pello, A.C. and Hirano, T. (1983) Controlling Mechanisms of Flame Spread. *Combustion Science and Technology*, **32**, 1-31. http://dx.doi.org/10.1080/00102208308923650

[2] De Ris, J. (1969) Spread of a Laminar Diffusion Flame. *Twelfth Symposium (International) on Combustion*, 241-252.

[3] Kazuho, S., Kazumasa, M. and Toshisuke, H. (1984) Flame Spread over Paper in an Air Stream with a Velocity Change. *Journal of Heat Transfer*, **106**, 707-712. http://dx.doi.org/10.1115/1.3246742

[4] Takashi, S. (1976) Summaries of Technical Papers of Annual Meeting. Architectural Institute of Japan, 1803-1804. (in Japanese)

[5] Toshiyuki, K., Takaaki, K. and Osamu, K. (2001) The Basic Study on the Evaluation of Disaster Mitigation Performance of the Local Area Road Network Based on Percolation Theory. *Journal of Architecture and Planning*, **549**, 231-237. (in Japanese)

[6] Takaaki, K. (2002) Operations Research as a Management Science. *Communications of the Operations Research Society of Japan*, **47**, 441-446. (in Japanese)

[7] Hiroshi, O., Hisashi, S., Masato, M. and Naoya, K. (2009) Flame-Spread Probability and Local Interactive Effects in Randomly Arranged Fuel-Droplet Arrays in Microgravity. *Combustion and Flame*, **156**, 763-770. http://dx.doi.org/10.1016/j.combustflame.2008.12.013

[8] Akira, U. (2011) Application of Percolation Theory to Combustion Problems. *Journal of the Combustion Society of Japan*, **53**, 145-152.

[9] Yukito, W., Akihiko, I. and Hiroyuki, T. (2009) Effect of Porosity on Flame Spread along a Thin Combustible Solid Randomly Distributed Pores. *Sixth International Symposium on Scale Modeling, Proceedings (CD-Rom)*, Kauai, Hawaii, 13-16 September 2009.

[10] Yukito, W., Akihiko, I. and Hiroyuki, T. (2010) Flame Spread along a Thin Randomly Distributed Combustible and Noncombustible Areas. *Proceedings of the Combustion Institute*, **33**, 2449-2455.

[11] Syuhei, A., Akihiko, I. and Hiroyuki, T. (2012) Flame Spread along a Thin Combustible Solid with Randomly Distributed Square Pores of Two Different Sizes. *Modern Applied Science*, **6**, 11-19.

[12] Stauffer, D. (1985) Introduction to Percolation Theory. Taylor and Francis, London

[13] Stauffer, D. and Aharony, A. (1992) Introduction to Percolation Theory. 2nd Edition, Taylor and Francis, London.

# 12

# Role Expansion: Implications of the Security Role Expanding into the EHS Function

**Wanda D. Minnick, Jonathan D. Cunkelman**

Safety Sciences Department, Indiana University of Pennsylvania, Indiana, USA
Email: wanda.minnick@iup.edu, J.D.Cunkelman@live.iup.edu

## Abstract

**The purpose of this study was to explore the perception of safety professionals regarding the expansion of their role into security functions. One hundred eleven returned surveys from practicing safety professionals in the manufacturing, oil and gas, and construction industries were examined. Fifty seven percent of respondents agree with role expansion into security and twenty three percent of participants do not. Thematic patterns emerged from their comments and are provided. The remaining twenty percent of respondents fell into a category of "it depends" and the provided reasons suggest other variables that could have influence. A Chi square test for independence indicated that role expansion is independent of industry type; suggesting there is agreement for role expansion into security functions regardless of industry type. Another primary finding is that although the majority of safety professionals agree with role expansion, security coursework is currently not required in a review of accredited occupational safety and health program curricula. This suggests recent graduates may not be formerly prepared for the security elements of their role and should seek or be offered additional training.**

## Keywords

**Role Expansion, Safety Professional, Security, Role Overload**

## 1. Introduction

The role of the safety professional can be defined as a person that engages in the recognition of hazards, the evaluation of risk, and provider of guidance and counsel on controlling or minimizing risk. The American National Standard Institute (ANSI), an organization that promulgates guidelines used in various business sectors, along with the American Society of Safety Engineers (ASSE) defines the scope and function of the professional safety position in guideline ANSI/ASSE Z590.2-2003 and is described on the ASSE body of knowledge website.

The primary tenets of the role being to identify, evaluate, and develop hazard controls. In addition the role includes warning and advising others in prevention strategies and overall risk reduction as well as measuring and auditing effectiveness of hazard controls [1]. The Board of Certified Safety Professionals further defines today's safety role as identifying what hazardous events can occur and the likelihood of that occurrence, the severity of the results, the overall risk (a combination of probability and severity), and cost [2]. In comparison, the American Society for Industrial Security (ASIS) defines the term security and describes an overall scope for the security function. It notes that security is a similar concept to safety with one clear distinction; within security added emphasis is placed on dangers originating from the outside [3]. ASIS also explains that the term security means that something is not only secure but it has been secured. The scope of the security function is to reduce the risk of unintentionally and naturally-caused crises and disasters that disrupt and have consequences on the products and services, operation, critical assets, and continuity of the organization and its stakeholders [4].

Many identify safety and security as two separate fields, but recent research has indicated that the safety professional's role is beginning to expand to include security functions. Specifically 25% of safety professionals from the manufacturing industry indicated they currently have security roles within their job function [5].Within that study, those that indicated the word "security" as a job descriptor explained their roles in some of the following ways: the development of security plans, conducting facility vulnerability assessments, being a member of the crisis team, designing and implementing various security systems to reduce vulnerabilities, working with classified documents and conducting personnel training in security awareness. Consideration of how the two disciplines were formed provides a foundation in which to examine the similarities and the potential merging of the two roles.

## 2. History of the Safety and Security Disciplines

The ASSE was found in 1911 and currently has 36,000 members. It is the oldest professional safety society and is committed to protecting people, property and the environment [6]. Hence, from the very beginning organized safety professionals have played a role in security. The American Society for Industrial Security (ASIS) was founded in 1955 and has over 38,000 members, each of whom has a role in the protection of assets-people, property, and/or information. It is currently known as ASIS international and has 234 worldwide chapters [4].

Both safety and security have relied on tremendous tragedy to bring about legislation that requires industry and others to maintain minimum standards. The triangle shirtwaist factory, BP in Texas and the Union Carbide plant in Bhopal, India are some of the tragic examples that led to legislation in safety. Likewise, the increased attention following the first World Trade Center attack in 1993 (truck Bomb), attacks on several US embassies and overseas military installations, as well as the events of 9/11/2001 heightened the awareness for the value of the security function. Another similarity exists in the reporting structure of each discipline, typically both functions reporting directly to upper management. In addition, both disciplines have the very difficult task of proving to management that their plans and actions are effective; meaning it can be hard to prove the loss you have prevented.

Safety and security professionals have a common goal of wanting to protect people and property, therefore it is important that their efforts, plans, and policies work together. Safety and security should have an understanding and knowledge about what the other is doing, as the very barriers designed to protect the employees from a security standpoint could also create a hazard. For example, the locks and doors that are installed to prevent unauthorized entry can fail or be misused and trap personnel and prevent evacuation in an emergency situation. Vehicle access barriers designed to prevent potential truck bombs from entering a facility could prevent emergency responders from timely access.

### Justification

A recent study of safety professionals in the manufacturing industry asked participants to describe the functions within their role. The intent of this study was to consider the number of roles in which safety professionals are engaged and if an increase in roles correlated with an increase in role conflict, a precursor to negative job states. Of the 356 qualitative responses, 25% ($N = 88$) participants listed the word "security" when providing a description of their role [5]. This finding suggests a growth trend (or baseline to mark potential growth) into security functions. In the same aforementioned study, it was concluded that as the number of roles increases so does the average role conflict score (a measure of job stress). Specifically, there was a significant difference in the average role conflict score by number of roles group membership ($F = 2.69$, $df = 5411$, $p = 0.021$), suggesting that further expansion of the safety professional's role (into security) could contribute to role overload. Brown, Jones and Leigh [7] explain that role overload is experienced "when role demands create the perception that available resources are inadequate to deal with them, resulting in distraction and stress" (p. 972). This purpose of this study is to explore the safety professional's perception of role expansion into the security function to discern not only if a study on role overload is warranted but also if role expansion into the security function is welcomed by safety professionals.

## 3. Method

Data for this study was obtained from administering an on-line survey to members of the American Society of Safety Engineers (ASSE), specifically members of the three largest practice specialty groups. To be included in this random sample, the following conditions had to be met: be a member of an academic specialty in the manufacturing, oil and gas, or construction industry and be employed as a safety professional; not be retired, not be employed as a full-time consultant, and not be employed as a full-time trainer. Justification for inclusion criteria was to limit participation to those members engrained in the daily operations of their company and those with an elevated expertise in a particular industry, as indicated by their membership in a specialty practice group. In addition, only those participants that answered the open-ended questions regarding role expansion were included, resulting in 111 applicable returned surveys. The survey instrument consisted of the following two questions: 1) Do you agree with the expansion of the safety role into security functions?: Yes or No and 2) describe why you do or why you do not agree with the expansion. An inductive approach was used in coding the qualitative data and enabled the identification of over-arching categories in which the reasons for agreement or disagreement could be sorted. Descriptive information considered in this study included industry type. Chi square test for independence was conducted to test the hypotheses below. If the $p$ value was $<0.05$, an association existed between the variables.

*Ho* = role expansion agreement is independent of industry type.

*Ha* = role expansion agreement is associated with industry type.

## 4. Results

**Frequencies:** There were 64 participants (57%) that agreed with the expansion of the safety professional's role into security functions as indicated in **Table 1**. There were 25 participants (23%) that did not agree with the expansion of the safety professional's role into security as indicated in **Table 1**. There were 20% ($N = 22$) of respondents that neither agreed or disagreed with role expansion, rather wrote the word "depends" in the open-ended question that asked "why do you agree or disagree".

Forty and a half percent ($N = 45$) of the participants were from the manufacturing industry, 40.5% ($N = 45$) were from the construction industry and 19% ($N = 18$) are from the oil and gas industry. There was no significant relationship found between industry type and role expansion agreement [$\chi^2$ (4, $N = 111$) = 1.71, $p = 0.79$]; resulting in a failure to reject the null hypothesis (see **Table 2**).

### 4.1. Agree with Role Expansion into Security Functions.

There were 64 participants (57%) that agree with the expansion of the safety professional's role into security. When asked "why", participants responded with three over-arching categories: "obvious overlap with current safety role" ($N = 38$), "pro generalist" ($N = 10$), meaning additional responsibilities adds value to the role and offers another skill set, and "yes, if appropriate education and resources are granted" ($N = 8$). Frequency information for these categories are provided in **Table 3**. Examples of responses coded as "*obvious overlap*" with current safety role include: "Safety is not limited to personal safety but has always had aspects of site and physical safety in it. There is a natural progression to security," "Safety, Env and Security are inter-related in many aspects. Joining of forces seems to be cost effective," "Everything is all inter-related. Adding Facilities pushed me into a security role but even before that I was involved in many aspects regarding chemical security, workplace violence, emergency response and scene control to name a few. So yes it is appropriate," and "I feel that the expansion into Security makes sense. Security of a facility has a direct impact on personal safety and is related to programs such as emergency preparedness (such as bomb threats, workplace violence, etc.). These two areas

**Table 1.** Descriptive Statistics for the question: "do you agree with expansion of the safety role into the security function?"

| Variable | Frequency (N) | Percent (%) |
|---|---|---|
| Respondents that answered "Yes" | 64 | 57% |
| Respondents that answered "No" | 25 | 23% |
| *Missing | 22 | 20% |
| Total | 111 | 100% |

*Respondents that did not answer yes or no, but wrote "depends" on open-ended question.

**Table 2.** Industry type by role expansion into security cross tabulation.

| | | | Role Expansion into Security | | | |
|---|---|---|---|---|---|---|
| | | | Missing | Agree with role expansion | Do not agree with role expansion | Total |
| Industry Type | Manufacturing | Count | 9 | 27 | 9 | 45 |
| | | Expected Count | 8.9 | 25.9 | 10.1 | 45.0 |
| | Construction | Count | 10 | 23 | 12 | 45 |
| | | Expected Count | 8.9 | 25.9 | 10 | 45.0 |
| | Oil and Gas | Count | 3 | 14 | 4 | 21 |
| | | Expected Count | 4.2 | 12.1 | 4.7 | 21.0 |
| Total | | Count | 22 | 64 | 25 | 111 |
| | | Expected Count | 22.0 | 64.0 | 25.0 | 111.0 |

**Table 3.** Descriptive statistics for the reasons participants answered "YES" for expansion into security functions.

| Variable | Frequency (N) | Percent (%) |
|---|---|---|
| Obvious Overlap | 38 | 59 |
| Pro-generalist viewpoint, expansion = job security | 10 | 16 |
| If appropriate education and resources are granted | 8 | 12.5 |
| Missing | 8 | 12.5 |
| Total | 64 | 100 |

seem to mesh well in my experience." Examples of response coded "*pro-generalist*" include: "We must always be adding value and if we can do more, then we become more valuable," "One cannot specialize and be ensured of future work. Security has become much more important now, even if security is contracted out the function is overseen by the Safety Dept," "They all overlap each other and if we pigeon hole ourselves strictly to safety the demand for us is very narrow," and "There are synergies in these roles, and in our industry all are doing more with less. I have security now at my request, but it also makes managing contractor safety easier because the guards report to me for all functions." Examples of responses coded "*yes, if appropriate education and resources are granted*" include: "Expansion can be appropriate if the company is willing to provide the necessary training. I don't believe the average safety professional has the background initially to cover multiple roles," "Yes, but my recent thrust into a larger role with Security reminded me that this was not an emphasis during college or any previous workplaces", and "Yes, I agree with the expansion; however, I don't agree with the methods. Environment and security do not simply fall in-line with what the typical Health & Safety professional's knowledge base is. Training must be provided or additional resources are necessary".

## 4.2. Do Not Agree with Role Expansion into Security Functions

There were 25 participants (23%) that did not agree with the expansion of the safety professional's role into security. When asked "why", participants responded with five over-arching categories: "separate discipline" ($N$ = 15), "not educated, do not have the skill set" ($N$ = 3), "distraction to primary safety role" ($N$ = 3), "role overload" ($N$ = 2), and "cop persona" ($N$ = 2). Frequency information for these categories are provided in **Table 4**. Exam-

**Table 4.** Descriptive statistics for the reasons participants answered "NO" for expansion into the security function.

| Variable | Frequency (N) | Percent (%) |
|---|---|---|
| Separate Discipline | 15 | 60 |
| Distraction to primary safety role | 3 | 12 |
| Not educated, do not have the skill set | 3 | 12 |
| Role Overload | 2 | 8 |
| Cop Persona | 2 | 8 |
| Missing | 0 | 0 |
| Total | 25 | 100 |

ples of responses coded "*separate discipline*" include: "I would not agree to expand into security as it is a separate and unrelated field," "no, security is a stand alone issue becoming more complex and time consuming, especially for Global operating companies," "No. I consider that safety is a career and job all by itself. Safety and security are not good partners as they do not complement each other in any way," and "In my opinion Security is it's own entity and is a different discipline." Examples of responses coded "*distraction to primary safety role*" include: "with role expansion you no longer become a practitioner and you get a peanut butter spread of experience but nothing real deep in any one area. The safety profession is becoming so wide it is losing its expertise and hands on experience to be credible," "No I don't agree, if the safety pro stays focused on eliminating injury rather than other distractions the safety pro and the organization benefit," and "No, I believe the safety role should remain specific to safety. Individuals who have multiple roles/responsibilities often find themselves unable to perform any one role as effectively as they feel they should. Often times performance morphs into a role of "Do what you have to do to get by and move on". Quality level suffers," Examples of responses coded "*Not educated, do not have skill set*" include: "Most safety professionals are ill-equipped either by training, knowledge or temperament to assume broader roles within their organizations" and "many people may have this in their title but not the qualifications necessary to effectively manage the additional roles." Examples of response coded "*role overload*" include: "no, I have enough to worry about as it is," and "no, I am trying to narrow my focus and avoid mission creep". Examples of responses coded "*cop persona*" include: "Pairing security with H&S can give the perception that H&S is also still a "policing function" and "Expansion into the security field is not appropriate because of the enforcement requirements. Safety professionals do not need to be known as cops also."

## 4.3. Depends

There were 20% ($N = 22$) of respondents that neither agreed or disagreed with role expansion, rather wrote the word "depends" in the open-ended question that asked "why do you agree or disagree". In expanding upon this answer, respondents most commonly wrote that it depends on complexity and size of the company ($N = 5$), depends whether resources such as time, employee to safety professional ratio is reasonable, and if additional training are available ($N = 8$), depends on the overall reporting structure or management style of the organization ($N = 5$) and lastly depends on industry type ($N = 4$).

# 5. Discussion

The majority of participants agree with the expansion (57%), however it was indicated in previous studies that additional roles can increase the level of role conflict in safety professionals. Regardless of the increased stress that additional roles bring to safety professionals, the majority still agree with expansion into security. In addition, there was no significant relationship found between industry type and role expansion agreement suggesting the growth trend or willingness to perform security functions is consistent across all industries. Considering that the majority of safety professionals in this study agree with the expansion and that 25% of safety professionals are currently performing security functions in manufacturing, it is obligatory to examine the education required to perform these additional roles.

It is important for the general safety practitioner to have a working knowledge of security. Many safety professionals are willing to accept an expanded role, but do not currently possess the skill set needed to move forward in that direction. Although there may be several ways in which to make this determination, one approach is to examine the coursework required within occupational safety degrees. There are currently 14 occupational safety and health programs that are listed as being accredited [8]. A review of the coursework and program requirements listed on the website for each program indicated that 2 programs have security-related coursework on a list of possible electives (non-mandated courses), the three courses are titled: *Introduction to Criminal Justice, Communication and Conflict, and Threat and Vulnerability Assessment and Management.* There were no programs that currently require a course with "security" in the title as part of the degree program. However, a more thorough review should be conducted since elements of the security function could be housed within the courses and just not listed as part of the course title.

It may be difficult for an occupational safety program to justify an entire course on security, however there are several options to consider. One option is to incorporate security elements into existing coursework. Modules on conflict resolution, homeland security, workplace violence, and managing lock down or shelter in place scenarios would serve as starting point. In developing these modules, reaching out to organizations for assistance can be productive. Jeff Schoepf, a Certified Protection Professional (CPP) and the Certification Coordinator at the ASIS International Pittsburgh chapter, explained that ASIS offers a multitude of educational information and is available to the public upon request (personal communication, March, 18, 2015). ASIS offers webinars, conferences, books, and classes that can be accessed by non-members for a fee. Interested parties may also attend a chapter meeting by contacting their local chapter, via www.*asis*online.org for meeting dates and times.

Additional resources include accessing information from the Department of Homeland Security (DHS). Not only does the DHS help protect and secure our national assets, they are concerned with the American presence overseas as well. For that reason they are very important to US based companies doing business or manufacturing in other countries and offer educational material on their website. In addition, FEMA offers online independent study courses and certifications that are related to disaster management. Many of these courses are based on fundamental safety and security training and it can be applied to overall security and safety at a facility. Additional information on the FEMA courses can be obtained at http://training.fema.gov/is/. Individual security companies offer education material as well. Lastly, using established professional organizations and their associated social media is also a resource.

Implications to the security function include the streamlining of managing risk within the safety role, both nationally and abroad. When asking a safety professional to describe a typical loss, he or she may suggest a fatality or accident. When a security professional is asked the same question he/she may say a breach of entry through a guarded gate, a theft, or a workplace shooting. A third party may say a loss is a loss regardless of where the box is located on the org chart and that the associated risks must be managed to prevent the loss. *Managing the risk* to prevent the loss is the key and the common denominator between safety and security. In terms of global implications, many organizations have been able to streamline efforts through the use of safety, environmental and quality management system correspondences [9]. The pending ISO 45001 Occupational Health and Safety management system slated for release in October, 2016 is a risk-based framework [10] [11]. Considering the safety role and security role are both risk-based, it could offer potential pathways in which to merge functions.

## 6. Limitations

Limitations of this study include sample size of each individual industry type and lack of additional descriptor variables such as industry size, complexity, and safety to employee ratio. In addition, it is possible that each safety professional may define or perceive "security" roles or functions differently, a common definition of the security role should be provided when creating a new survey or inventory. It would also be of importance to consider the percentage of time dedicated to the security functions.

## 7. Conclusion

Safety and security have a lot of similar duties, history, problems, and goals. Safety professionals are seeing an increase in responsibilities relating to security. Previous studies indicate that 25% of current practicing safety professionals include security in their job description; however there are no accredited occupational safety and health programs that require coursework on security. To properly address the gap between the current training that is offered to our formally educated professionals, it is important that we look to other sources. As part of our ethical obligations we are to be aware of our limitations and not reach beyond our skill set. In order to meet the demands of employers, safety professionals may need to reach out to other sources to expand our training. Although role expansion can increase role conflict and stress [5], the majority of participants (57%) in this study agree with role expansion into security. Additional research in this area is merited to discern if the addition of security function to the safety professional is contributing to role overload. Additional variables of importance to include in a future study on role overload is to clearly identify how participants are defining their security function, quantify how many separate roles (*i.e.* environmental, quality) they are performing, the size of the company, and the number of employees the safety professional is supporting in their current role. In addition, a study on the perceptions from traditional security personnel on "role creep" of safety professionals in to their roles would be warranted.

## References

[1] ANSI/ASSE Z590.2-2003 Standard (2015) Criteria for Establishing the Scope and Functions of the Professional Safety Position. http://www.safetybok.org/resources/criteria_for_establishing_the_scope_and_functions_of_the_professional_safety_position/

[2] BSCP (2015) The Safety Professional Today. http://www.bcsp.org/SH-E-Practice

[3] ASIS (2015) Security Glossary. https://www.asisonline.org/Membership/Library/Security-Glossary/Pages/Security-Glossary-A.aspx

[4] ASIS (2015) About ASIS. https://www.asisonline.org/About-ASIS/Pages/default.aspx

[5] Minnick, W.D. (2012) Organizational and Personal Characteristics That Influence Role Conflict and Role Ambiguity in the Safety Professional. Doctoral Dissertation.

[6] ASSE (2015) About ASSE. http://www.asse.org/about/

[7] Brown, S.P., Jones, E. and Leigh, T.W. (2005) The Attenuating Effect of Role Overload on Relationships Linking Self-Efficacy and Goal Level to Work Performance. *Journal of Applied Psychology*, **90**, 972. http://dx.doi.org/10.1037/0021-9010.90.5.972

[8] ASSE (2015) 2014-2015 Safety-Related Programs Accredited by ABET. http://www.asse.org/assets/1/7/2014-2015_Safety-Related_Programs_Accredited_by_ABET.pdf

[9] BSI (2007) BS OHSAS 18001:2007: Occupational Health and Safety Management Systems-Requirements.

[10] BSI (2015) ISO 45001: New International Standard for Occupational Health & Safety Management Systems. http://www.bsigroup.com/en-US/OHSAS-18001-Occupational-Health-and-Safety/ISO-45001/

[11] ISO (2015) ISO 45001: Briefing Notes. http://www.iso.org/iso/iso_45001_briefing_note.pdf

# 13

# Assessment of Safety Practices and Injuries Associated with Wood Processing in a Timber Company in Ghana

**Stephen J. Mitchual*, Mark Donkoh, Francis Bih**

Department of Construction and Wood Technology Education, University of Education, Winneba, Kumasi Campus, Kumasi, Ghana

Email: *Stephen.mitchual@yahoo.com

## Abstract

**This paper reports the assessment of safety practices and injuries in a timber company in Ghana. A cross-sectional survey design was used for the study. Data were collected from 300 respondents at a sawmill in Ghana using a 5-point Likert-type scale questionnaire. Descriptive and inferential statistics were used for the data analyses. The result indicates that the mean rating of the respondents on their use of personal protective equipment ranged from 1.09 to 2.51. This implies that the workers of the sawmill rarely or never wear: gloves, overall, goggles, face shield, nose and mouth mask, earplugs or helmet during wood processing. Additionally, the mean ratings of the respondents regarding safety practices, machine and maintenance ranged from 2.42 to 4.03 with three out of the eight items having mean ratings higher than the theoretical mean of 3.0. The ratings of the respondents of the saw doctoring department on their practice of safety on machine and maintenance were significantly higher than those of the other departments. The result of the study further suggests that the workers of this sawmill do rarely report of: skin irritation/dermatitis, eye irritation, lack of appetite, arm/leg amputation, neck pains, hearing loss and poor eye sight problems. However, they usually report of headaches, nausea, respiratory problems, small cuts, back pains, and hip and leg pains. The outcome of this study therefore suggests that the management of the sawmill studied needs to do more to enforce practice of safety, especially the use of personal protective equipment, to reduce hazards and injuries associated with wood processing in the firm.**

## Keywords

**Safety Practices, Wood Processing, Wood Processing Injuries, Woodworkers**

*Corresponding author.

## 1. Introduction

Keeping the workplace safe must not be the concern of only workers and companies but also national and global economies whose productivity and competitiveness play a major role on safe working environment. The International Labour Organisation (ILO) considers issues relating to occupational health and safety (OHS) to be of much importance to the extent that it has devoted about 80% of its standards and instruments either wholly or partly to it [1]. This notwithstanding, currently around 160 million people are estimated to suffer from occupational diseases. Additionally, two million people die every year as a result of occupational accidents and work-related diseases and injuries [2]. These numbers may translate into an estimate of nearly 5 percent loss of the world's GDP [1]. Furthermore, according to the World Health Organisation (WHO) as cited in [3], poor occupational health and reduced working capacity of workers may cause economic loss up to 10% - 20% of the Gross National Product of many countries for which Ghana is not an exception.

In Ghana the timber industry has been making substantial contribution to its economy over the years. As indicated by [4], the wood processing industry contributes about 11 percent of Ghana's foreign earnings, employs over 100,000 workers and provides a livelihood for over 2.5 million Ghanaians. However, the operations of the timber industry have been found to be associated with high levels of occupational hazards which result in illness and injuries. Records from the past two decades show that about 50 percent of fatal accidents in Ghana's industrial sector came from the wood working industry [5] [6]. According to [1], these figures may have been underestimated as nearly 50% of workplace injuries are not reported, especially in developing countries like Ghana. What the above suggests is that a lot of pragmatic steps need to be put in place to reduce occupational hazards so as to create a safe working environment and also improve the well-being of the workers in the wood processing industry.

Issues regarding OHS in Ghana are regulated by the Department of Factories Inspectorate. This department is responsible for promotion and enforcement of regulatory measures on OHS. In the absence of a national OHS policy in Ghana, the Factories, Offices and Shops Act of 1970, Workmen's Compensation Law, Act 187 (1987) and Labour Act 651 (2003) are the main legislative documents on OHS (Ghana News Agency, 2003) as cited in [7]. In spite of this, OHS has not attained a high profile in the sustainable development agenda of Ghana, where most of the working class is engaged in primary industries such as agriculture, logging, and mining [8]. This has resulted in high industrial-related accidents, injuries and diseases that cost Ghana about 7 percent of its GDP. Even though according to the Factories, Offices and Shop Act of 1970, all the stakeholders namely, the state, the employer and the employee have certain responsibilities, duties and obligations in reducing occupational hazards, the ultimate responsibility of safety practice rests on the worker/employee who directly experiences the effects of occupational hazards. Workers must follow established safety procedures and practice safety to ensure their own safety and good health as well as those of their co-workers. However, much is not known about the extent to which workers in the timber industry in Ghana consciously practice safety. Therefore, this paper seeks to assess the safety practices and injuries associated with wood processing in a timber company in Ghana so as to ascertain the safety culture in this company.

## 2. Research Questions

This study, which assessed the conformity to safety practices and injuries associated with wood processing at a timber processing firm in Ghana is guided by the following research questions:

1) How do woodworkers rate their practice of safety?
2) How does the department of woodworkers affect their rating of practice of safety?
3) How do woodworkers rate the effect of wood processing on their health?
4) How does the department of woodworkers affect their rating of effect of wood processing on their health?

## 3. Research Methodology

### 3.1. Research Design

A cross-sectional survey design was used for this study. According to [9], cross-sectional surveys are appropriate for situations where the data to be collected are about self-reported beliefs or behavior. Besides, it enables the researcher to collect data and compare many different variables at the same time without manipulating the study environment.

### 3.2. Population and Sample

The population for this study comprised 702 workers of a selected sawmill in the Ashanti region of Ghana. This includes: 372 workers at the sawmill department, 221 workers at the veneering department, 92 workers at the moulding department, and 17 workers at the sawdoctoring department. Stratified random sampling technique was used to select respondents from the various departments of the selected company. The sample size was 426 and was determined in accordance with the mathematical formula: $n = N/\left(1+N(\alpha)^2\right)$ where n = sample size; N = sampling frame; $\alpha$ = confidence level [7].

### 3.3. Instruments

The instrument used for the larger study was a questionnaire which was adapted from safety management perception questionnaire prepared by [10]. The questionnaire was made up of five (5) parts. The first part dealt with the demographic data of the respondents namely: gender, length of service, level of education, and department. The second part addressed issues on awareness of OHS by woodworkers of the selected company. The other three parts of the questionnaire assessed: 1) the willingness of the workers to use safety practice equipment 2) conformity to safety practices by the workers of the wood processing company, and 3) the effect of wood processing on the health of the woodworkers. The focus of this paper is on the conformity to safety practices by the workers of the wood processing company, and the effect of wood processing on the health of the woodworkers. With respect to conformity to safety practices, respondents were asked to indicate whether they never, rarely, sometimes, usually or always: "do wear gloves/mittens when working", "do wear overall when working", "do wear goggles when working", "do wear face shield when working", "do wear nose and mouth mask when working", "do wear earplugs or ear muffs when working", "do wear helmet when working" etc. On the issue of the effect of wood processing on the health of the woodworkers, respondents were asked to indicate whether they never, rarely, sometimes, usually or always report of: skin irritation/dermatitis, headaches, nausea, eye irritation, respiratory problems, lack of appetite etc. A pilot study was conducted to assess the validity (internal consistency) and reliability of the questionnaire in order to enhance its accuracy for assessment and evaluation. Participants for the pilot study were selected from four departments of the same company namely: veneering, sawmilling, moulding and sawdoctoring. They completed the sample questionnaire and in addition provided suggestions for its modification to help remove any ambiguity. The reliability of the questionnaire, which is concerned with its ability to measure consistently, was determined using the Cronbach's alpha. A Cronbach's alpha value of 0.78 which was obtained for the questionnaire was considered adequate.

### 3.4. Data Collection Procedure

The data for this part of the study was collected using a five-point Likert scale questionnaire at a single point in time. The questionnaire was administered by the researchers to the participants directly. A total number of about 426 questionnaires were distributed. The number of questionnaires successfully completed and returned were 300 (Sawmilling = 135; Veneering = 102; Moulding = 52; Sawdoctoring = 11). This represents a return rate of about 70%. According to Dillman (2000) as cited in [11], a return rate of 70% is a representative sample of the population. Participants were made to indicate their willingness to participate in the study. Directives on the questionnaires ensured respondents anonymity and confidentiality.

### 3.5. Method of Data Analysis

The data were analysed using descriptive and inferential statistics. Software used for the analysis was Statistical Package for Social Scientists (SPSS). The mean and standard deviation of the ratings for each of the items were computed, and the mean compared to the theoretical mean rating of 3.0 (assuming normal distribution of responses) to ascertain the respondents' perception on the themes studied. Additionally, the effects of department on the respondents' adherence to safety practices as well as injuries associated with wood processing were determined. An item-by-item one-way analysis of variance (ANOVA) at 5% level of significance was performed to establish possible significant difference in the respondents' ratings of the factors of this study. P-values lower than 0.05 were deemed significant. In most of such situations Scheffe's post hoc test was used to make pair wise comparison of the means.

# 4. Results and Discussion

## 4.1. Woodworkers Practice of Safety

Occupational health and safety can be important for several reasons, in that, good OHS practice can reduce employee injury and illness-related costs. Additionally, cost associated with sick leave and disability benefit can be reduced. Even though most of these costs could be prevented or significantly reduced through implementation of sound prevention, reporting and inspection practices they persist at various workplaces. **Table 1** indicates the mean and the resultant mean ratings as well as their corresponding standard deviations of practice of safety by the respondents. The mean rating of the 300 respondents on the use of personal protective equipment ranged from 1.09 (*SD* = 0.548) for item # 7 to 2.51 (*SD* = 1.013) for item # 5.

That of safety practices, machine and maintenance, ranged from 2.42 (*SD* = 1.826) for item # 14 to 4.03 (*SD* = 1.549) for item # 10. The resultant mean rating of the respondents' practice of safety was 2.17 (*SD* = 1.343). Apart from three items ("do you ensure that guards and fences are in place when working", "do you ensure that workers put off electrical gargets", "do you ensure that workers maintain and repair machines") for which the respondents' ratings were higher, all the other items had mean ratings lower than the theoretical mean of 3.0. This suggests that the workers of the sawmill studied rarely or never wear: gloves, overall, goggles, face shield, nose and mouth mask, earplugs or helmet during wood processing. Furthermore, the workers would rarely or never: insist that only trained personnel operate machines, ensure that workers adhere to safety rules, ensure that saws are adequately conditioned, and ensure that worn out chains and ropes are changed.

The rating of the respondents, especially on the use of personal protective equipment was a true reflection of what pertains at the sawmill used for the study. Indeed, most of the workers were either in "slippers" or "canvass" which do not protect their foot/toes (**Figure 1**). Additionally, almost all the workers were not in goggles, helmet, nose and mouth mask, overall (**Figure 2**), gloves/mittens or earplugs. This result is consistent with the outcome of a study conducted by [12] and [13]. In a study conducted on the theme "Awareness of Occupational Hazards, Health Problems and Safety Measures among Sawmill Workers in North Central Nigeria", [12] concluded that less than 20% of the sawmill workers wore protective devices/clothing and this was due to the fact that health and safety standards were neither practiced nor enforced.

**Table 1.** Descriptive statistics on ratings of respondents on practice of safety.

| Item # | Elements of safety practice | Mean rating (n = 300) | Standard deviation |
|---|---|---|---|
| | **Personal protective equipment** | | |
| 1 | Do you wear gloves/mittens when working? | 1.72 | 1.082 |
| 2 | Do you wear overall when working? | 1.10 | 0.520 |
| 3 | Do you wear goggles when working? | 1.19 | 0.754 |
| 4 | Do you wear face shield when working? | 1.14 | 0.651 |
| 5 | Do you wear nose and mouth mask when working? | 2.51 | 1.013 |
| 6 | Do you wear earplugs or ear muffs when working? | 1.22 | 0.720 |
| 7 | Do you wear helmet when working? | 1.09 | 0.548 |
| | **Machine and maintenance** | | |
| 8 | Do you ensure that guards and fences are in place when working? | 3.03 | 1.890 |
| 9 | Do you ensure that trained personnel operate the machines? | 2.56 | 1.528 |
| 10 | Do you ensure that workers put off electrical gargets? | 4.03 | 1.549 |
| 11 | Do you ensure that workers adhere to safety rules? | 2.58 | 1.570 |
| 12 | Do you ensure that workers maintain and repair machines? | 3.31 | 1.854 |
| 13 | Do you ensure that saws are adequately conditioned? | 2.43 | 1.854 |
| 14 | Do you ensure that worn out chains and ropes are changed? | 2.42 | 1.826 |
| | Resultant mean for elements of safety practice | 2.17 | 1.343 |

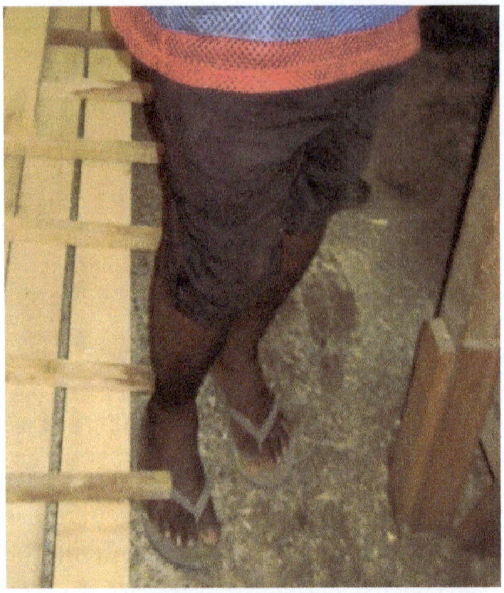

**Figure 1.** Picture of woodworker wearing bath room slippers whilst working.

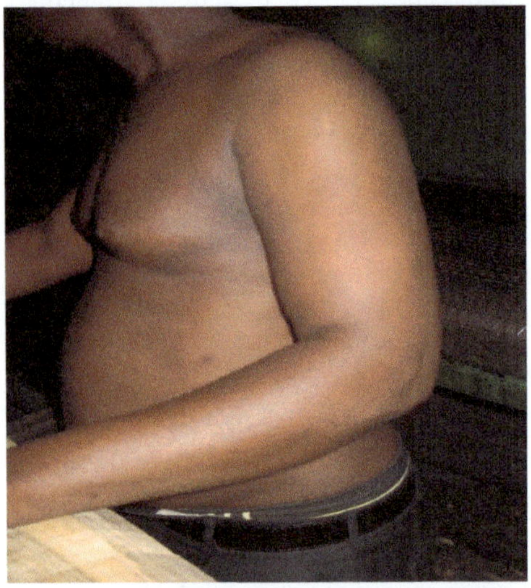

**Figure 2.** Picture of a worker neither in shirt or overall. He complained the environment was too warm.

A study by [13] on small scale sawmilling industries in Tamale Metropolis in Ghana also indicated that a significant number of the workers in the study area did not use personal protective equipment when operating machines or performing jobs that require their use. The non adherence to safety practices by the woodworkers would expose them to various degrees of hazards and injuries. For example, according to [14], many timber species contain chemicals that can irritate the eyes, nose and throat, causing shortness of breath, dryness and soreness of the throat and sneezing. Therefore, if workers do not adopt the use of the appropriate personal protective equipment they will expose themselves to various degrees of health problems and injuries.

## 4.2. Comparison of Ratings of Departments on Practice of Safety by Woodworkers

The effect of department on respondents' rating of their practice of safety is shown in **Table 2**. With the exception of the sawdoctoring department for which the mean rating of the respondents on item # 3 was equal to the

**Table 2.** *ANOVA* for the ratings of respondents' on practice of safety.

| Item # | Elements of safety practice | Veneering ($n_1$ = 102) | | Sawmilling ($n_2$ = 135) | | Moulding ($n_3$ = 52) | | Sawdoctoring ($n_4$ = 11) | | F-value | p-value |
|---|---|---|---|---|---|---|---|---|---|---|---|
| | | Mean | SD | Mean | SD | Mean | SD | Mean | SD | | |
| | **Personal protective equipment** | | | | | | | | | | |
| 1 | Do you wear gloves/mittens when working? | 1.53 | 1.01 | 1.80 | 1.04 | 1.65 | 1.01 | 2.91 | 1.70 | 6.09 | 0.000* |
| 2 | Do you wear overall when working? | 1.12 | 0.55 | 1.10 | 0.56 | 1.00 | 0.00 | 1.36 | 0.81 | 1.63 | 0.182† |
| 3 | Do you wear goggles when working? | 1.21 | 0.80 | 1.10 | 0.56 | 1.00 | 0.00 | 3.00 | 1.55 | 29.28 | 0.000* |
| 4 | Do you wear face shield when working? | 1.16 | 0.73 | 1.01 | 0.17 | 1.12 | 0.62 | 2.73 | 1.42 | 30.42 | 0.000* |
| 5 | Do you wear nose and mouth mask when working? | 2.48 | 1.08 | 2.58 | 0.93 | 2.46 | 1.06 | 2.18 | 1.17 | 0.65 | 0.582† |
| 6 | Do you wear earplugs or ear muffs when working? | 1.20 | 0.77 | 1.24 | 0.74 | 1.15 | 0.54 | 1.45 | 0.82 | 0.59 | 0.619† |
| 7 | Do you wear helmet when working? | 1.10 | 0.59 | 1.10 | 0.61 | 1.00 | 0.00 | 1.18 | 0.60 | 0.60 | 0.615† |
| | **Machine and maintenance** | | | | | | | | | | |
| 8 | Do you ensure that guards and fences are in place when working? | 3.09 | 1.80 | 2.84 | 1.92 | 3.08 | 1.98 | 4.55 | 1.21 | 2.93 | 0.034* |
| 9 | Do you ensure that trained personnel operate the machines? | 2.85 | 1.42 | 2.33 | 1.53 | 2.46 | 1.64 | 3.09 | 1.58 | 2.81 | 0.040* |
| 10 | Do you ensure that workers put off electrical gargets? | 4.48 | 1.20 | 3.83 | 1.61 | 3.69 | 1.81 | 4.00 | 1.61 | 4.61 | 0.004* |
| 11 | Do you ensure that workers adhere to safety rules? | 2.75 | 1.34 | 2.40 | 1.65 | 2.54 | 1.71 | 3.55 | 1.57 | 2.40 | 0.068† |
| 12 | Do you ensure that workers maintain and repair machines? | 3.91 | 1.55 | 3.00 | 1.92 | 2.81 | 1.91 | 3.91 | 1.87 | 6.88 | 0.000* |
| 13 | Do you ensure that saws are adequately conditioned? | 1.65 | 1.38 | 2.84 | 1.92 | 2.46 | 1.90 | 4.64 | 1.21 | 6.26 | 0.000* |
| 14 | Do you ensure that worn out chains and ropes are changed? | 1.82 | 1.55 | 2.74 | 1.89 | 2.58 | 1.91 | 3.27 | 1.68 | 15.37 | 0.000* |

*Statistically significant at 0.05 level of significance; †Not statistically significant at 0.05 level of significance.

theoretical mean of 3.0, all the other items under the use of personal protective equipment were rated less than the theoretical mean. It is also worth noting that all the item-by-item mean rating of the respondents in the department of sawdoctoring for practice of safety, that is, machine and maintenance was higher than the theoretical mean of 3.0. The rating of the respondents in this department for the practice of safety, machine and maintenance, ranged from 3.09 for item # 9 to 4.64 for item # 13. This result implies that the workers at the sawdoctoring department are more likely to usually or always enforce safety practices in respect of machine and maintenance as compared to those of the other departments who would never or rarely enforce such safety practices. The item-by-item one-way analysis of variance (*ANOVA*) shows that at 5% level of significance, the department for which respondents belongs to significantly had effect on the rating of their use of personal protective equipment for the following items: "do you wear gloves/mittens when working", "do you wear goggles when working" and "do you wear face shield when working".

Even though there were significant differences in the respondents' rating for the use of personal protective equipment for the items mentioned above, with the exception of the item "do you wear goggles when working" all the other items were rated lower than the theoretical mean of 3.0. This means that irrespective of the differences in the ratings of the respondents of departments, they would rarely or never wear gloves/mittens and face shield when working. On the contrary, the rating of the respondents on the following items: "do you wear overall when working", "do you wear nose and mouth mask when working", "do you wear earplugs or ear muffs when working" and "do you wear helmet when working" did not differ significantly.

With respect to practice of safety, machine and maintenance, apart from the item "do you ensure that workers adhere to safety rules", the respondents' rating for all the other items differed significantly. To determine the items whose mean ratings differed significantly, an item-by-item Scheffe's post hoc multiple comparison of means was performed especially for the items whose mean ratings were higher than the theoretical mean of 3.0 (Item # 8, 10, 12). The result shows that for item # 8 there was a significant difference between respondents rating for sawmilling and sawdoctoring departments, with that of sawdoctoring being significantly higher than that of sawmilling department. In addition there was significant difference in the ratings of respondents in the veneering and sawmilling departments, and, veneering and moulding departments for items # 10 and # 12 respectively.

## 4.3. Health Hazards and Injuries Associated with Wood Processing

The result shown in **Table 3** indicates the means of the respondents' ratings of health hazards and injuries associated with wood processing as well as their corresponding standard deviations. **Table 3** also indicates the resultant mean ratings of the respondents. Seven out of the total number of 13 items were rated lower than the theoretical mean value of 3.0 and the ratings ranged from 1.10 to 2.99. This implies that the respondents do never or rarely report of: skin irritation/dermatitis, eye irritation, lack of appetite, arm/leg amputation, neck pains, hearing loss, and poor eye sight. On the contrary, six out of the total number of the 13 items were rated higher than the theoretical mean value of 3.0 and the ratings ranged from 3.08 to 3.40. This also suggest that on the average, the respondents would usually or always report of headaches, nausea, respiratory problems, small cuts, back pains, and hip and leg pains. **Figure 3** and **Figure 4** for instance are samples of pictures showing two workers who sustained various degrees of cut injuries during wood processing in the timber processing firm studied. Health hazards and injuries associated with wood processing have been a major problem in the timber industry worldwide.

Analysis of data collected from 60 respondents from three constituencies in the Tamale Metropolis in Ghana indicated: cut-type of injuries, fractures, sprains, catarrh, waist pains, eye problems and dizziness as the range of injuries and illness associated with wood processing [13]. Similar study by [12] also indicated that minor accidents constituted about 40.5% of the health problems that the workers experienced. Furthermore, [12] indicated that this was followed by stress and exhaustion after work (25.7%) and eye irritation (14.0%). What the above suggest is that the timber industry needs to do more to protect the health and safety of its employees since the human and economic costs of occupational accidents and diseases remain high and it requires concerted efforts to control them.

## 4.4. Comparison of Ratings of Respondents in Departments Regarding Effect of Wood Processing on the Health of Woodworker

The comparison of the respondents' ratings of health hazards and injuries associated with wood processing for the four departments is shown in **Table 4**. The result indicates that at 5% level of significance, there was a significant difference in the ratings of the respondents on the occurrence of health problems and injuries related to wood processing for six of the items. Significant differences between the respondents' ratings occurred for the following items: "I do report of eye irritation", "I do report of lack of appetite", "have you ever had arm/leg amputation", "I do report of back pains", "I do report of hip and leg pains", and "I do report of hearing loss". For four of these items ("I do report of eye irritation", "I do report of lack of appetite", "have you ever had arm/leg amputation", "I do report of hearing loss") the mean ratings were generally lower than the theoretical mean of 3.0. What this means is that even though there were significant differences in the ratings of the departments, those health hazards would never or will rarely occur. In view of that Scheffe's post hoc multiple comparisons of means was performed for the items: "I do report of back pains" and "I do report of hip and leg pains".

The result for pair-wise comparison of means for the item # 10 (I do report of back pains) indicated that the mean rating of the workers in the veneering ($\bar{x} = 3.39$), sawmilling ($\bar{x} = 3.40$) and the moulding ($\bar{x} = 3.62$) departments were significantly higher than that of the sawdoctoring department ($\bar{x} = 2.45$). This may be due to the fact that the nature of work in the three departments (veneering, sawmilling, moulding) requires a lot of standing and bending that could give rise to frequent pains at their back. The result for pair-wise comparison of means for item # 11 also indicated that the mean rating of the workers in the moulding department ($\bar{x} = 3.46$) was significantly higher than that of the sawdoctoring department ($\bar{x} = 2.64$). This may also be due to the same reason that the nature of work in the moulding department requires a lot of standing and bending that could result in frequent hip and leg pains.

**Table 3.** Descriptive statistics on ratings of respondents' on effect of wood processing on the health of woodworkers.

| Item # | Elements of health hazards of wood processing on woodworkers | Mean rating (n = 300) | Standard deviation |
|---|---|---|---|
| | **Exposure to sawdust** | | |
| 1 | I do report of skin irritation/dermatitis | 2.25 | 1.122 |
| 2 | I do report of headaches | 3.38 | 1.252 |
| 3 | I do report of nausea | 3.13 | 0.846 |
| 4 | I do report of eye irritation | 2.57 | 1.346 |
| 5 | I do report of respiratory problems | 3.08 | 1.030 |
| 6 | I do report of lack of appetite | 1.49 | 0.909 |
| | **Exposure to cutting sharp edge** | | |
| 7 | I do report of small cuts | 3.27 | 0.981 |
| 8 | Have you ever had arm/leg amputation | 1.10 | 0.514 |
| | **Heavy lifting injury** | | |
| 9 | I do report of neck pains | 2.99 | 1.076 |
| 10 | I do report of back pains | 3.40 | 0.888 |
| 11 | I do report of hip and leg pains | 3.35 | 0.862 |
| | **Noise/poor ventilation injuries** | | |
| 12 | I do report of hearing loss | 1.35 | 0.810 |
| 13 | I do report of poor eye sight | 1.77 | 1.003 |
| | Resultant mean for the elements of health hazard of wood processing on woodworkers | 2.55 | 0.993 |

**Table 4.** *ANOVA* for ratings of respondents' of wood processing effect on the health of woodworkers.

| Item # | Wood processing effect on the health of woodworkers | Veneering ($n_1 = 102$) | | Sawmilling ($n_2 = 135$) | | Moulding ($n_3 = 52$) | | Sawdoctoring ($n_4 = 11$) | | F-value | p-value |
|---|---|---|---|---|---|---|---|---|---|---|---|
| | | Mean | SD | Mean | SD | Mean | SD | Mean | SD | | |
| | **Exposure to sawdust** | | | | | | | | | | |
| 1 | I do report of skin irritation | 2.31 | 1.08 | 2.21 | 1.11 | 2.27 | 1.21 | 2.09 | 1.30 | 0.252 | 0.860† |
| 2 | I do report of headaches | 3.25 | 1.06 | 3.52 | 1.30 | 3.37 | 1.39 | 2.91 | 1.58 | 1.417 | 0.238† |
| 3 | I do report of nausea | 3.19 | 0.73 | 3.16 | 0.88 | 3.06 | 0.85 | 2.55 | 1.21 | 2.120 | 0.098† |
| 4 | I do report of eye irritation | 2.77 | 1.31 | 2.43 | 1.35 | 2.35 | 1.34 | 3.45 | 1.13 | 3.419 | 0.018* |
| 5 | I do report of respiratory problems | 3.10 | 0.91 | 3.19 | 1.06 | 2.79 | 1.16 | 2.91 | 0.94 | 1.988 | 0.116† |
| 6 | I do report of lack of appetite | 1.53 | 0.93 | 1.39 | 0.85 | 1.46 | 0.85 | 2.36 | 1.29 | 4.109 | 0.007* |
| | **Exposure to cutting sharp edge** | | | | | | | | | | |
| 7 | I do report of small cuts | 3.22 | 0.75 | 3.38 | 1.08 | 3.15 | 1.07 | 3.09 | 1.14 | 1.012 | 0.388† |
| 8 | Have you ever had arm/leg amputation | 1.07 | 0.35 | 1.07 | 0.43 | 1.15 | 0.67 | 1.55 | 1.29 | 3.336 | 0.020* |
| | **Heavy lifting injury** | | | | | | | | | | |
| 9 | I do report of neck pains | 3.14 | 0.90 | 2.88 | 1.17 | 3.06 | 1.07 | 2.55 | 1.21 | 1.804 | 0.147† |
| 10 | I do report of back pains | 3.39 | 0.80 | 3.40 | 0.87 | 3.62 | 0.91 | 2.45 | 1.29 | 5.402 | 0.001* |
| 11 | I do report of hip and leg pains | 3.36 | 0.79 | 3.35 | 0.82 | 3.46 | 0.98 | 2.64 | 1.12 | 2.865 | 0.037* |
| | **Noise/poor ventilation injuries** | | | | | | | | | | |
| 12 | I do report of hearing loss | 1.38 | 0.80 | 1.21 | 0.66 | 1.46 | 0.94 | 2.09 | 1.38 | 4.892 | 0.002* |
| 13 | I do report of poor eye sight | 1.82 | 1.01 | 1.69 | 0.94 | 1.77 | 1.06 | 2.36 | 1.29 | 1.687 | 0.170† |

*Statistically significant at 0.05 level of significance; †Not statistically significant at 0.05 level of significance.

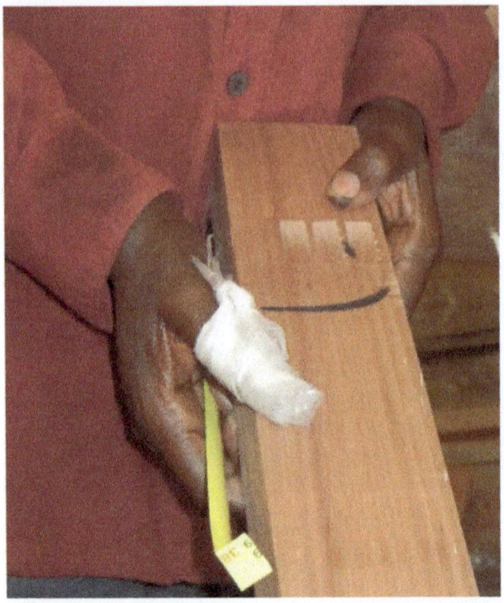

**Figure 3.** Picture of a woodworker with a minor cut.

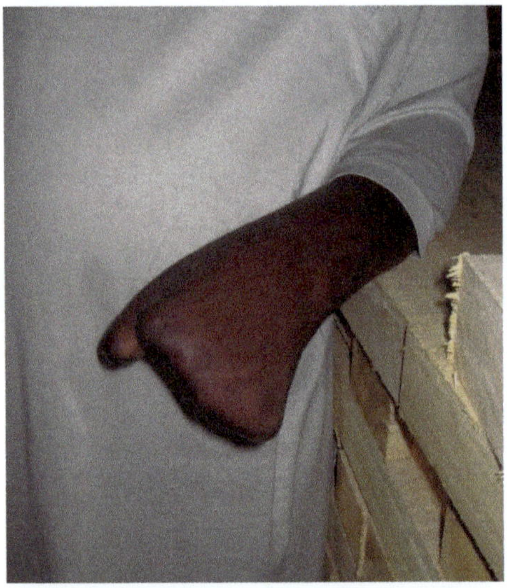

**Figure 4.** Picture of a woodworker with all the fingers cut-off during operation of a machine without prior training.

## 5. Conclusion and Recommendation

Studies conducted in other parts of the world suggest that the wood industry is exposed to high levels of occupation hazards. This in no doubt affects the social well-being of the workers, their families as well as the economies of this industry. In view of the above, this study was conducted to assess the safety practices and injuries associated with a wood processing company in Ghana. The study found that the workers of the timber firm studied would rarely or never wear: gloves, overall, goggles, face shield, nose and mouth mask, earplugs or helmet during wood processing. Additionally, the workers would rarely or never: insist that only trained personnel operate machines; ensure that workers adhere to safety rules; ensure that saws are adequately conditioned; and ensure that worn out chains and ropes are changed. The workers at the saw doctoring department are also more likely to enforce safety practices on machine and maintenance as compared to those of the other departments. It could further be concluded from the study that the workers of this sawmill do rarely report of: skin irritation/ dermatitis, eye irritation, lack of appetite, arm/leg amputation, neck pains, hearing loss and poor eye sight. Nevertheless, they usually report of headaches, nausea, respiratory problems, small cuts, back pains, and hip and leg pains. It is therefore recommended that occupational health and safety practices should be promoted in the sawmill studied through adoption of good health and safety practice as a personal and organisational value.

## Acknowledgements

The authors are grateful to the management of the timber firm used for the study. Our appreciation also goes to the workers for willingly accepting to be part of this study. Special thanks also go to Mr. Francis Donkor of the Department of Mechanical Technology Education, University of Education for his professional advice during the preparation of the instrument for the study.

## References

[1] Alli, B.O. (2008) Fundamental Principles of Occupational Health and Safety. 2nd Edition, International Labour Office, Geneva.

[2] ILO (2013) Health and Safety at Work: Facts and Figures. http://www.ilo.org/global/about-the-ilo/media-centre/issue-briefs/WCMS_206117/lang--en/index.htm

[3] Amponsah-Tawiah, K. and Dartey-Baah, K. (2011) Occupational Health and Safety: Key Issues and Concerns in Ghana. *International Journal of Business and Social Science*, **2**, 119-126.

[4] Acquah-Moses, E.K. (2002) Supply to the Local Market: From Policy to Chain Saw Lumber Production Production a Necessary Evil? *Troperbs International-Ghana Workshop*, *Proceeding* 2, Wood Industries Training Centre, Akyawkrom, Ejisu, Kumasi.

[5] Boateng, M. and Nimako, M. (2000) A Situational Analysis of Occupational Health and Safety in Ghana. Unpublished Draft Report, Ministry of Employment and Social Welfare, Accra, 15-22.

[6] Ministry of Health, Ghana Health Services (2002) A Handbook of Occupational Health. Occupational and Environmental Health Unit, Accra.

[7] Adei, D. and Kunfaa E.Y. (2007) Occupational Health and Safety Policy in the Operation of the Wood Processing Industry in Kumasi, Ghana. *Journal of Science and Technology*, **27**, 161-173.

[8] Amponsah-Tawiah, K. (2013) Occupational Health and Safety and Sustainable Development in Ghana. *International Journal of Business Administration*, **4**, 74-78. http://dx.doi.org/10.5430/ijba.v4n2p74

[9] Neuman, W.L. (2000) Social Research Methods: Qualitative and Quantitative Approaches. Allynn & Bacon, Boston.

[10] Lapidus, R.A. and Waite, M.J. (2001) Safety Management Perception Questionnaire. http://www.sipeonlinetraining.com/A-ZResources/Safety%20Management%20Perception%20Survey.pdf

[11] Bright Ideas (2009) How Many Is Enough? The Quest for an Acceptable Survey Response Rate. https://kkbiersdorff.wordpress.com/2009/09/16/how-many-is-enough/

[12] Osagbemi, G.K., La-Kadri, R.T. and Aderibigbe, S.A. (2010) Awareness of Occupational Hazards, Health Problems and Safety Measures among Sawmill Workers in North Central Nigeria. *TAF Preventive Medicine Bulletin*, **9**, 325-328.

[13] Ochire-Boadu, K., Kusi, E. and Lawer, E.A. (2014) Occupational Hazards and Safety Practices: A Concern among Small Scale Sawmilling Industries in Tamale Metropolis, Ghana. *International Journal of Scientific & Technology Research*, **10**, 234-236.

[14] Ahman, M., Holmstrom, M., Cynkier, I. and Soderman, E. (1996) Work Related Impairment of Nasal Function in Swedish Woodwork Teachers. *Occupational and Environmental Medicine*, **53**, 112-117. http://dx.doi.org/10.1136/oem.53.2.112

# 14

# OSHA's Enforcement of Asbestos Standards in the Construction Industry

**Christopher Allen Janicak**

Department of Safety Sciences, Indiana University of Pennsylvania, Indiana, USA
Email: cjanicak@iup.edu

## Abstract

**Background: Exposure to asbestos continues to be a concern for workers in the construction industry. Asbestos can still be found in many construction products and in hundreds of thousands of buildings in the United States. Methods: Data from OSHA's Integrated Management Information System (IMIS) were used to identify inspections in which violations of OSHA's asbestos standards were cited from 2010 to 2012. Employers selected for analysis had NAICS codes in the construction industry sector. Descriptive statistics were used to describe the characteristics of OSHA's enforcement approach to asbestos standards violations in the construction industry. Nonparametric statistics were used to identify significant differences in penalties assessed for violating asbestos standards based upon the type of violation and the type of inspection. Results: This study identified 4017 violations from 846 inspections in which the asbestos standards were cited in the construction industry. Employee complaints and referrals resulted in the largest number enforcement activities. Significant differences were identified in the fines assessed for different types of violations and inspections. Site preparation contractors, residential construction, and commercial and institutional building construction trades experienced the greatest number of violations. Most frequently cited standards included employees performing work in areas that were not properly regulated, personal protective equipment not meeting standards, and employee training not meeting standards. Conclusions: Recommended control measures include conducting targeted inspections in construction industry trades with a greater potential exposure to asbestos, improving worker awareness of asbestos and its hazards, strengthening the fining structure for asbestos violations, and conducting further research to determine underlying reasons for employers' inability to comply with OSHA's asbestos standards.**

## Keywords

**Asbestos, Construction Industry, OSHA, Enforcement**

# 1. Introduction

Over the years, the use of asbestos in the United States has declined substantially and mining of asbestos in the United States ceased in 2002 [1]. However, many asbestos products remain in use and new asbestos-containing products continue to be manufactured in or imported into the United States [1]. Exposure to asbestos is a great concern especially for people working in the construction industry. The presence of asbestos in construction materials from use decades ago, the continued use of asbestos containing materials not banned today in the United States, and work activities that require workers to disturb asbestos containing materials through demolition, alterations, and construction activities add up to increased chances for exposure to airborne asbestos.

To protect workers from asbestos exposure, the Occupational Safety and Health Administration (OSHA) has adopted the asbestos standards for the construction industry, 29 CFR 1926.1101 [2]. The OSHA standards apply to asbestos family of materials of chrysotile, amosite, crocidolite, tremolite asbestos, anthophyllite asbestos, actinolite asbestos, and any of these minerals that have been chemically treated and/or altered [2]. The standards do not apply to asbestos-containing asphalt roof coatings, cements and mastics [2]. The OSH Act is enforced by the federal OSHA in 29 states while 21 states have state plans which meet or exceed the requirements of federal OSHA [3].

The OSHA asbestos standards establish requirements for monitoring, measuring, and protecting workers who may potentially be exposed. Most OSHA state plan states adopted the 29 CRF 1926.1101 federal OSHA standards. Two states, California (CAL/OSHA) and Washington State (WISHA) have their own standards for asbestos in the construction industry [4].

In addition to promulgating standards, OSHA has the responsibility of ensuring safe and healthful workplaces through enforcement efforts. These workplace inspections can be initiated through a number of avenues ranging from planned inspections in which the site falls under a targeted OSHA inspection priority to employee complaints and referrals. OSHA also has established inspection priorities with their number one priority being imminent dangers followed by fatalities, catastrophes, and accidents [5]. Employers on the other hand, have the responsibility to protect their employees from the hazards of asbestos and federal OSHA and state OSHA plans have the responsibility of enforcing the asbestos safety standards. The purpose of this study was to examine the enforcement of occupational safety and health standards for asbestos in the construction industry. With a better understanding of how these standards are being enforced, recommend actions can be developed which, when implemented, can better protect workers.

# 2. Methodology

## 2.1. Data

The focus of this study was on the safety inspection and enforcement activities related to asbestos in the construction industry. Data from OSHA's Integrated Management Information System (IMIS) was analyzed for inspections conducted by the federal Occupational Safety and Health Administration and the agencies enforcing state plan occupational safety and health standards. The IMIS system contained data from both federal OSHA states and state plan states. Inspection data and violation data in which violations of OSHA's 29 CFR 1926.1101, CAL/OSHA's, Title 8 Section 1529, or WISHA's WAC 296-62-07705 were cited were included for analysis. Inspections were conducted between January 1, 2010 and December 31, 2012 involving employers with NAICS codes falling in the construction industry sector.

The following variables from the IMIS records were examined in this study:

*Asbestos Standard Violated*: The specific asbestos standard number that was violated was identified. The standard numbers selected for this study included OSHA's 29 CFR 1926.1101, CAL/OSHA's Title 8 Section 1529, and WISHA's WAC 296-62-077.

*Penalties*: The current penalty assigned to each standard violated was identified. Current penalties are the penalties assessed following any reductions or deletions from administrative actions such as informal conferences.

*Type of Inspection*: The type of inspection was classified as follows [6]:

*Fatalities and Catastrophes*—Incidents that involved a death or the hospitalization of three or more employees.

*Complaints*—Allegations of hazards or violations by employees.

*Referrals*—Referrals of hazard information from other federal, state or local agencies, individuals, organizations or the media receive consideration for inspection.

*Follow-Ups*—Checks for abatement of violations cited during previous inspections conducted by OSHA in certain circumstances.

*Planned or Programmed Investigations*—Inspections aimed at specific high-hazard industries or individual workplaces that have experienced high rates of injuries and illnesses.

*Unprogrammed Related Investigations*—Inspections of employers at multi-employer work sites whose operations are not directly affected by the subject of the conditions identified in the complaint, accident, or referral.

*Type of Violation*: The violations were classified according to their severity according to the following OSHA classification system [5]:

*Willful Violation*: A *willful* violation exists where the evidence shows either an intentional violation of the Act or plain indifference to its requirements.

*Repeat Violation*: A repeated violation exists if that employer has been cited previously for a substantially similar condition and the citation has become a final order.

*Serious Violation*: A serious violation exists when a hazard causes or was likely to cause death or serious physical harm.

*Other-Than-Serious Violation*: An other-than-serious violation is cited in situations where the most serious injury or illness that would be likely to result from a hazardous condition cannot reasonably be predicted to cause death or serious *physical* harm to exposed employees but does have a direct and immediate relationship to their safety and health.

*NAICS Code*: The North American Industrial Classification System (NAICS) code for the establishment site as determined by OSHA was used to classify establishments by industry sector. The construction industry consists of industry sectors with NAICS codes that range from 230000 to 238990.

*Number of Employed Persons*: The number of employed persons over the three year period was obtained using the Bureau of Labor Statistics Current Population Survey.

### 2.2. Statistical Analyses

Data was analyzed descriptively to summarize the number of asbestos standard violations by industry, the type of inspection conducted, the standard number that was violated, and the type of violation. Penalties assessed for asbestos violations were summarized by type of inspection, standard number, and type of violation.

Because the distribution of penalties was not normally distributed, use of non-parametric statistical procedures was required. The Kruskal-Wallis test examines ranks across the categorical independent groups [7]. If all of the samples were from the same population, then the mixture of high, medium and low ranks should be evenly distributed across the groups [7].

The Kruskal-Wallis test was used in this study to identify the presence of significant differences in the mean rankings of penalties assessed across the different types of inspections. A second Kruskal-Wallis test was used to identify significant differences in the mean rankings of penalties assessed between the different asbestos violation categories. *Post hoc* Mann-Whitney U tests were used to identify significant differences between each of the pairwise groups [8].

## 3. Results

### 3.1. Type of Inspections

Data from the IMIS system identified 4017 violations from 846 inspections in which the asbestos standards were cited in construction industry establishments during the analysis period of January 1, 2010 to December 31, 2012. Employee complaints accounted for approximately 34.0 percent of all asbestos violations, referrals accounted for 22.8 percent, and planned inspections accounted for approximately 20.3 percent (see **Table 1**).

Overall, the average penalty assessed for a violation of an asbestos standard was $809. The average penalties ranged from $239 for violations classified as "Other-than-serious" to $9205 for violations classified as "Willful". A summary of the results appears in **Table 2**.

**Table 1.** Frequency, percentage, and average penalty of asbestos violations by inspection type (2010-2012).

| Inspection Type | Descriptive Statistics | | |
|---|---|---|---|
| | N | Percent | Average Penalty (USD) per Violation |
| Accident | 5 | 0.1 | 343 |
| Complaint | 1365 | 34.0 | 1090 |
| Referral | 914 | 22.8 | 915 |
| Follow Up | 5 | .1 | 1040 |
| Unprogrammed-Related | 744 | 18.5 | 558 |
| Planned | 815 | 20.3 | 537 |
| Programmed-Related | 156 | 3.9 | 402 |
| Programmed-Other | 6 | .1 | 214 |
| Unprogrammed-Other | 7 | .2 | 167 |
| Total | 4017 | 100.0 | $809 |

**Table 2.** Frequency, percentage, and average penalty of asbestos violations by violation type (2010-2012).

| Inspection Type | Descriptive Statistics | | |
|---|---|---|---|
| | N | Percent | Average Penalty (USD) per Violation |
| Serious | 3196 | 79.6 | 637 |
| Other-Than-Serious | 692 | 17.2 | 239 |
| Repeat | 18 | 0.4 | 1372 |
| Willful | 111 | 2.8 | 9,205 |
| Total | 4017 | 100.0 | $809 |

## 3.2. Violations

An analysis of asbestos related inspections and violations identified the Site Preparation Contractor sector as having the largest percentage of asbestos inspections (20.2%) and the largest percentage of asbestos violations (24.8%) (see **Table 3**). However, this industry only accounted for less than 5 percent of all asbestos inspections and had an inspection rate of 8.29 inspections per 1000 employees which is below the overall inspection rate of 9.04 inspections per 100,000 employees. This industry comprises establishments primarily engaged in site preparation activities, such as excavating and grading, demolition of buildings and other structures, septic system installation, and house moving (BLS, 2014).

## 3.3. Industry Sector

The Commercial and Institutional Building Construction industry accounted for approximately 15.4 percent of all inspections and 19.3 percent of all inspections resulting in asbestos violations. This industry was the second highest in terms of the frequency of both all inspections and asbestos related inspections. Residential Remodelers were the third most frequently cited industry for asbestos violations accounting for approximately 12.9 percent of all asbestos inspections and 13.3 percent of all asbestos violations. However, this industry accounted for less than 3 percent of all inspections. An analysis of all inspections in construction industry NAICS code found the roofing industry accounted the most inspections overall with approximately 16.2 percent of all inspections but less than 9 percent of all asbestos violations and less than 9 percent of inspections resulting in asbestos violations.

**Table 3.** Frequency, percentage, and average penalty of asbestos violations by violation type (2010-2012).

| Industry Name and NAICS Code | Inspections per 1000 | All Inspections | | Asbestos Inspections | | Asbestos Violations | |
|---|---|---|---|---|---|---|---|
| | Employees | N | % | N | % | N | % |
| New Single-Family Housing Construction (236,115) | 6.45 | 5531 | 3.8 | 37 | 4.4 | 110 | 2.7 |
| New Multifamily Housing Construction (236,116) | 44.23 | 2864 | 2.0 | 10 | 1.2 | 44 | 1.1 |
| New Housing For-Sale Builders (236,117) | 3.26 | 225 | 0.2 | 2 | 0.2 | 8 | .2 |
| Residential Remodelers (236,118) | 4.71 | 3427 | 2.4 | 109 | 12.9 | 533 | 13.3 |
| Industrial Building Construction (236,210) | 5.49 | 2467 | 1.7 | 11 | 1.3 | 54 | 1.3 |
| Commercial and Institutional Building Construction (236,220) | 15.02 | 22,524 | 15.7 | 163 | 19.3 | 752 | 18.7 |
| Water and Sewer Line and Related Structures Construction (237,110) | 12.53 | 5725 | 4.0 | 9 | 1.1 | 47 | 1.2 |
| Power and Communication Line and Related Structures Const. (237,130) | 2.09 | 873 | 0.6 | 1 | 0.1 | 3 | 0.1 |
| Land Subdivision (237,210) | 0.73 | 102 | 0.1 | 4 | 0.5 | 13 | 0.3 |
| Highway, Street, and Bridge Construction (237,310) | 8.54 | 7299 | 5.1 | 3 | 0.4 | 9 | 0.2 |
| Other Heavy and Civil Engineering Construction (237,990) | 6.12 | 1766 | 1.2 | 1 | 0.1 | 1 | 0.0 |
| Poured Concrete Foundation and Structure Contractors (238,110) | 8.52 | 2617 | 1.8 | 3 | 0.4 | 12 | 0.3 |
| Structural Steel and Precast Concrete Contractors (238,120) | 22.64 | 3152 | 2.2 | 1 | .1 | 2 | 0.0 |
| Framing Contractors (238,130) | 57.06 | 9065 | 6.3 | 16 | 1.9 | 50 | 1.2 |
| Masonry Contractors (238,140) | 23.63 | 9182 | 6.4 | 7 | 0.8 | 38 | 0.9 |
| Glass and Glazing Contractors (238,150) | 6.86 | 979 | 0.7 | 6 | 0.7 | 29 | 0.7 |
| Roofing Contractors (238,160) | 47.93 | 23,411 | 16.3 | 73 | 8.6 | 351 | 8.7 |
| Siding Contractors (238,170) | 29.00 | 2701 | 1.9 | 5 | 0.6 | 63 | 1.6 |
| Other Foundation, Structure, and Building Exterior Contractors (238,190) | 8.65 | 936 | 0.7 | 3 | 0.4 | 11 | 0.3 |
| Electrical Contractors and Other Wiring Installation Contractors (238,210) | 3.66 | 7909 | 5.5 | 38 | 4.5 | 161 | 4.0 |
| Plumbing, Heating, and Air-Conditioning Contractors (238,220) | 2.87 | 6938 | 4.8 | 59 | 7.0 | 283 | 7.0 |
| Other Building Equipment Contractors (238,290) | 2.74 | 974 | 0.7 | 4 | 0.5 | 21 | 0.5 |
| Drywall and Insulation Contractors (238,310) | 6.88 | 4046 | 2.8 | 19 | 2.2 | 35 | 0.9 |
| Painting and Wall Covering Contractors (238,320) | 7.16 | 3543 | 2.5 | 13 | 1.5 | 69 | 1.7 |
| Flooring Contractors (238,330) | 3.79 | 660 | 0.5 | 22 | 2.6 | 148 | 3.7 |
| Finish Carpentry Contractors (238,350) | 5.08 | 1659 | 1.2 | 5 | 0.6 | 17 | 0.4 |
| Other Building Finishing Contractors (238,390) | 7.82 | 1439 | 1.0 | 6 | 0.7 | 41 | 1.0 |
| Site Preparation Contractors (238,910) | 8.29 | 6566 | 4.6 | 171 | 20.2 | 997 | 24.8 |
| All Other Specialty Trade Contractors (238,990) | 6.81 | 5234 | 3.6 | 45 | 5.3 | 115 | 2.9 |
| Total | 9.04 | 143,814 | 100.0 | 846 | 100.0 | 4017 | 100.0 |

## 3.4. Asbestos Standard Violated

An analysis of violations by asbestos standard was performed for the three groups of standards; OSHA's 1926.1101 standards, CAL/OSHA's 1529 asbestos standards, and WISHA's 296-62-077 asbestos standards. During the analysis period, there were 3360 violations of the 1926.1101 asbestos standards cited. The most frequently cited sections the standards included violations of the requirements for asbestos work to be conducted in regulated areas, lack of employee training, personal protective equipment (PPE) violations, and a lack of monitoring and exposure assessment (see **Table 4**).

Similar results were identified for the leading CAL/OSHA 1529 standards cited. Overall, there was 123 violations of the 1529 standards with the most frequently cited sections including violations of the requirements for exposure assessment, regulated work areas, employee training, PPE, and control measures. Frequently cited violations of the WISHA standards included violations of standards pertaining to the communication of hazards, monitoring criteria, PPE, and employee training. Overall, there were 534 violations of the asbestos standards identified for the analysis period. Because of the degree of hazards associated with willful violations and the penalties they can carry, an analysis of willful violations of the asbestos standards was conducted. Most frequently cited standards in which willful violations were identified include standards pertaining to work in regulated areas, employee training, monitoring, and respiratory protection.

## 3.5. Inferential Tests

Kruskal-Wallis tests were conducted to determine if significant differences in mean rankings of penalties assessed for asbestos violations exist between the types of violations. The data met the assumptions of the test procedure as described in the Methods section. There were 4,017 cases in the analysis. The Kruskal-Wallis test identified significant differences in the mean ranks of the penalties assessed by type of violation ($X^2 = 133.15$, d.f. = 3, $p < 0.000$) (see **Table 5**). *Post hoc* tests using Mann-Whitney U tests identified significant differences between pairwise mean rankings (see **Table 6**). These *post hoc* tests identified significant differences between the mean rankings for all pairs except for the mean ranking of penalties assessed for repeat violations compared to willful violations. Significant results indicate the penalties differ significantly between groups.

Kruskal-Wallis tests also identified significant differences in the mean rankings of the penalties assessed by type of inspection ($X^2 = 48.49$, d.f. = 8, $p < 0.001$) (see **Table 7**). Mann-Whitney U *post hoc* tests identified a number of significant differences between pairs of mean rankings (see **Table 8**). Mean rankings for penalties assessed were significantly different for follow up inspections compared to all other types of inspections. Complaint and referral inspections penalties were significantly different. Other pairwise comparisons had mixed results.

## 4. Discussion

Exposure to asbestos poses a major health hazard to workers in the construction industry despite efforts to eliminate its use. While there have been a number of bans on the use asbestos enacted in the United States, it is still

**Table 4.** Most frequently cited 1926.1101 standards (2010-2012).

| Standard | N | Percent (Total N = 3360) |
|---|---|---|
| Class I, II, and III asbestos work conducted in regulated areas (1926.1101 E01) | 226 | 6.7 |
| Employee training (1926.1101 K09 I) | 167 | 5.0 |
| Protective clothing (1926.1101 I01) | 160 | 4.8 |
| Initial exposure assessment (1926.1101 F02I) | 146 | 4.4 |
| Duties of employers, identify ACM materials (1926.1101 K03 I) | 118 | 3.5 |
| Competent person (1926.1101 E06) | 109 | 3.2 |
| Methods of compliance, engineering controls and work practices (1926.1101 G01) | 98 | 2.9 |
| Class III controls (1926.1101 G08III) | 92 | 2.7 |
| Class I work, less than 25 linear feet or 10 square feet (1926.1101 J02I) | 90 | 2.7 |
| Respiratory protection (1926.1101 H01) | 89 | 2.7 |
| Total | 1,295 | 38.5 |

**Table 5.** Mean rankings of penalties assessed by type of violation.

| Type of Violation | N | Percent Mean Rank |
|---|---|---|
| Serious | 3196 | 2079.36 |
| Other-Than-Serious | 692 | 1607.84 |
| Repeat | 18 | 2877.75 |
| Willful | 111 | 243.10 |
| Total | 4017 | |

**Table 6.** Mann-whitney u pairswise comparisons: significance of mean rankings of penalties by type of violation.

| Type of Violation | Serious | Other-Than-Serious | Repeat |
|---|---|---|---|
| Serious | - | | |
| Other-than-Serious | 0.002 | - | |
| Repeat | 0.000 | 0.000 | - |
| Willful | 0.000 | 0.003 | 0.571 |

**Table 7.** Mean rankings of penalties assessed by type of violation.

| Type of Inspection | N | Percent Mean Rank |
|---|---|---|
| Accident | 5 | 2614.30 |
| Complaint | 1365 | 2058.07 |
| Referral | 914 | 2124.90 |
| Follow Up | 5 | 3108.30 |
| Unprogrammed-Related | 744 | 2002.21 |
| Planned | 815 | 1838.08 |
| Programmed-Related | 156 | 1802.79 |
| Programmed-Other | 6 | 1639.67 |
| Unprogrammed-Other | 7 | 1623.43 |
| Total | 4017 | |

**Table 8.** Mann-whitney u pairswise comparisons: significance of mean rankings of penalities by type of inspection.

| Type of Inspection | Accident | Complaint | Referral | Follow Up | Unprogrammed-Related | Planned | Programmed-Related | Programmed-Other |
|---|---|---|---|---|---|---|---|---|
| Accident | - | | | | | | | |
| Complaint | 0.296 | - | | | | | | |
| Referral | 0.352 | 0.200 | - | | | | | |
| Follow Up | 0.108 | 0.052 | 0.041 | - | | | | |
| Unprogrammed-Related | 0.151 | 0.242 | 0.041 | 0.014 | - | | | |
| Planned | 0.085 | 0.000 | 0.000 | 0.004 | 0.001 | - | | |
| Programmed-Related | 0.045 | 0.006 | 0.001 | 0.004 | 0.027 | 0.623 | - | |
| Programmed-Other | 0.202 | 0.304 | 0.207 | 0.022 | 0.348 | 0.578 | 0.717 | - |
| Unprogrammed-Other | 0.186 | 0.340 | 0.283 | 0.022 | 0.407 | 0.631 | 0.778 | 0.861 |

present in many forms of construction materials found in hundreds of thousands of existing buildings. This study examined OSHA's enforcement of the asbestos standards in the construction industry.

Over the three year analysis period this study covered, OSHA cited violations of the asbestos standards over 4000 times in 809 inspections. Employers engaged in site preparation trades, which includes demolition-related activities, were cited most often. However, the inspection rate based upon the number of workers was far less than many other construction industries. It appears there is a focus by OSHA on conducting inspections in industries such as the roofing industry, framing industry, and multi-family construction industry and less of a focus on the industries identified with the higher numbers of asbestos violations. To better protect workers, industries with greater exposures to asbestos should receive a higher inspection priority.

The extent to which OSHA conducts planned inspections and enforces the asbestos standards does not appear to be representative of the potential exposure in the construction industry. It appears the majority of OSHA's enforcement activities related to the asbestos standards were the result of employee complaints and referrals. These complaints and referrals accounted for more than half of all asbestos-related inspections and violations.

For OSHA's complaint inspection process to be more effective, employees must be knowledgeable of the asbestos standards, know the hazards of asbestos exposure, have skills for recognizing asbestos sources in the workplace, and be knowledgeable of OSHA's role in enforcing standards. Education and awareness programs could strengthen the enforcement of the asbestos standards by increasing workers' awareness and action resulting in higher numbers of worker complaints and corresponding enforcement activities.

Planned inspections on the other hand, while accounting for the largest percentage of willful violations, yielded less than one quarter of all asbestos violations. To strengthen the inspection process, targeting the industry groups with planned inspections focused on asbestos could have a positive impact on protecting workers. The site preparation industry group was identified as one potential focus industry due to asbestos exposure from demolition activities. The residential construction and the commercial and institutional building construction also had a disproportionate number of inspections compared to the number of asbestos-related inspections and should also be included as a sector for increased planned inspections focused on asbestos.

An examination of penalties determined OSHA's citation and penalty process resulted in the largest penalties assessed for willful violations. The higher fines can be expected for the willful violations because OSHA's penalty system establishes this practice. Of concern, however, is the fact that willfully exposing workers to asbestos in the construction industry resulted in an average penalty that was less than $10,000 per violation. Serious violations of asbestos standards resulted in average penalties less than $1000. A criticism of OSHA for decades has been the penalties assessed for violating OSHA standards that do not appear to match the potential severity of the long-term effects of exposure and do not serve as a true deterrent.

To protect workers, employers are required to comply with the asbestos standards. Whether it is a federal OSHA site or a state plan site, the requirements are, for the most part, the same. In fact, for many state plan states, their standards are identical to the federal standards. An analysis of the asbestos standards cited yielded similar results across these federal and state plans. Common violations included exposing workers to asbestos because regulated areas were not established, exposure monitoring was not properly conducted, and appropriate control measures were not used. The more serious violations classified as being willful included work being conducted in areas that were not regulated, a lack of exposure monitoring, a lack of PPE, and inadequate employee training. The reasons why employers failed to comply with these standards were not within the scope of this study but should be the focus of future research.

## References


[1] Centers for Disease Control and Prevention (CDC) (2011) Current Intelligence Bulletin 62: Asbestos Fibers and Other Elongate Mineral Particles: State of the Science and Roadmap for Research. Centers for Disease Control and Prevention, Washington DC. http://writingcenter.waldenu.edu/706.htm

[2] Occupational Safety and Health Administration (OSHA) (2014) 29 CRF 1926.1101: Asbestos. Washington DC.

[3] Occupational Safety and Health Administration (OSHA) (2014) Frequently Asked Questions about State Occupational Safety and Health Plans. Washington DC. https://www.osha.gov/dcsp/osp/faq.html

[4] California Division of Occupational Safety and Health (CAL/OSHA) (2014) Title 8, Section 1529: Asbestos. California Division of Occupational Safety and Health, Oakland, CA.

[5] Occupational Safety and Health Administration (OSHA) (2011) CPL 02-00-150: Field Operations Manual (FOM).

Washington DC.

[6] U.S. Department of Labor: Occupational Safety and Health Administration (OSHA) (2014) Fact Sheet, OSHA Inspections. Washington DC.

[7] Sprent, P. and Smeeton, N. (2007) Applied Nonparametric Statistical Methods. Chapman & Hall, Boca Raton, FL.

[8] Hayes, W. (1988) Statistics. Holt, Reinhart, and Winston, Inc., Chicago, IL.

# 15

# Public Perception of Physical Risks: Effect of the Experience of Repeated Explosion Accidents at a Chemical Plant

**Yusuke Koshiba[1*], Hideo Ohtani[2]**

[1]Department of Materials Science and Chemical Engineering, Yokohama National University, Yokohama, Japan
[2]Department of Safety Management, Yokohama National University, Yokohama, Japan
Email: *ykoshiba@ynu.ac.jp

## Abstract

**To explore the influence of experiencing accidental explosions, a questionnaire-based survey was conducted. This paper compares the risk perception characteristics of inhabitants living near a chemical plant at which accidental explosions occurred (group A) and of those living near a plant where no serious accidents have occurred (group B). The survey forms contain the following items: demographics, knowledge about causes of accidental explosion, dread, frequency, and risk rating variables. In addition, participants of group A are asked about their level of participation in physical risk communication. The survey reveals that 1) participants of group A have a greater sense of dread than those of group B; 2) participants of group A are more likely to perceive accidental explosions in chemical plants that occur more frequently; 3) risk rating of physical risks is independent of that of health risks; 4) the "frequency" variable is a key factor that influences decision making whether they participate in physical risk communication or not.**

## Keywords

**Physical Risk, Explosion Accident, Risk Perception, Experience, Chemical Plant**

## 1. Introduction

Physical risk is the risk of death, injury and property damage caused by a physical effect (e.g., explosion, fire, radiant heat, projection) [1]. Industrial accidents involving hazardous materials can result from explosions, fires, chemical spills and leaks, leading to human injury and environmental damage. Industrial accidents may cause both casualties and economic loss; thus, accidental explosions and fires at a chemical plant pose a social prob-

*Corresponding author.

lem in many countries [2]-[5]. Predictably, any industrial plant dealing with hazardous materials poses a physical risk. In Japan, explosion and fire accidents involving dangerous goods have increased over the previous two decades (1994-2011) [6]. In particular, serious accidental explosions and fires have received much attention owing to the fact that such accidents tend to be highly publicized. Further, the social and political impacts of accidental explosions and fires are often intensified by public perception of risks [7]. Previously, many risk perception studies have focused on natural hazards [8], and health and environmental problems [9]. Reportedly, the experience of a hazard has a greater effect on risk perception and risk-related behavior [8]. The effect of the experience of a hazard event on risk perceptions may vary depending on the type of risk event or activity involved. However, few studies have been reported on the effects of the experience of a physical hazard. Thus, the development of effective risk management and communication about physical risks requires an understanding of the public's perception of physical risk.

In the present study, we compare the risk perception of people living near a chemical plant at which accidental explosions/fires occurred and that of people living near a chemical plant where no other serious explosion/fire accidents happened nearby. This study employed a questionnaire survey to investigate the influence of the direct explosion experience. Risk perception of the two groups is characterized by five parameters: demographics, knowledge about causes of accidental explosions, dread, frequency of accidents, and risk rating variables. The reasons why these parameters are employed in this study are as follows. Demographic variables have been found to be an important predictor of risk perception [10]. Many studies on risk perception have suggested that "dread" is the most important factor [11] [12]. In this study, because a majority of people living near a chemical plant have twice experienced accidental explosions of the chemical plant, "frequency" is included as a variable. In addition, these people who live near a chemical plant where accidental explosions/fires occurred are asked whether they want to participate in physical risk communication.

## 2. Accident Description

Yokohama is the second largest city in Japan by population and an important industrial city in the Keihin industrial region (**Figure 1**). Yokohama's major industries include petrochemical and other chemicals, automobiles, and machinery. While many facilities are located in lightly populated areas, some chemical plants manufacturing hazardous materials are located in close proximity to residential areas as well.

On January 7, 2010, 5:47 p.m., a strong explosion caused by a runaway reaction occurred at a chemical plant (hereafter referred to as Plant A) near a residential area [13]. This plant is located in an industrial park in Yokohama city, which borders Tokyo Bay to the east and a residential area to the west. In the residential area, there are several schools and stations. The explosion injured eight operators at the chemical plant and four people outside of it. The blast also damaged roughly 17 buildings at the plant including four that were completely destroyed and three that subsequently burned down. In addition, 82 buildings and 69 cars were damaged outside of

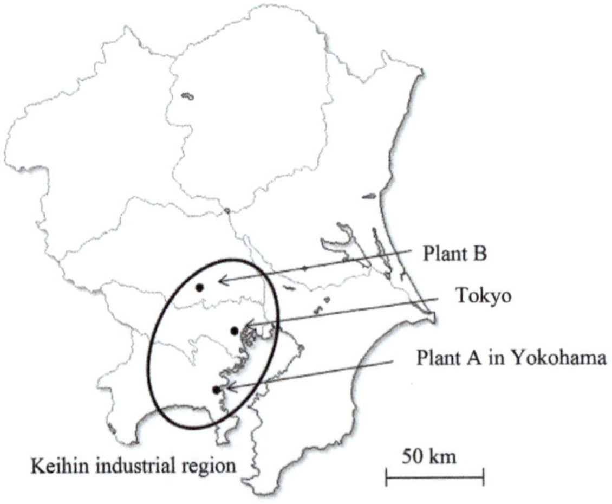


**Figure 1.** Map of Keihin industrial region.

the plant. The explosion blasted several pieces of debris: a 10-ton autoclave used as a reactor vessel, a stirring motor, and a part of the autoclave were found 38 m, 48 m, and 739 m away from the plant, respectively. Newspapers reported that many nearby inhabitants heard a huge blast and witnessed fireball, smoke plume, and flying debris.

Another explosion accident occurred at the same chemical plant A on April 7, 2008 [13]. This explosion caused one death and one injury. A fire triggered by this explosion burned down one building and damaged others. To be precise, two huge accidental explosions have occurred at this chemical plant A in a short period. Hence, when an explosion occurred in 2010, the event was highly publicized by local media.

## 3. Methods

### 3.1. Participants and Data Collection

The main objective of this study is to elucidate how the experiences of explosion accidents influence risk perception characteristics. Perceived risk can be directly affected by the experiences of repeated explosion and fire accidents. In this paper, we surveyed the differences in the perceived physical risk between inhabitants living near plant A and those living near a chemical plant manufacturing similar hazardous materials but where no serious industrial accident has occurred (hereafter referred to as Plant B).

Because one piece of debris, which was believed to be a part of the autoclave, was discovered approximately 740 m away by the blasts, a total of 52 adult inhabitants (29 males and 23 females, see **Table 1**) who lived within 700 m from plant A (hereafter referred to as group A) were randomly sampled. All the participants were interviewed using a face-to-face interview to achieve a high response rate and to provide assistance in answering. The survey was completed from January 2010 to February 2010, and all subjects were unpaid. Note that no industrial accident occurred at the area during the survey.

Aside from this survey, face-to-face interviews were conducted with 87 inhabitants (group B: 43 males and 44 females) who lived near plant B from February 2010 to March 2010 employing the same method as group A. The chemical plant B is also located in the Keihin industrial region. As mentioned above, no serious fire/explosion incident had occurred at plant B for a long period. Plant B was selected as it was in the same industry (hazardous material manufacture) and had nearly equal staffing and site area. However, there may be slight differences in their safety management system and the amount of chemicals stored and used. Although a simple comparison is not accurate, we do not address this problem here because this study focuses simply on the impact of repeated industrial accidents such as explosion and fire on risk perception.

### 3.2. Instrument

In the questionnaire, the participants had to answer utilizing five-point scales, yes/no, multiple choice, and open response. The questionnaire included 11 questions and took approximately eight minutes to complete.

The participants were asked two basic demographic questions: gender and age. In general, various perceived

**Table 1.** Demographic data of participants (group A: n = 52; group B: n = 87).

| Variables | Group A | Group B |
|---|---|---|
| | Number of respondents, n (%) | |
| Gender | | |
| Male | 29 (55.8%) | 43 (49.4%) |
| Female | 23 (44.2%) | 44 (50.6%) |
| Experience | | |
| Yes | 47 (90.4%) | 44 (50.6%) |
| No | 5 (9.6%) | 43 (49.4%) |
| Relationship | | |
| Yes | 8 (15.4%) | — |
| No | 44 (84.6%) | — |

risks (e.g., health risk) are influenced by age [14]. Perhaps the variable affects the explanatory variables. Unfortunately, this study cannot use the date of the age variable because there were a number of missing values. Group A was asked about the experience as follows: "Have you ever heard, felt, or seen fire and/or explosion at chemical plant A?" On the other hand, the question to group B was "Have you ever heard, felt, or seen fire and/or explosion at a chemical plant in Japan?" The next question for group A was about the relationship between the respondents (or their family) and the industrial park including chemical plant A: "Do/Did you have someone who works/worked at the industry park in your household or family?" The respondents had to choose 1) Yes (self/family member) or 2) No. Those who directly benefit from the industry park (*i.e.*, themselves or their family member employed in the industry park) tend to understand risks involved [15] to characterize risks in terms of probability [16] or to rate risks as low [17]. The relationship, in any case, can significantly affect the explanatory variables.

The next question asked about the specific knowledge of the cause of fire/explosion accidents. The question to group A was "Do you know the cause of the fire/explosion accident at the chemical plant A?", and the question to group B was "Do you know the cause of the fire/explosion accident at a chemical plant in Japan?".

Participants were then asked about dread and frequency on five-point scales: "How much do you dread the physical hazard from chemical plants?", and "How likely do you think fire/explosion accidents at chemical plants in Japan are occurring a year ranging from 1 (not at all) to 5 (very often)?".

The next questions asked were about rating the size of the physical risk related to the physical hazard compared with that related to other hazards such as nuclear power, typhoon, earthquake, genetically modified foods (GMO), new influenza, automobiles, airplanes, and health problems resulting from drinking alcohol and smoking. The first five hazards can be considered to be involuntary risk activities, whereas the others are voluntary risk activities. The reason for selecting these hazards is that all of them are usually present in Japan. In particular, the 2009 flu pandemic caused approximately 200 deaths in Japan [18]. Such a comparison makes it possible to measure the relative magnitude of perceived risk associated with physical hazards.

Finally, the participants were asked whether to participate in physical risk communication. When the participant responded yes, the desire for additional information in physical risk communication was evaluated by multiple-choice questions.

### 3.3. Quantification Method Type II

Hayashi's quantification method type II (hereafter referred to as quantification method II) [19] [20] is a linear discriminant analysis for categorical data. This method pairs a categorical variable with a dummy variable (*i.e.*, 0 or 1), allowing the analysis of a quantitative relationship between an objective variable and an explanatory variable. The quantification method type II has been used to quantitatively analyze categorical data in a number of studies. For instance, Ju and Sohn analyzed injury phenomena in traffic accidents [21], and Saito *et al.* revealed risk factors regarding forest road collapse [22]. In this study, we applied the quantification method II to the data of group A in order to clarify an important factor affecting the level of interest in participating in physical risk communication, *i.e.*, the presence or absence of interest in participation in the risk communication was selected as an objective variable and other variables were employed as explanatory variables.

## 4. Results

Differences were considered to be significant at $p < 0.05$ in the statistical analyses. An economical relationship with the industrial park (*i.e.*, the participants themselves or their households who worked at the industrial park) is shown in **Table 1**. 15.6% of group A was or had been economically tied to the industrial park. While only 50.6% of group B had heard and/or saw a fire/explosion accident at chemical plants, a majority (90.4%) of group A reported that they had felt the fire/explosion at plant A (see **Table 1**). Fisher's exact test comparing the difference of the proportions revealed that the difference was statistically significant ($p < 0.001$). Of the participants who answered yes to the question, more than one-third (34.0%) in group A and 40.9% of group B answered that they knew the accident cause. There were no significant differences in the proportions between groups A and B ($p = 0.52$).

### 4.1. Ratings of Dread, Frequency, and Perceived Risk Variables

The means of rated dread for group A and group B were M = 4.69 ($SD = 0.46$) and M = 3.91 ($SD = 1.18$), respectively (see **Table 2**). M and *SD* represent mean value and one standard deviation of the data, respectively. It is important to note that 1) in group B, approximately 5% and 13% of respondents rated their dread as "Not at all" and "A little" respectively; 2) in contrast, in group A, all the participants indicated that their dread were "Somewhat" (30.8%) or "Very" (69.2%, mode). Owing to lack of normality in these data (Kolmogorov-Smirnov test, $p < 0.001$), the Mann-Whitney *U* test was applied to nonparametric analysis. The test demonstrated a significant difference in the perception of dread between group A and group B ($p < 0.001$), thereby indicating that participants in group A had a greater sense of dread than those in group B.

The mean value of rated frequency for group B was M = 2.57 (*SD* = 0.84). In group B, only a few participants answered "Not at all" (n = 4, 4.5%). The mode value was found to be "A little" (n = 46, 52.3%). On the other hand, in group A, the mean value was M = 3.04 (*SD* = 0.68), and no participant rated their frequency as "Not at all." Because the Kolmogorov-Smirnov test showed that the distribution of the data was not normal ($p < 0.001$), the Mann-Whitney *U* test was used. As a result, the means of frequency rated were significantly different ($p < 0.001$). This indicated that experienced participants were more likely to perceive fire/explosion accidents in chemical plants as something that will occur more frequently.

**Table 3** presents the risk ratings of a variety of hazards by comparing them with the physical risk from chemical plants. When the value is three, the risk rating of the hazard is rated as equivalent to the physical risk. Higher values indicate greater perceived risk ratings than the physical risks. In group A, "Nuclear power," "Typhoon," "Earthquake," "Genetically modified foods (GMO)," and "Airplanes" were perceived as a greater risk activities/event. Conversely, "New influenza," "Automobiles," "Alcohol," and "Smoking" were less fearsome to group A. In group B, the risk ratings of "Nuclear power," "Earthquake," "GMO," and "Airplanes" were higher than three. Because the scale type used was a rank scale, the Mann-Whitney *U* test was applied to compare the risk ratings, allowing it to roughly divide the nine hazards into three categories: 1) hazards that the physical risk rating relatively increased; 2) hazards that the physical risk rating relatively decreased; and 3) hazards that the physical risk rating remained unchanged after the experience of the accidents. Category 1 included "Nuclear power". That is, the participants of group A tended to rate the physical risk as something smaller than the risk of nuclear power ($p < 0.05$). "Typhoon" and "Earthquake" fell into Category 2. This means that participants of group A tended to perceive the physical risk as greater than the risk of the two natural disasters (Typhoon, $p < 0.05$; Earthquake, $p < 0.001$). Category 3 contained "GMO," "New influenza," "Automobiles," "Airplanes," "Alcohol", and "Smoking". These activities/events are hazards for which there are no significant differences in the risk ratings between group A and group B (all $p > 0.05$). Most of those are health risks. The prominent effect of individuals' experiences of the accidental explosions/fires on their risk judgments was not found: it was possible that the risk rating of physical risk was not strongly associated with that of health risks.

## 4.2. Level of Interest in Risk Communication and Desire for Information

While approximately half of the respondents of group B (51.1%) were concerned about participation in risk communication of physical risks, a majority of the respondents of group A (80.8%) did so (see **Table 4**). There were significant differences in these data between group A and group B (Fisher's exact test, $p < 0.001$).

**Figure 2** showed proportions of group A participants who desired information concerning fire/explosion accidents. A total of 38.5% of participants wished information on the names and amounts of hazardous chemicals handled at chemical plant A. Those who desired information on an emergency manual were less than 40%. These represented desire for safety information. In contrast, most respondents desired information on frequency of fire/explosion accident (88.5%), anticipated extent of damage (75.0%), and riskiness of the plant (67.3%). This indicated that participants of group A preferred risk information rather than safety information.

**Table 2.** Means of ratings dread and frequency variables.

| | Mean (*SD*) | |
|---|---|---|
| | Dread | Frequency |
| Group A | 4.69 (0.46) | 3.04 (0.68) |
| Group B | 3.91 (1.18) | 2.57 (0.84) |
| *p* value | 0.000*** | 0.000*** |

*$p < 0.05$; **$p < 0.01$; ***$p < 0.001$.

**Table 3.** Means of risk ratings.

| Hazards | Mean (*SD*) | | *p* value |
|---|---|---|---|
| | Group A | Group B | |
| Nuclear power | 4.35 (1.27) | 4.76 (0.80) | 0.018* |
| Typhoon | 3.42 (1.47) | 2.91 (1.22) | 0.029* |
| Earthquake | 4.49 (1.06) | 3.04 (1.48) | 0.000*** |
| Genetically modified foods | 3.09 (1.36) | 3.40 (1.57) | 0.209 |
| New influenza | 2.91 (1.55) | 2.55 (1.34) | 0.197 |
| Automobile | 2.35 (1.31) | 2.62 (1.22) | 0.132 |
| Airplane | 3.70 (1.49) | 4.11 (1.36) | 0.087 |
| Alcohol | 1.79 (1.30) | 2.13 (1.19) | 0.055 |
| Smoking | 2.16 (1.63) | 2.33 (1.30) | 0.242 |

*$p < 0.05$; **$p < 0.01$; ***$p < 0.001$.

**Table 4.** Proportions of participants who were concerned about participation in the risk communication.

| | Proportion (%) |
|---|---|
| Group A | 80.8 |
| Group B | 51.1 |

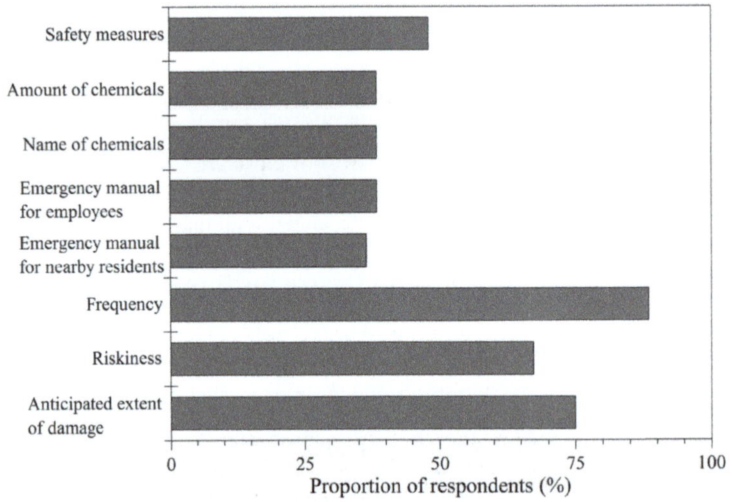


**Figure 2.** Proportions of participants of group A who desired information concerning fire/explosion accidents.

### 4.3. Identification of Factors Affecting the Level of Interest for Risk Communication

To elucidate an important factor affecting the level of interest in participating in physical risk communication, we applied quantification method II to the data of group A. Because this method required independence among the objective variables, the chi-squared ($\chi^2$) test was conducted for each objective variable. Because the "Knowledge" variable was not independent of other variables, five variables, *i.e.*, "Gender," "Experience," "Relationship," "Dread," and "Frequency" were finally determined to be explanatory variables. The results obtained from the $\chi^2$ test are summarized in **Table 5** (all $p > 0.05$).

**Figure 4** shows the category scores of the three variables. In this figure, the positive category score has a positive effect on the objective variable. Conversely, the negative category score exerts a negative effect, meaning

**Table 5.** Results of $\chi^2$ test for the explanatory variables.

| | 1 | 2 | 3 | 4 | 5 |
|---|---|---|---|---|---|
| 1. Gender | | 0.976[a] | 0.111 | 0.963 | 0.457[a] |
| 2. Experience | | | 0.316 | 0.423[a] | 0.774[a] |
| 3. Frequency | | | | 0.444 | 0.416 |
| 4. Dread | | | | | 0.423[a] |
| 5. Relationship | | | | | |

[a]: Yate's continuity correction.

that the category score tends to positively correlate with the desire to participate in risk communication. Regarding "Frequency" category scores were in order of "Very (CS = +0.88)", "Somewhat (CS = +0.55)" > "Neither nor (CS = −0.04)" > "A little (CS = −0.74)" > "Not at all (CS = −4.26)", indicating that there was a positive correlation between the "Frequency" variable and the objective variable. That is, respondents who perceived fire/explosion accidents as something that occurred more frequently tended to have an interest in participating in risk communication processes. In addition, those who answered "Not at all" were found to express a very low level of concern. Furthermore, "Relationship" those with an economical relationship with the industrial park (CS = +0.68) wanted to participate in risk communication. Concerning "Dread" category scores of "Very" and "Somewhat" were +0.18 and −0.43, respectively. Thus, those who had a greater sense of dread tended to desire participation in risk communication.

## 5. Discussion

Our findings in this study revealed that there were no significant differences between groups A and B in the proportion of respondents who knew the accident cause, notwithstanding the fact that most participants of group A directly experienced the explosion accidents twice in a short period. The experience of a hazard has been considered and identified as an important factor governing people's emotion [23]. Kuhar *et al.* have also shown that people influenced by disasters tended to seek out information regarding the risk [24]. Conversely, people who have constantly experienced extreme risks are sometimes less concerned [25]. As a result, the participants of groups A and B did not significantly differ in the percent of respondents who knew the accident cause.

Our survey showed that group A tended to rate that likelihood of the accident occurring was more frequent and they were more fearful of it (see **Table 2**). Ho *et al.* studied risk perception of victims concerning natural hazards of flood and landslide [26]. This study indicated that people with more disaster experiences had a greater dread of these disasters and rated them as something that occurred more frequently. Although it should be noted that the perceived risk may depend on the type of hazard, the fact agrees well with our results of perception of physical risk. According to the availability heuristic [27], people tend to overestimate the likelihood that an event will occur if they have experienced serious consequence of hazards. Thus, the results of the perception of physical risk in this study can be explained by the availability heuristic.

As shown in **Table 3**, nuclear power was assigned to Category 1. This meant that the perceived physical risk was relatively amplified; in other words, the risk rating of the nuclear power was relatively reduced. The reason could be that most respondents perceived the likelihood that nuclear accidents would occur to be very low. However, this result is not in agreement with the findings reported by Grasa *et al.* [17], who reported that the risk ratings of chemical plants at which industrial chemical accidents [28] have occurred were lower than those of nuclear power plants by surveying inhabitants near the plants. This difference was probably due to the difference in the number of experience: most participants of group A experienced the repeated accidents in a short period. Note that our questionnaire survey was conducted before the 2011 Tohoku earthquake [29] that caused the Fukushima Daiichi nuclear disaster [30]. Thus, it is now possible that the risk rating of "Nuclear power" has increased. **Table 3** also revealed that Category 2 contained typhoon and earthquake, which meant that the physical risk rating relatively decreased. To put it in another way, the perceived risks of typhoon and earthquake were relatively enhanced. In Japan, these two natural disasters occur very frequently and often cause loss of life and property damage. Although most respondents of group A have undoubtedly experienced the accidental explosions/fires, they also more often have experiences of typhoon and earthquake. Unfortunately, although our survey does not elucidate the loss and damage of group A by explosion/fire accidents, the effect on the frequency of the event occurring perhaps resulted in the higher risk rating of the two natural disasters compared with that of an explosion/fire accident in the chemical plant.

As shown in **Figure 3**, the item range of "Frequency" variable was largest among the five explanatory variables, indicating that it remarkably dominated the objective variable (*i.e.*, level of interest in participating physical risk communication). In addition, as can be seen from **Figure 4**, participants who witnessed the explosion/fire accidents tended to have an interest more frequently. It is known that greater frequency of involuntary experiences has a positive relationship with greater concern [31]. It was possible that the participants of group A who had experienced an accident twice were strongly concerned about explosion/fire accidents. As a result, the frequency factor governed the objective variable. Interestingly, the item range of "Gender" had a low item range of 0.31. In general, the gender variable was found to play an important role in the perception of health and environmental risks [10]; however, the gender variable hardly influenced the objective variable in the present study. This was probably because there were no significant differences in "Dread" and "Frequency" ratings between

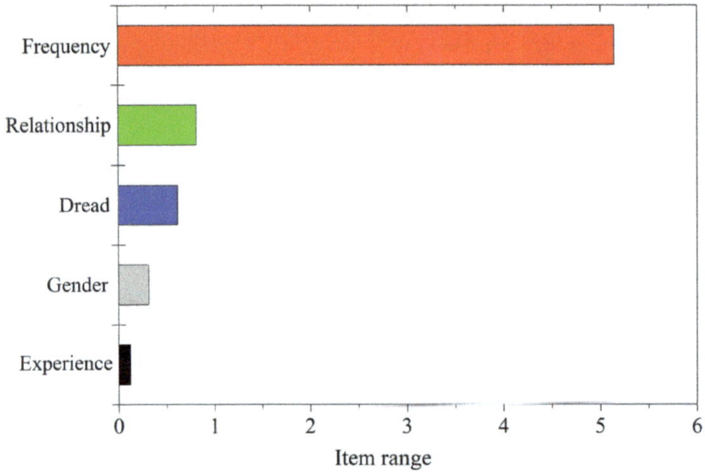


**Figure 3.** Results of item ranges (IR). As the value of the IR increases, the explanatory variable greatly affects the objective variable.

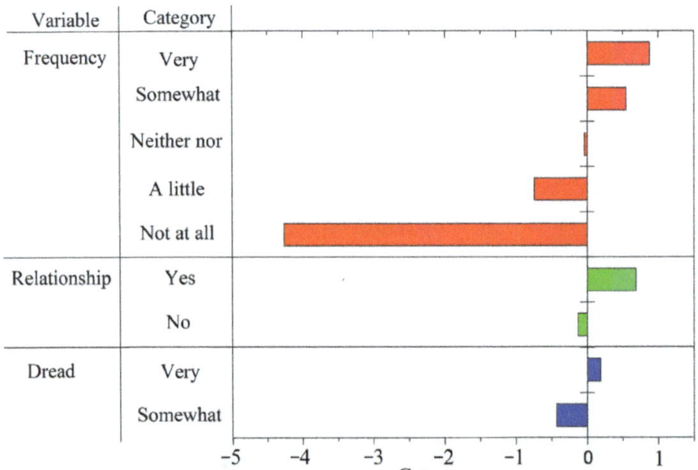


**Figure 4.** Results of category scores (CS). The positive CS has a positive effect on the objective variable. Conversely, the negative CS exerts a negative effect.

males and females in group A (Dread: $M_{male} = 4.96$, $M_{female} = 4.70$, $U$ test, $p = 0.96$; Freq.: $M_{male} = 3.10$, $M_{female} = 3.00$, $U$ test, $p = 0.61$).

Here some limitations of the present study should be acknowledged. The first limitation is that the questionnaire survey was restricted to inhabitants living near the two chemical plants: plants A and B. The second concerns the relatively small sample size (n = 52 and 87). That said few studies had been reported on perceptions of physical risks; thus, this research provides new insights into the influence of experiencing repeated explosion accidents on the public perception of physical risks. Our findings will be useful for designing effective physical risk communication processes.

## 6. Conclusions

On the basis of the questionnaire survey presented in this paper, the following main conclusions can be obtained:

1) People living near a chemical plant where accidental fires and explosions have previously occurred (group A) have a greater sense of dread than those living near a chemical plant where no fire/explosion accidents have occurred (group B) ($M_{group\ A} = 4.69$, $M_{group\ B} = 3.91$, $U$ test, $p < 0.001$).

2) Participants of group A are more likely to perceive accidental fires and explosions at chemical plants as something that occur more frequently than those of group B ($M_{group\ A} = 3.04$, $M_{group\ B} = 2.57$, $U$ test, $p < 0.001$).

3) The experience of fire/explosion accidents does not influence risk ratings of health risks such as genetically modified foods, new influenza, health problems resulting from drinking alcohol and smoking. That is, the risk rating of physical risks is probably independent of that of health risks.

4) The experience of accidental fires and explosions remarkably increases the level of interest in participating in physical risk communication (group A: 80.8%, group B: 51.1%, Fisher's exact test, $p < 0.001$).

5) In physical risk communication, a majority of participants of group A requested information concerning the frequency of fire/explosion accidents (88.5%), extent of damage (75.0%), and accident risks (67.3%).

6) Quantification method II reveals that "frequency" is a dominant factor affecting the level of interest in participating physical risk communication. In addition, participants who perceived fire/explosion accidents as something that occurs more frequently tend to show higher levels.

On the basis of the results presented in this study, we conclude that explosion accidents have a marked impact on risk perception characteristics of inhabitants living near a chemical plant at which explosion accidents occurred.

## References


[1] United Nations (2011) Globally Harmonized System of Classification and Labelling of Chemicals. 4th Edition, United Nations Publication, New York and Geneva.

[2] Ettouney, R.S. and EL-Rafai, M.A. (2012) Explosion of Ammonium Nitrate Solutions, Two Case Studies. *Process Safety and Environmental Protection*, **90**, 1-7.

[3] Taveau, J. (2010) Risk Assessment and Land-Use Planning Regulations in France Following the AZF Disaster. *Journal of Loss Prevention in the Process Industries*, **23**, 813-823. http://dx.doi.org/10.1016/j.jlp.2010.04.003

[4] Chen, J.R., Tsai, H.Y., Chen, S.K., Pan, H.R., Hu, S.C., Shen, C.C., Kuan, C.M., Lee, Y.C. and Wu, C.C. (2006) Analysis of a Silane Explosion in a Photovoltaic Fabrication Plant. *Process Safety Progress*, **25**, 237-244. http://dx.doi.org/10.1002/prs.10136

[5] Lacoursiere, J.P. (2005) An Explosion Caused by Mixing Incompatible Liquids. *Process Safety Progress*, **24**, 115-119. http://dx.doi.org/10.1002/prs.10066

[6] Fire and Disaster Management Agency (2012) Dangerous Goods Accidents in 2011. *Safety & Tomorrow*, **144**, 19-21 (in Japanese).

[7] Burns, W.J. and Slovic, P. (2012) Risk Perception and Behaviors: Anticipating and Responding to Crises. *Risk Analysis*, **32**, 579-582. http://dx.doi.org/10.1111/j.1539-6924.2012.01791.x

[8] Kung, Y.W. and Chen, S.H. (2012) Perception of Earthquake Risk in Taiwan: Effects of Gender and Past Earthquake Experience. *Risk Analysis*, **32**, 1535-1546. http://dx.doi.org/10.1111/j.1539-6924.2011.01760.x

[9] Grasmück, D. and Scholz, R.W. (2005) Risk Perception of Heavy Metal Soil Contamination by High-Exposed and Low-Exposed Inhabitants: The Role of Knowledge and Emotional Concerns. *Risk Analysis*, **25**, 611-622. http://dx.doi.org/10.1111/j.1539-6924.2005.00628.x

[10] Flynn, J., Slovic, P. and Mertz, C.K. (1994) Gender, Race, and Perception of Environmental Health Risks. *Risk Analysis*, **14**, 1101-1108. http://dx.doi.org/10.1111/j.1539-6924.1994.tb00082.x

[11] Slovic, P. (2000) The Perception of Risk. Earthscan Ltd., Washington DC.

[12] Bronfman, N.C. and Cifuentes, L.A. (2003) Risk Perception in a Developing Country: The Case of Chile. *Risk Analysis*, **23**, 1271-1285. http://dx.doi.org/10.1111/j.0272-4332.2003.00400.x

[13] Aoki, T. (2010) Explosion Accidents at a Chemical Factory in Yokohama. *Safety & Tomorrow*, **131**, 34-38 (in Japanese).

[14] Armaş, I. (2006) Earthquake Risk Perception in Bucharest, Romania. *Risk Analysis*, **26**, 1223-1234. http://dx.doi.org/10.1111/j.1539-6924.2006.00810.x

[15] Cutchin, M.P., Martin, K.R., Owen, S.V. and Goodwin, J.S. (2008) Concern about Petrochemical Health Risk before and after a Refinery Explosion. *Risk Analysis*, **28**, 589-601. http://dx.doi.org/10.1111/j.1539-6924.2008.01050.x

[16] Sjöberg, L. and Drottz-Sjöberg, B.M. (1991) Knowledge and Risk Perception among Nuclear Power Plant Employees. *Risk Analysis*, **11**, 607-618. http://dx.doi.org/10.1111/j.1539-6924.1991.tb00650.x

[17] Grasa, G., Navarro, V., Rubio, O., Peña, J.A. and Santamaria, J. (2002). A Comparative Study of Risk Perception in Small and Large Communities Where Chemical Industries Are Present. *Journal of Loss Prevention in the Process Industries*, **15**, 525-530. http://dx.doi.org/10.1016/S0950-4230(02)00033-5

[18] Ministry of Health, Labour and Welfare (2010) Information about Influenza A (H1N1) Pdm. http://www.mhlw.go.jp/english/topics/influenza_a/situation_update/dl/100331_the_situation_of_H1N1_in_Japan.pdf

[19] Hayashi, C. (1952) On the Prediction of Phenomena from Qualitative Data and the Quantification of Qualitative Data from the Mathematico-Statistical Point of View. *Annals of the Institute of Statistical Mathematics*, **3**, 69-98. http://dx.doi.org/10.1007/BF02949778

[20] Hayashi, C. (1950) On the Quantification of Qualitative Data from the Mathematico-Statistical Point of View. *Annals of the Institute of Statistical Mathematics*, **2**, 35-47. http://dx.doi.org/10.1007/BF02919500

[21] Ju, Y.H. and Sohn, S.Y. (2011) Quantification Method Analysis of the Relationship between Occupant Injury and Environmental Factors in Traffic Accidents. *Accident Analysis and Prevention*, **43**, 342-351. http://dx.doi.org/10.1016/j.aap.2010.09.002

[22] Saito, M., Shindo, D., Kinoshita, W., Ueki, T. and Inoue, H. (2015) Extraction of Collapse Risk Factors for Operation-Road in Terasawayama. *Formath*, **14**, 60-72. http://dx.doi.org/10.15684/formath.14.007

[23] Richardson, B., Sorensen, J. and Soderstrom, E.J. (1987) Explaining the Social and Psychological Impacts of a Nuclear Power Plant Accident. *Journal of Applied Social Psychology*, **17**, 16-36. http://dx.doi.org/10.1111/j.1559-1816.1987.tb00290.x

[24] Kuhar, S.E., Nierenberg, K., Kirkpatrick, B. and Tobin, G.A. (2009) Public Perceptions of Florida Red Tide Risks. *Risk Analysis*, **29**, 963-969. http://dx.doi.org/10.1111/j.1539-6924.2009.01228.x

[25] MacGregor, D., Slovic, P., Mason, R.G., Detweiler, J., Binney, S.E. and Dodd, B. (1994) Perceived Risks of Radioactive Waste Transport through Oregon: Results of a Statewide Survey. *Risk Analysis*, **14**, 5-14. http://dx.doi.org/10.1111/j.1539-6924.1994.tb00022.x

[26] Ho, M.-C., Shaw, D., Lin, S. and Chiu, Y.-C. (2008) How Do Disaster Characteristics Influence Risk Perception? *Risk Analysis*, **28**, 635-643. http://dx.doi.org/10.1111/j.1539-6924.2008.01040.x

[27] Tversky, A. and Kahneman, D. (1974) Judgment under Uncertainty: Heuristics and Biases. *Science*, **185**, 1124-1131. http://dx.doi.org/10.1126/science.185.4157.1124

[28] Marco, E., Peña, J.A. and Santamaría, J. (1998) The Chlorine Release at Flix (Spain) on January 21st 1996: A Case Study. *Journal of Loss Prevention in the Process Industries*, **11**, 153-160. http://dx.doi.org/10.1016/S0950-4230(97)00014-4

[29] Japan Meteorological Agency (2011) The 2011 off the Pacific Coast of Tohoku Earthquake—First Report. http://www.jma.go.jp/jma/en/News/2011_Earthquake_01.html

[30] IAEA (2011) Fukushima Nuclear Accident Update Log. http://www.iaea.org/newscenter/news/2011/fukushimafull.html

[31] Barnett, J. and Breakwell, G.M. (2001) Risk Perception and Experience: Hazard Personality Profiles and Individual Differences. *Risk Analysis*, **21**, 171-177. http://dx.doi.org/10.1111/0272-4332.211099

# 16

# A Novel Application of Inertial Measurement Units (IMUs) as Vehicular Technologies for Drowsy Driving Detection via Steering Wheel Movement

**Samuel A. Lawoyin, Ding-Yu Fei, Ou Bai**

Department of Biomedical Engineering, Virginia Commonwealth University, Richmond, VA, USA
Email: lawoyins@vcu.edu, fei@vcu.edu,obai@vcu.edu

## Abstract

**Introduction: Vehicular technologies intended for the improvement of driver safety are especially critical today in the view of the thousands of deaths that occur annually due to drowsy driving. Current technologies include physiological methods like electroencephalography (EEG), behavioral methods including driver video monitoring, and vehicle measures which include lane and steering wheel tracking. These current technologies are impractical in their current implementations as they cannot readily be used outside of laboratory settings due to their requirements for intrusive electrodes, expensive cameras, and complex equipment. An earlier article demonstrated an effective method for wheel tracking using only an accelerometer; however the introduction of integrated gyroscopes and accelerometers has afforded further opportunities. Objective: This paper introduces a novel, low-cost, and easy to implement an approach to address this unmet problem. Method: Through the use of an Inertial Measurement Unit (IMU) combining a gyroscope and an accelerometer, measurements of steering wheel behavior were recorded in both simulator and real world driving while compared against a standard potentiometer. Results: The excellent agreement between potentiometer recorded angles and IMU estimated angles ($R^2 = 0.98$, $P < 0.001$) suggests that the complicated installation of potentiometers in vehicle steering columns is no longer a necessary step for steering wheel monitoring. Conclusion: This paper presents an IMU based method for drowsy steering-wheel behavioral tracking which is cost-effective, easy to implement, and accurately estimates steering behaviors. The results suggest that this novel vehicle technology offers hope for improving road safety.**

## Keywords

**Driver Safety, Drowsy and Fatigued Driving, Accidents, Accident Prevention, Inertial Measurement**

**Units**

## 1. Introduction

Many methods have been proposed for the detection of drowsy or fatigued driving. Researchers have tried physiological measures which include heart rate, breathing rate, and brain activity via electroencephalography (EEG). Behavioral measures including eye blinks and eye closures have also been used. Because these measures have obtrusive properties, they are impractical for daily use with commuters and researchers have turned to vehicle-based measures instead. Vehicle-based measures are embedded into the vehicle and its systems, making them unobtrusive to the driver. These measures include monitoring of the drivers steering patterns for signs of declining alertness and lapses in wakefulness which contribute to road accidents.

Drowsy and fatigued driving is a significant problem. It is responsible for about 1200 deaths and 76,000 injuries every year in the United States alone [1] and the annual numbers worldwide are even higher. Greater than 14% of respondents who polled in Ontario Canada admitted having fallen asleep or nodded off while driving [2]. Sleep-deprived individuals become undetected drowsy drivers on the roads and highways [3]. The other types of drowsy drivers are those who initially received sufficient sleep but engaged in prolonged driving which caused their awareness to deteriorate over time [4]. Prolonged wakefulness is just as dangerous to driver safety as alcohol intoxication [5].

Physiological measures of drowsiness including EEG are not a practical measure of drowsiness due to the complexity of setup and the non-portability of equipment. Commuters would be unable or unwilling to prepare and apply EEG to themselves daily. In addition research has shown that engine vibrations affect EEG outcomes [6].

Eyelid closures are behaviors that can be observed for drowsiness detection. Eye closures which last for more than half a second are especially indicative of sleepiness [7]. PERcentage of eye CLOSure (PERCLOS) refers to the percent of time that a driver's eyelids are closed over a given interval. It has shown good promise as a means for drowsiness detection [8]. Video-based eye closure monitoring methods are ineffective if the driver is wearing eyeglasses [9] and give false positive readings if the driver looks down and around him [10]. Electrooculography (EOG) methods for PERCLOS are too intrusive for widespread daily use due to their use of electrodes and wires.

An important vehicle-based method for determining drowsy driving is the monitoring of a driver's Steering Wheel Movements (SWM) for drowsy patterns. The relationship between SWM and driver drowsiness has been well documented. The correlation between a driver's intervals of steering adjustments and their level of drowsiness has been consistently seen by researchers [11] [12]. Drivers make fewer regular maneuvers when drowsy than when alert. Despite the decline in overall steering inputs, the fewer inputs made are sudden and larger in degree [13] [14]. Steering inputs in fatigued drivers are shown to have fewer micro corrections and more macro-corrections, with sleeping drivers making no corrections [13]-[15]. It has been demonstrated that as the majority of sampled drivers become drowsy, they tend to increasingly trend towards faster and larger steering corrections [13].

With the knowledge that SWM is a highly effective and highly accurate measure of drowsy driving, there have been several methods used by researchers to monitor SWM. Sayed and Eskandarian [16] measured SWM using built-in equipment found in complex vehicle simulators. This approach is cost-prohibitive to the average user and excessively powered for users requiring only SWM monitoring without extra options. Thiffault and Bergeron [13] placed potentiometers along the axis of the steering column to directly measure the turn angle. This would require users to have the technical knowledge and dexterity to install a potentiometer into the steering column of their vehicle or vehicle simulator. Potentiometer use would also require the dismantling of the current vehicle setup to install the potentiometer. Despite the numerous progresses made by researchers towards an effective method of deploying the SWM method on a wide scale, the use of a gyroscope has not been explored despite its ubiquitous use as the standard for rotational measurements in ships, airplanes, submarines and space-crafts. Despite the large amounts of linear vibrations, noise, jolts and accelerations in the gyroscopes routine applications, it remains a highly accurate and widely proliferated standard of rotation measurement. Even though gyroscopes are highly proliferated measures of rotation, their positional measurements have a tendency to drift with time [17]. Luinge [17] used gyroscopes to measure human orientation. To address the drift problem, a 3D accelerometer which served as a tilt sensor was used to continuously recalibrate the gyroscope. Lee *et al.* [18] used a gyroscope to estimate the posture of a mobile inverted pendulum. For drift correction in that case, a tilt sensor was fused with the gyroscope. In spaceflight, NASA uses accelerometers to detect and null errors generated in IMU devices. The accelerations caused by the earth are definitely known, and any discrepancies are considered errors. In terrestrial flights, vacuum gyroscope drifts cause errors in directional readings and pilots have to manually recalibrate the readings by hand, referencing a compass. Ring Laser Gyroscopes (RLGs) are less prone to drift, but they cost upwards of $300,000 for a tri-axial unit [19]. Greene [19] proposed the use of a piezoelectric gyroscope in place of an RLG. Due to drift, an accelerometer was used to stabilize the gyroscopes. In our case, the use of regular automatic recalibration from an accelerometer would ensure seamless operation.

In this study, an Inertial Measurement Unit (IMU) based approach for monitoring the SWM is proposed. IMU devices include gyroscopes and accelerometers. Due to modern Micro Electro Mechanical (MEMS) technologies, gyroscopes and accelerometers are now very affordable to obtain and have very compact form factors. They are no longer restricted by cost or complexity to advanced navigation devices. The proposed use of gyroscopes for SWM monitoring requires a minimal setup that is easy to install and uninstall. The only requirement is that the gyroscope should be affixed to a surface of the steering wheel that would allow the device to be perpendicular to the steering column axis.

An earlier study conducted showed the high accuracy and efficacy of using an accelerometer for the monitoring of drowsiness via SWM methods [20]. The method proposed in this study depends upon the accurate accelerometer-based method from the earlier study for solving the gyroscopes drift problems. In low noise or simulator based environments, the accelerometer-based method might be adequate or preferable if an experimenter has access to cheaper accelerometers. Although there has come to be widespread use of analog accelerometers, there has since come about the availability of 6-axis digital MEMS sensors which incorporate a gyroscope and an accelerometer in a tiny footprint ($4 \times 4 \times 0.9$ mm) such as the one used in this study. The benefit of this newer sensor type is that the combined setup helps to simultaneously improve the accuracy of both the accelerometer and the gyroscope. Because accelerometers are prone to linear vibration noise and gyroscopes are prone to slow drifts, the combination of the two sensors has provided new opportunities for SWM monitoring that were originally not available in discrete analog accelerometers. Further, these opportunities are in a very tiny, unobtrusive, and inexpensive package.

An algorithm was developed in this study for monitoring SWM that utilizes a gyroscope's proficiency in detecting angular velocities. The gyroscope outputs the rate of angular change, and then the proposed algorithm interprets this data into SWM position angles with the assistance of an integrated accelerometer which accounts for drift and is located on the same single-DIEmems.

This solution can meet the unmet problem to curb drowsy driving. Despite the safety efforts of the NHSTA and The Federal Highway Administration (FHWA), drowsy drivers continue to take to the wheel and deaths and injuries continue to yield consistently high annual fatality figures. The knowledge for detecting drowsy driving exists but remains within the positive results of numerous successful driving trials. Manufacturers have found no adequate method to deploy these known techniques. The necessity for a practical and inexpensive means for drowsy driving monitoring is especially pertinent as an effective counter to the high fatality numbers. The proposed method is simple, cost-effective and provides not only for drowsy monitoring on new vehicles, but also allows for retrofitting on older vehicles and current model vehicles which on average continue to be manufactured with no drowsy driving detection mechanisms.

Section 2 describes the method and materials for using a gyroscope-accelerometer fusion for SWM detection, Section 3 describes the various tests performed to evaluate the efficacy of the method, Section 4 lists the results obtained from the design and testing of the method, while Sections 5 and 6 discuss the results and list conclusions.

## 2. Material and Method

### 2.1. The Basic Concept and Algorithm to Calculate SWM Using a Gyroscope-Accelerometer Fusion

Gyroscopes detect angular velocity and they can be used to derive information about the angular orientation of the steering wheel. An equation for real time monitoring of the rotational position of a gyroscope is given by

[21]:

$$\theta_{gyro}[n] = \theta_{gyro}[n-1] + \Delta n\dot{\theta}[n] \tag{1}$$

where the gyroscope positional angle $\theta_{gyro}[n]$ is based upon knowledge of the last positional sample $\theta_{gyro}[n-1]$ as well as knowledge of the angular displacement since the last sample, which is the product of the rate of angular change $\dot{\theta}[n]$, and the sample interval, $\Delta n$. The new position of gyroscope orientation can be determined as $\theta_{gyro}[n]$ with the above Equation (1). The equation was designed especially for capturing rota tional movements originating from human motion [21]. SWM is a product of human motion, and will be served well by this method.

A drawback to using gyroscopes for detection of angular rotation is the tendency for gyroscope positional values to drift [17] [21].

The second part of the proposed IMU device is the accelerometer. An equation for extracting SWM angle solely via an accelerometer is given as [20]:

$$\hat{\theta}[n] = atan\left(\frac{A_x[n]}{A_y[n]}\right) \tag{2}$$

where $\hat{\theta}$ was the SWM angle being estimated from the accelerometer readings of $A_x[n]$ and $A_y[n]$, $\hat{\theta}[n]$ had a strong positive correlation with the steering wheels SWM angle $\theta_w[n]$ [20].

The addition of an accelerometer to the gyroscope compensated for gyroscope drift via the accelerometers perpetual ability for gravitational alignment. This is predicated upon the fact that the operation of Equation (2) depends upon relative readings of gravity on the accelerometers separate axes. The accelerometers tendency to pick up linear vibrations was in turn countered by the gyroscope which has sensitivity to angular velocity. The IMU fusion led to a highly effective combination. When the steering wheel was in a neutral position as shown in **Figure 1**, the main sensor was fastened to the steering wheel surface such the accelerometer gave a neutral angular reading of $\hat{\theta}[n] = 0°$ and the gyroscope Z-axis ($g_z$ marked as "$x$" was parallel to the steering column axis.

Combining Equations (1) and (2), a complimentary filter was designed to maximize the strengths of both IMU devices. An ideal relationship between $\theta_{\text{final}}$ and $\hat{\theta}$ which would be easy to update in real time was found to be the causal system:

$$\theta_{\text{final}}[n] = \left(\theta_{\text{final}}[n-1] + \left(\frac{\Delta n\dot{\theta}[n] + \Delta n\dot{\theta}[n-1]}{2}\right)\right) \times \beta_{gyro} + atan\left(\frac{A_x[n]}{A_y[n]}\right) \times \beta_{accel} \tag{3}$$

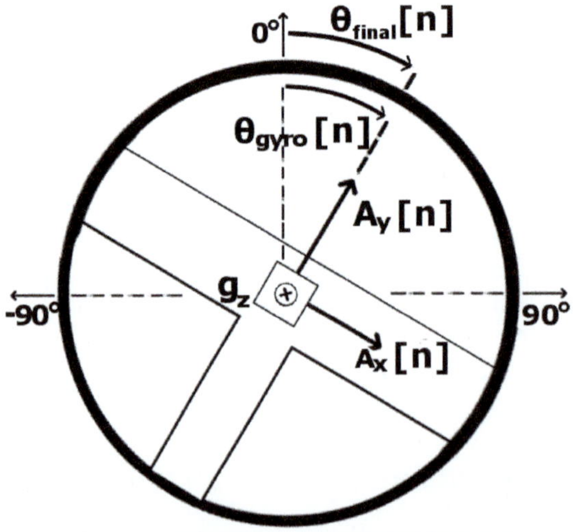


**Figure 1.** The mapping of the IMU device to the steering wheel.

which is effectively a weighted addition of Equation (1) and Equation (2) with a few slight modifications. The first modification was that $\theta_{gyro}[n]$ took over the role of $\theta_{\text{final}}[n]$. This consequentially resulted in the causal system referencing $\theta_{\text{final}}[n-1]$ rather than $\theta_{gyro}[n-1]$, which is incidentally more correct for the newly formed system in terms of accurately calculating angle based in part upon the last known position. The second difference was that the angular velocity output of the gyroscope was averaged over current and last known reading. This was intended to provide a smoother reading and to improve overall accuracy rates of the newly fused system. At 250 Hz of sampling frequency, which yields 250 samples each second, the averaging of only 2 samples will not adversely affect the overall signal even in the very short term.

Finally, $\beta_{gyro}$ and $\beta_{accel}$ were chosen as the coefficients for determining the percentage contribution of each element in Equation (3) to the overall IMU fusion reading of SWM. The summation case therefore must always hold that:

$$\beta_{gyro} + \beta_{accel} = 1 \qquad (4)$$

The full process for determination of the coefficient weights for Equation (3) is described in Section 3.

For comparison against current potentiometer based SWM angle recordings, a linear potentiometer in series with the steering axis was used as a reference. Linear potentiometer output voltages vary in linear proportion to their angle of rotation and can be modelled as a standard linear equation:

$$\theta_w = m \times V_p + b \qquad (5)$$

where $\theta_w$ was the steering wheel angle of rotation in degrees (°) and $V_p$ was the potentiometer voltage in volts (V). $m$ was the slope of the linear relationship and $b$ was the $y$-intercept of the linear relationship.

To customize our model, the parameters $m$ and $V_p$ were generated by sampling 90 data points per quadrant of the steering wheel, yielding approximately 1 sample of $V_p$ per 1°. Using these data points to generate a linear relationship gave values of $m = -93.409\,°/\text{V}$ and $b = 177.400°$. All further potentiometer readings of SWM angle were calculated by using these parameters with Equation (5) for derivation of $\theta_w$.

## 2.2. Equipment

The steering wheel used for simulator tasks was the Top Drive GT (Logic3, Hertfordshire, England). Simulator driving tasks were performed using the OpenDS driving simulation software.

An MPU-6050 (InvenSense, San Jose, California) which is a 6-axis combined MEMS gyroscope + accelerometer was the main sensor. The sensitivity of the gyroscope was set at $\pm 250°\cdot s^{-1}$ while the sensitivity of the accelerometer was set at ±2 g. At 4 mm × 4 mm × 0.9 mm and weighing less than a gram, the sensor lends itself to portability and non-intrusiveness in any SWM application

The IMU data was collected using an amplifier based on the TI-ADS1299 Analog Front-End (Texas Instruments, Dallas, TX). All data were sampled at 250 Hz.

Data were analyzed with MATLAB. For statistical analysis, linear correlations between data were determined through linear regression, Pearson's Linear Correlation coefficients, and Spearman's Rho. P-values were recorded at $\alpha = 0.05$ unless otherwise specified. The correlations between potentiometer measured SWM and SWM estimated via the gyroscope-accelerometer algorithm were determined using the cross correlation (xcorr) function of the MATLAB signal processing toolbox.

# 3. Testing

## 3.1. Test for Applicability of SWM Monitoring with an Accelerometer as the Sole IMU Input

For this test, Equation (3) was used for generating the SWM signal $\theta_{\text{final}}[n]$. However, the signal here was generated using IMU data collected from the accelerometer only during road tests. For this purpose, $\beta_{gyro}$ was set to 0. The resulting signal was compared against potentiometer readings during the same period to determine the usability of the signal and its correctness.

## 3.2. Test for Applicability of SWM Monitoring with a Gyroscope as the Sole IMU

Similar to the previous test, Equation (3) was used for generating the SWM signal $\theta_{\text{final}}[n]$ using IMU data collected from the gyroscope during road tests. In this case, $\beta_{accel}$ was set to 0. The resulting signal was compared against potentiometer readings during the same period to determine the usability of the signal and its correctness.

### 3.3 Test for Equal Weighting of Gyroscope: Accelerometer Coefficients $\beta_{gyro}$:$\beta_{accel}$

This test was intended to implement true signal combinations as described by Equation (3). Accelerometer and gyroscope input were initially combined at a ratio of 50:50 for $\beta_{gyro}$:$\beta_{accel}$.

### 3.4. Test to Determine Optimal Weights for Gyroscope: Accelerometer Coefficients $\beta_{gyro}$:$\beta_{accel}$

Combining the two inertial measures of SWM measurements into a single efficient unit required the optimal weight distribution of each component. It was intended that the shock resistant gyroscope which was sensitive to angular rotations inherent to steering behavior and less sensitive to linear or translational noise would provide the bulk of SWM monitoring data. It was also intended that the drift resistant accelerometer would contribute just enough orientation data to ensure that the gyroscope measurement was perpetually calibrated against gravity so that the angle did not drift with time.

Road tests on the high way were useful for making a determination of what ratio of $\beta_{gyro}$:$\beta_{accel}$ was most effective. The aim was to decide which weight ratio yielded the best data in relation to the potentiometer, since the method was to eventually be an efficient replacement of the potentiometer for steering behavior monitoring.

### 3.5. Test for Accuracy of SWM Readings against Linear Potentiometer Using Selected Weights

After the setup from 2.2 had been used to establish a relationship between $\theta_{gyro}$, $\theta_{\text{final}}$, and $\hat{\theta}$, as well as a standard for $\theta_w$, a participant was recruited to perform driving simulator activities for 45 minutes. The correlation between $\theta_{\text{final}}$ and $\theta_w$ over this prolonged period was calculated.

Once strong correlation was seen in a simulator environment, an actual road test was performed which involved the physical mounting of the simulator's steering wheel platform into the vehicle interior while driving tasks were performed by a passenger. This test involved about 20 minutes of driving tasks involving high speed highway driving and city driving in stop-and-go traffic.

## 4. Results

Data were analyzed using MATLAB. Correlations between the inertial unit's SWM data and the potentiometers SWM data were determined via Pearson's Linear Correlation coefficients, Spearman's Rho, and Kendall's tau. Signal cross correlation between the Inertial Measurement Units output signal and the potentiometers output signal were determined through the xcorr function of the MATLAB signal processing toolbox. P-values were recorded at $\alpha < 0.05$.

### 4.1. SWM Monitoring with an Accelerometer as the Sole IMU

For this test, Equation (3) was used to generate the SWM signal $\theta_{\text{final}}[n]$ from IMU data collected from the accelerometer during road tests. In this case, $\beta_{gyro}$ was set to 0.

The purely accelerometer signal demonstrated noticeable amounts of road noise during road tests (**Figure 2(a)**). Although a low pass filter conveniently removed the noise (**Figure 2(b)**), the use of gyroscope fusion demonstrated better results when optimal weighting eliminated the need for hardware or software filtering (Section 4.4). **Table 1** shows the results obtained when each IMU device was used as the sole SWM input. **Table 1** also shows the result of a 50:50 IMU device input ratio.

### 4.2. SWM Monitoring Using a Gyroscope as the Sole IMU Input

The SWM signal $\theta_{\text{final}}[n]$ was generated from IMU data collected solely from the gyroscope during road tests. In this case, $\beta_{accel}$ was set to 0. **Figure 2(c)** shows a 24 second section of the gyroscope output waveform after about 10 minutes of road driving. While the gyroscope output was of the correct waveform to match the potentiometer output, the drifting caused the signal to eventually center around 150° (**Figure 2(c)**) whereas it would

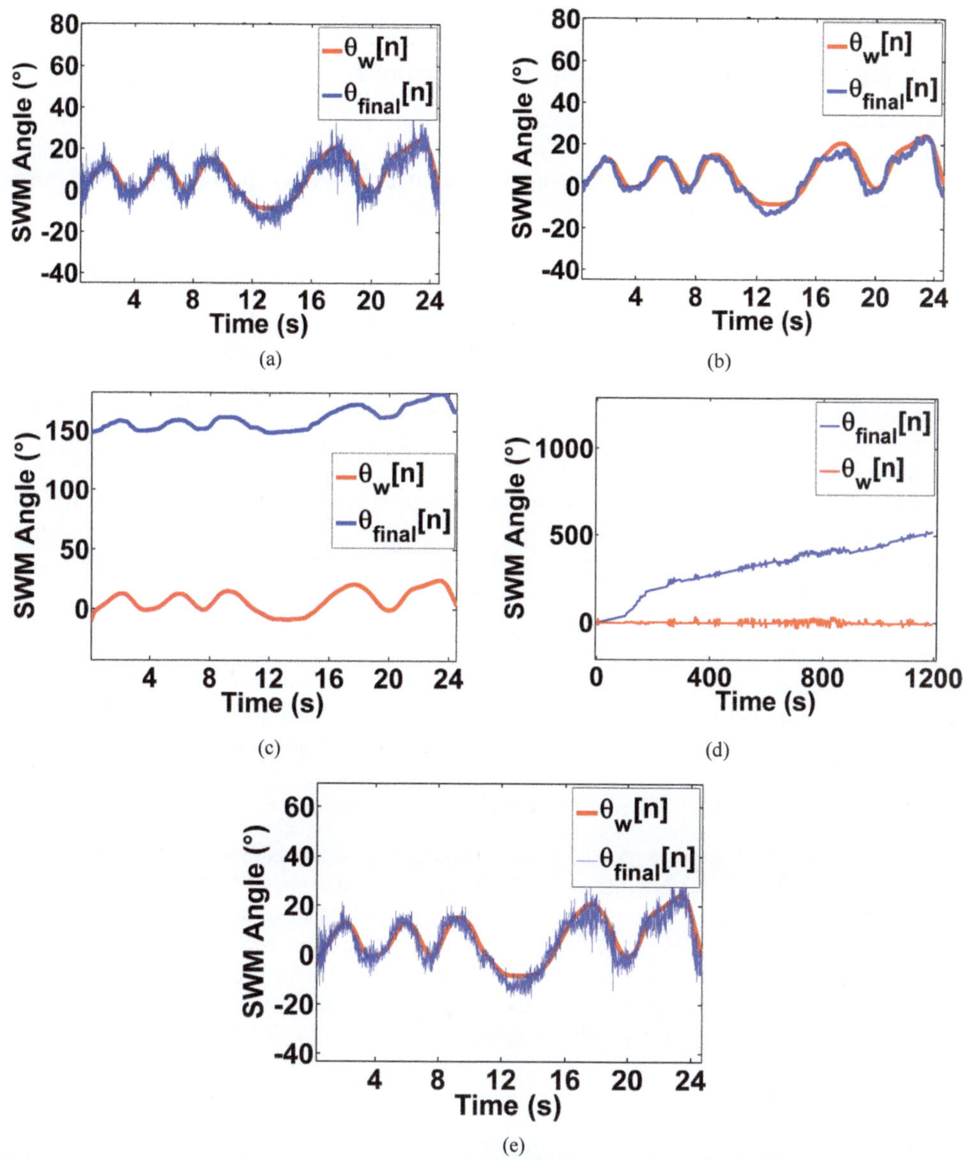


**Figure 2.** (a) Accelerometer only SWM signal; (b) Accelerometer only SWM signal passed through a 4[th] order low pass Butterworth filter; (c) Gyroscope only signal demonstrating slow drift; (d) Gyroscope only signal from road test demonstrating how the gyroscope signal would wander into a slow drift in the longer term; (e) A 50:50 distribution of accelerometer: gyroscope signals.

**Table 1.** Various IMU device ratios and their correlation to potentiometer.

| | Ratio | Correlation to potentiometer results | | | |
|---|---|---|---|---|---|
| | $\beta_{accel}$:$\beta_{gyro}$ | Xcorr | Spearman's | Pearson's | Kendall's |
| a | 100:0 | 0.91 | 0.88 | 0.89 | 0.70 |
| b | 100:0 (5Hz low-pass) | 0.96 | 0.95 | 0.96 | 0.80 |
| c & d | 0:100 (high gyro drift) | 0.63 | 0.85 | 0.85 | 0.69 |
| e | 50:50 | 0.93 | 0.90 | 0.91 | 0.73 |

always be centered at 0° when calibrated by the accelerometer complement.

### 4.3. SWM Monitoring Using Equal Weighting of Gyroscope and Accelerometer Inputs

This test was intended to implement true signal combinations as described by Equation (3). Accelerometer and gyroscope input were initially combined at a ratio of 50:50 for $\beta_{gyro}$:$\beta_{accel}$. The signal generated from this ratio yielded a fairly noisy signal (**Figure 2(d)**). However, the signal generated through this weighting was less noisy than the signal generated by the accelerometer only.

### 4.4. Optimal Weights Discovery for Gyroscope: Accelerometer Coefficients $\beta_{gyro}$:$\beta_{accel}$

Data used to optimize the weight ratio were collected during actual road driving to ensure a robust selection. Various ratios were tried during this analysis and a few of the important ratios are shown in **Table 2**.

The ratio which was finally chosen was 99:1 or $\beta_{gyro} = 0.99$, $\beta_{accel} = 0.01$ (**Figure 2(c)**) because it had a stronger correlation to potentiometer readings in all measures of correlation used, indicating signal correctness. Additionally, visual inspection revealed a good balance between noise reduction properties and better signal agreement with potentiometer readings.

SWM readings from cases in which the $\beta_{gyro}$:$\beta_{accel}$ ratio favored the accelerometer tended towards introducing linear vibrations. These are very easily removable using a low pass filter or an averaging filter. However, using the selected weight ratio of 99:1 as in this case, the accelerometer and gyroscope fusion yielded data that was not significantly affected by linear noises or vibrations, even during highway driving, and city driving on rough roads. It was unnecessary to filter the data. The gyroscope's design as a measure of angular velocity about clearly defined axes contributed greatly to the efficacy of this method for low-noise SWM monitoring. SWM readings from cases in which the $\beta_{gyro}$:$\beta_{accel}$ ratio heavily favored the gyroscope tended towards introducing slow signal drift, while SWM readings from cases in which the $\beta_{gyro}$:$\beta_{accel}$ ratio heavily favored the accelerometer tended towards introducing artifacts (**Figure 2(a)**). The selected ratio yielded optimal results.

### 4.5. Accuracy of SWM Readings as Compared to the Potentiometer

When subjected to prolonged SWM inputs over a 45 minute driving task, a strong cross correlation between the two signals $\theta_{\text{final}}$ and $\theta_w$ was discovered (xcorr: 0.99; $R^2$: 0.96; P = 0; Pearson: 0.99; $P < 0.05$).

As an extension of this test, the SWM readings derived from s $\theta_{\text{final}}$ data were found to be capable of keeping up with even the most rapid steering wheel movements that were tested. The steering wheel was subjected to greater than 20 sudden rotations at rates up to 150°$s^{-1}$ and the results during high speed rotations yielded a high signal cross correlation between $\theta_w$ and $\theta_{\text{final}}$ (xcorr: 0.98; $P < 0.05$; Pearson: 0.97; $P < 0.05$). A small section of this test is shown in **Figure 4(a)**. The MPU-6050 sensor has a range of up to ±2000°$s^{-1}$ and was configured in this test for use up to ±250°$\cdot s^{-1}$.

To plot **Figure 4** shown below, Equation (4) was used to convert the potentiometer voltages recorded from its amplifier channel directly into $\theta_w$. Over the same sample period, Equation (3) was used to convert readings collected from the gyroscope and accelerometer channels directly into $\theta_{\text{final}}$. $\theta_w$ and $\theta_{\text{final}}$ were then overlaid in the resulting plots.

The road test using the simulator steering wheel showed very positive results for the proposed method. The signal $\theta_{\text{final}}$ was highly correlated to the potentiometer output $\theta_w$ (xcorr: 0.992; Pearson: 0.988, P = 0; $R^2$: 0.976, P

**Table 2.** Various IMU device ratios and their correlation to potentiometer.

| | Ratio | Correlation to potentiometer results | | | |
|---|---|---|---|---|---|
| | $\beta_{accel}$:$\beta_{gyro}$ | XCorr | | $\beta_{accel}$:$\beta_{gyro}$ | XCorr |
| a | 10:90 | 0.92 | 0.89 | 0.89 | 0.71 |
| b | 90:10 | 0.97 | 0.96 | 0.95 | 0.81 |
| **c** | **99:1** | **0.98** | **0.97** | **0.98** | **0.88** |
| d | 99.5:0.5 | 0.94 | 0.90 | 0.90 | 0.73 |

= 0) and the signals overlapped each other for the majority of the recorded time (**Figure 3(b)**).

## 5. Discussion

The method was tested for efficacy during real road driving. The described method allows for an inexpensive, non-intrusive, and very easy to implement drowsiness detection system without the requirement for complex equipment or major modifications to the current steering system. Although some minor vibrations were seen during the mounting of the device in road tests, these vibrations affected angular signal at less than 0.1° angular displacement when unfiltered. However it is important to know that SWM assessment of driver drowsiness is a vehicle based behavioral measure which relies upon detection of trends slowly increasing towards drowsiness and not necessarily upon precision within 0.1°. Further assessments of the method through the creation of a mobile phone application were able to utilize the mobile devices internal gyroscope and accelerometer for accurate SWM monitoring for drowsiness detection.

$\beta_{gyro}$ was eventually chosen to be 0.99 and $\beta_{accel}$ was chosen as 0.01. The output $\theta_{\text{final}}$ was the positional angle result of the combined IMU setup in degrees. This output indicated the current wheel orientation in units of degrees (°). The shock resistant gyroscope provided most of the SWM monitoring while the drift resistant accelerometer contributed only the minimum amount of orientation data to ensure that the gyroscope measurement was perpetually calibrated against gravity and did not drift with time.

These findings are important because the method does not require extensive modifications to existing vehicle setups. The high affordability of this primarily gyroscope-based method also improves the feasibility of wide scale deployment. Many individual researchers and federal regulators have invested large amounts of time and manpower to stem the thousands of highway fatalities and injuries that occur worldwide each year as a result of drowsy driving. Although these efforts have yielded reliable methods such as SWM, which has been touted by researchers and government agencies as a potential lifesaver, there has still been no widespread practical means to apply this method. As a result, the vast majority of highway vehicles continue to operate without drowsy driving detection mechanisms, and thousands of fatalities and injuries continue to occur annually. With this method, the well documented SWM method of drowsy driving detection can be applied to curb highway accidents

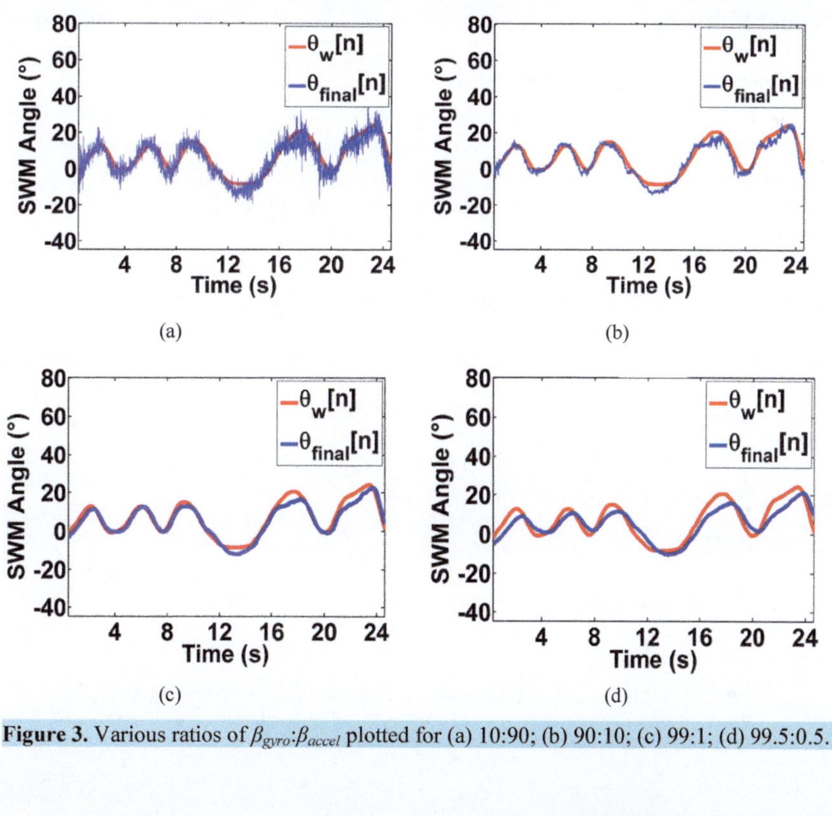


(a) (b) (c) (d)

**Figure 3.** Various ratios of $\beta_{gyro}$:$\beta_{accel}$ plotted for (a) 10:90; (b) 90:10; (c) 99:1; (d) 99.5:0.5.

and deaths with minimal cost to drivers and car manufacturers.

Future work will involve embedding this technology into vehicle steering wheels to further the potential for its eventual integration into vehicular systems. In addition, the development of a steering mount implementation will be researched to provide an easy alternative for drivers whose vehicles do not come with such technologies. Other future work includes the investigation of alternate inertial components by manufacturer to further optimize cost/performance output for the end user.

### 5.1. Slight Variations between $\theta_w$ and $\theta_{\text{final}}$

Slight variations existed between $\theta_w$ and $\theta_{\text{final}}$ which could be observed in the high velocity rotation testing performed in 3.5. The $\theta_{\text{final}}$ signal which comprised mostly of gyroscope data exhibited "ears" at the beginning and end of very sudden turns during periods of high and rapidly changing angular velocities (**Figure 4(a)**). Overall, the $\theta_{\text{final}}$ readings maintained high accuracy, even during sudden movements. Although such unusually high velocity steering activities are not expected to occur except during the most extreme driving cases, possibly involving road accidents. Despite this, the readings of $\theta_{\text{final}}$ maintained high accuracy.

### 5.2. Rotations beyond 360°

Because most steering wheels rotate through more than 360°, predictive readings can be used to compensate for this effect. A tiny microcontroller for example can be used to adjust for this. If a range of 361° to 720° are the base readings of the sensor, any measurements beyond a full counter-clockwise rotation would adaptively read between 0° to 360° and any measurements beyond a full clockwise rotation would adaptively read between 721° to 1080°.

### 5.3. Comparison of the Proposed Method with the Earlier Accelerometer-Based Method

The proposed method yielded a more noise resistant method of SWM monitoring when compared to the previous accelerometer-based method [20]. The use of a low pass filter is effective against vehicle and road noises using the previous method, however, if a practical application of SWM monitoring calls for no phase shifting-margin, then the current fusion method might be better suited. Phase shifted signals retain their accurate waveforms, however time delay could occur if filtered improperly. The currently proposed method did not demonstrate any need for filtering, even during real road trials.

The unfiltered accelerometer-based method, while more prone to linear vibration noise than the current method, is very effective in drowsy driving simulation tasks especially as it is dangerous to place sleep deprived subjects on the highway.

The benefit of the currently proposed system is the enhancement of the strengths and weaknesses of two completely different sensors in a method whereby they both work more effectively. The use of a gyroscope for the majority of the SWM data eliminates the problem of linear vibrations due to the gyroscopes insensitivity to such data. It is seen that both methods are effective and accurate for their individually specific tasks. The current method was not prone to road noise, engine noises, and other vehicle noises.

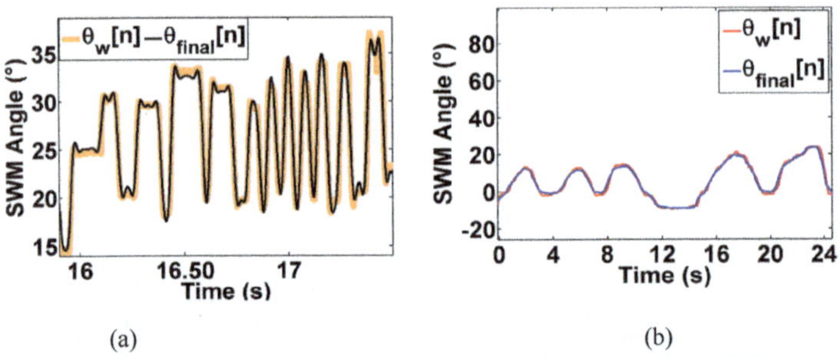


**Figure 4.** (a) High speed SWM outputs remained highly accurate representations of ground truth steering movements; (b) The final signal (right) matches the potentiometer signal.

## 6. Conclusion

This study demonstrated that the effective fusion of a simple gyroscope and accelerometer can be used to accurately monitor SWM for drowsy driving activities including sudden corrections and wide angle corrections. The efficacy of the method was confirmed by comparing the SWM estimates generated by the method with actual SWM readings collected from the steering-wheel potentiometer which yielded high correlation. The high correlation suggests that the method could be used as a direct replacement of other SWM measures for the implementation of SWM-based drowsy detection algorithms.

## References

[1] Rau, P. (1996) NHTSA's Drowsy Driver Research Program. National Highway Traffic Safety Administration. Washington DC.

[2] Vanlaar, W., Simpson, H., Mayhew, D. and Robertson, R. (2008) Fatigued and Drowsy Driving: A Survey of Attitudes, Opinions and Behaviors. *Journal of Safety Research*, **39**, 303-309. http://dx.doi.org/10.1016/j.jsr.2007.12.007

[3] Connor, J., Norton, R., Ameratunga, S., Robinson, E., Civil, I., Dunn, R., Bailey, J. and Jackson, R. (2002) Driver Sleepiness and Risk of Serious Injury to Car Occupants: Population Based Case Control Study. *BMJ*, **324**, 1125. http://dx.doi.org/10.1136/bmj.324.7346.1125

[4] Hamblin, P. (1987) Lorry Driver's Time Habits in Work and Their Involvement in Traffic Accidents. *Ergonomics*, **30**, 1323-1333. http://dx.doi.org/10.1080/00140138708966026

[5] Maclean, A.W., Davies, D.R. and Thiele, K. (2003) The Hazards and Prevention of Driving While Sleepy. *Sleep Medicine Reviews*, **7**, 507-521. http://dx.doi.org/10.1016/S1087-0792(03)90004-9

[6] Kim, J.Y., Jeong, C.H., Jung, M.J., Park, J.H. and Jung, D.H. (2013) Highly Reliable Driving Workload Analysis Using Driver Electroencephalogram (EEG) Activities during Driving. *International Journal of Automotive Technology*, **14**, 965-970. http://dx.doi.org/10.1007/s12239-013-0106-z

[7] Ogawa, K. and Shimotani, M. (1997) A Drowsiness Detection system. *Mitsubishi Electric Advance*, **78**, 13-16.

[8] Wierwille, W. (1999) Historical Perspective on Slow Eyelid Closure: Whence PERCLOS? In: *Technical Proceedings of Ocular Measures of Driver Alertness Conference*, Herndon, VA (FHWA Technical Report No. MC-99-136), Federal Highway Administration, Office of Motor Carrier and Highway Safety, Washington DC, 31-53.

[9] Bowman, D.S., Schaudt, W.A. and Hanowski, R.J. (2012) Advances in Drowsy Driver Assistance Systems through Data Fusion. In: *Handbook of Intelligent Vehicles*, Springer, Berlin, 895-912.

[10] Wierwille, W., Hanowski, R., Olson, R., Dinges, D., Price, N., Maislin, G., Powell IV, J., Ecker, A., Mallis, M. and Szuba, M. (2003) NHTSA Drowsy Driver Detection and Interface Project-Final Report. Contract No. DTNH22-D-00-07007, Task Order, 1.

[11] Fukuda, J., Akutsu, E. and Aoki, K. (1995) An Estimation of Driver's Drowsiness Level Using Interval of Steering Adjustment for Lane Keeping. *JSAE Review*, **16**, 197-199. http://dx.doi.org/10.1016/0389-4304(94)00070-A

[12] Elling, M. and Sherman, P. (1994) Evaluation of Steering Wheel Measures for Drowsy Drivers. *Proceedings of the 27th ISATA*, Aachen, 31 October-4 November 1994, 207-214.

[13] Thiffault, P. and Bergeron, J. (2003) Monotony of Road Environment and Driver Fatigue: A Simulator Study. *Accident Analysis & Prevention*, **35**, 381-391. http://dx.doi.org/10.1016/S0001-4575(02)00014-3

[14] Eskandarian, A. and Mortazavi, A. (2007) Evaluation of a Smart Algorithm for Commercial Vehicle Driver Drowsiness Detection. *Proceedings of IEEE Conference on Intelligent Vehicles Symposium*, Istanbul, 13-15 June 2007, 553-559.

[15] Yabuta, K., Iizuka, H., Yanagishima, T., Kataoka, Y. and Seno, T. (1985) The Development of Drowsiness Warning Devices. *Proceedings of the* 10*th International Technical Conference on Experimental Safety Vehicles* (*ESV*), *Oxford, England*, **10**, 282-288.

[16] Sayed, R. and Eskandarian, A. (2001) Unobtrusive Drowsiness Detection by Neural Network Learning of Driver Steering. *Proceedings of the Institution of Mechanical Engineers, Part D*: *Journal of Automobile Engineering*, **215**, 969-975. http://dx.doi.org/10.1243/0954407011528536

[17] Luinge, H., Veltink, P. and Baten, C. (1999) Estimating Orientation with Gyroscopes and Accelerometers. *Technology and Health Care*, **7**, 455-459.

[18] Lee, H., Ryu, S. and Lee, J. (2009) Optimal Posture of Mobile Inverted Pendulum Using a Single Gyroscope and Tilt Sensor. *Proceedings of ICCAS-SICE*, Fukuoka, 18-21 August 2009, 865-870.

[19] Greene, M. (1996) A Solid State Attitude Heading Reference System for General Aviation. *Proceedings of IEEE Con-*

*ference on Emerging Technologies and Factory Automation*, Kauai, 18-21 November 1996, 413-417.

[20] Lawoyin, S.A., Fei, D.Y. and Bai, O. (2014) Accelerometer-Based Steering Wheel Movement Monitoring for Drowsy Driving Detection. *Proceedings of the Institution of Mechanical Engineers, Part D: Journal of Automobile Engineering*, Published Online 31 October 2014. http://dx.doi.org/10.1177/0954407014536148

[21] Sakaguchi, T., Kanamori, T. and Katayose, H. (1996) Human Motion Capture by Integrating Gyroscopes and Accelerometers. *Proceedings of IEEE/SICE/RSJ International Conference on Multisensor Fusion and Integration for Intelligent Systems*, Washington DC, 8-11 December 1996, 470-475.

17

# Whole-Body Vibration in Automated Residential Solid Waste Collection

**Helmut Paschold**

Department of Safety Sciences, College of Health and Human Services, Indiana University of Pennsylvania, Indiana, PA, USA.

Email: helmut.paschold@iup.edu

## Abstract

**Drivers of residential solid waste collection trucks are exposed to a wide variety of physical and health hazards. Automated robotic arm collection methods are intended to eliminate physical and ergonomic hazards associated with manual waste collection. However, whole-body vibration (WBV) exposure is a hazard that may be significant and greater than being found in semi-automated or manual methods. WBV is measured in a pilot field study of automated waste collection trucks during normal operation for nearly eight-hour periods on two consecutive days. All four trials are run under similar route conditions. Vibrations are measured at the seat-driver interface using a tri-axial accelerometer seat pad and portable vibration monitors. Measured WBV levels are compared with prior published data from waste collection trucks and health standards. The average WBV exposure value, corresponding to 0.99 $m \cdot s^{-2}$ for frequency-weighted r.m.s. acceleration, is above the ISO 2631-1 action value of 0.50 $m \cdot s^{-2}$ and below a limit value of 1.15 $m \cdot s^{-2}$ as given in the European Parliament Directive 2002/44/EC. This exposure level suggests the presence of potential health risks.**

## Keywords

**Whole-Body Vibration, Municipal Solid Waste, Automated MSW Trucks**

## 1. Introduction

Human exposure to whole-body vibration (WBV) is recognized as an occupational health risk factor, especially for lower back pain (LBP). Directive 2002/44/EC of the European Parliament recognizes worker risk from exposure to vibrations and the effects on the muscular/skeletal, neurological, and vascular systems [1]. Bovenzi and Hulshof [2] reported in a review of WBV epidemiological studies that there was clear evidence linking occupational WBV with increased risk of LBP. An epidemiological study of professional drivers found LBP frequency, intensity and disability increased significantly with increasing cumulative WBV exposure [3]. While acknowledging that LBP rates were higher for car drivers than non-drivers, and the contribution of WBV to car drivers' LBP was uncertain [4]. Schwarze [5] reported 27% to 35% of 388 vehicle operators' LBP was linked to WBV, with the group of drivers exposed to high vibration doses having an LBP risk 1.55 times greater than the control group experiencing low vibration doses. The U.S. National Institute of Occupational Safety and Health (NIOSH), in a review of literature, found 15 of 19 WBV and LBP studies consistent in showing positive associations [6]. NIOSH assigned their highest ranking of "strong evidence" to the association between WBV and LBP, concluding with:

Laboratory studies had demonstrated WBV effects on the vertebra, intervertebral discs, and supporting musculature. Both experimental and epidemiologic evidence suggested that WBV might act in combination with other work-related factors such as prolonged sitting, lifting, and awkward postures to cause increased risk of back disorder (p. 6-33).

Guidelines for human WBV exposure could be found in Annex B of ISO 2631-1, *Mechanical vibration and shock—Evaluation of human exposure to whole-body vibration* [7]. Based upon the parameters of ISO 2631-1 Equation B.2, the action level for an 8-hour exposure was 0.5 $m \cdot s^{-2}$ and the upper level boundary for likely health risk was approximately 0.8 $m \cdot s^{-2}$. Directive 2002/44/EC of the European Parliament presented a WBV daily exposure action value of 0.5 $m \cdot s^{-2}$ and a limit value of 1.15 $m \cdot s^{-2}$ for the frequency-weighted r.m.s. acceleration [1]. The U.S. standard ANSI S3.18-2002 ISO 2631-1-1997 [8] adopted the ISO WBV standard [9]. Similarly, the ACGIH Threshold Limit Values were also based on the ISO standard [10]. Both ANSI and ACGIH standards were voluntary standards in the U.S. The Occupational Safety and Health Administration (OSHA) or OSHA approved State programs were responsible for the enforcement of mandatory standards being adopted as law. The OSHA general industry, maritime and construction standards did not include WBV [11].

Solid waste collection was recognized as a hazardous occupation. The 2012 lost-time and restricted-work accident rate for U.S. solid waste collection was over 2.7 times greater than the combined private industry sector rates [12]. WBV had been specifically identified as one of the many hazards facing refuse collection employees [13].

Residential solid waste collection is typically performed with trucks possessing a haulage capacity sufficient to reduce frequent trips to a landfill for off-loading, yet small enough to be maneuverable in urban traffic. These vehicles are likely to expose operators and passengers to WBV. Residential solid waste collection methods range from manual to semi-automated to automated (or mechanized) loading of waste. Manual collection and loading are typically accomplished by a work crew of two to three persons physically moving waste containers from a collection point and then lifting and dumping these into the hoppers of trucks. Semi-automated systems still require the manual movement and placement of the waste containers at the truck hopper by a work crew; however, a hydraulic or mechanical device lifts and empties the waste containers into the hopper. In the fully automated system, a driver operating a joystick controls robotic arm collects solid waste from containers, without leaving the vehicle cab. In this situation, the collection container is positioned at the curbside by the resident [14].

Manual and semi-automated solid waste collection vehicles expose drivers to similar WBV as with other commercial vehicles, such as delivery trucks. In the automated method, operators drive the truck, **Figure 1**, to an approximate alignment with the container and extend the robotic arm. This arm, as shown in **Figure 2**, grasps, lifts, and dumps the filled container into a hopper at the top of the truck, returning the container back to its curbside location. Automated operators are exposed to the vibrations generated by and transmitted from the robotic arm lifter operating perpendicular to the forward direction. The automated vehicles tend to have more stops and starts, compared to manual or semi-automated collection methods, as the driver aligns with each individual container.

Solid waste collection management eliminated manual material handling, walking, and direct contact with the solid waste by the implementation of the robotic arm collection system; however, the change may have introduced new ergonomic stressors due to joystick manipulation, neck and trunk-turning posture, and possible increases in WBV. While ergonomic stressors due to continuous poor posture (neck and trunk turning) and joystick maneuvering are fairly well recognized, few safety and health professionals in the US have an adequate awareness or understanding of WBV to assess the potential risk to workers [15]. NIOSH [6] reported epidemiological studies linking neck pain and awkward work posture, and also between neck disorders and forceful movements. Bovenzi [16] reported a significant association between neck pain and WBV exposure in a cohort study of professional drivers. Further study of the postural component during WBV exposure for automated

**Figure 1.** Automated collection truck with close-up of robotic arm on mid-section right-hand side of the vehicle.

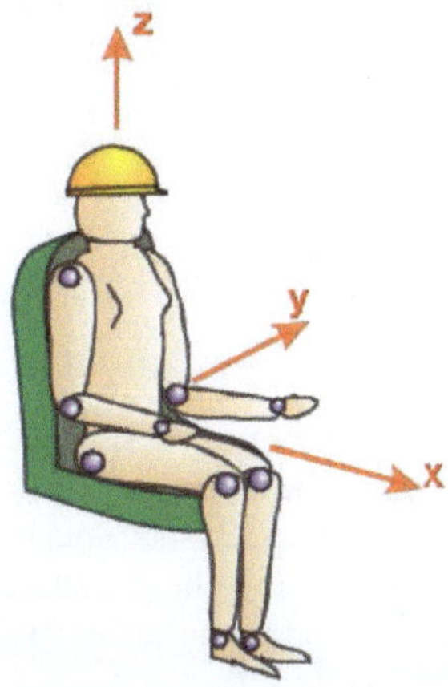


**Figure 2.** Depiction of seated X, Y, and Z axes (Source: Guide to good practice on Whole-Body Vibration, _11_08 WBV_Good_practice_Guide v6.7h English). http://resource.isvr.soton.ac.uk/HRV/VIBGUIDE/2008_11_08%20WBV_Good_practice_Guide%20v6.7h%20English.pdf

waste collection drivers is warranted.

Three prior studies evaluated solid waste collection trucks and WBV levels [3] [17] [18]. Direct correspondence with the authors was required to categorize the type of waste collection methodology as to manual, semi-automated or automated. The earliest study of manual residential waste collection found WBV exposure exceeding recommended limits in Japan [17]. Two semi-automated collection studies were located: one in Portugal reported WBV doses exceeding the action value in 5 of 13 garbage trucks [18] and another in Italy found WBV levels below the action value in all six solid waste collection trucks [3]. No published WBV study results were found for residential solid waste collection in the U.S. or for automated collection elsewhere.

The objective of this study was to obtain measurements of WBV exposure to operators of automated residential solid waste collection vehicles at the contact interface between the driver and seat and compare these measurements with health risk guidelines. If a WBV health risk is apparent, further study of WBV source generation will be conducted to propose remedial actions.

## 2. Methods

### 2.1. Solid Waste Collection Trucks and Routes

The trucks monitored in this study were representative of the types used for automated collection. On Day 1, a 2007 Sterling Condor (#1) and a 2003 Peterbilt (#2) truck were monitored. On Day 2, the test was repeated on two 2007 Sterling Condors—one was the same vehicle and driver (#1) as Day 1 and another 2007 Sterling

Condor with a different driver (#3). All trucks were equipped with the McNeilus 31-yard side loader compactor, **Figure 1**. Identical air ride seats were installed on the driver side of all trucks. The three drivers, of similar stature and weight, had been driving automated trucks for at least one year.

The trucks were run on their regular routes with no deviations from normal operation. The pre-trip inspections performed by the drivers indicated normal operating condition for all trucks. The four residential routes were within the same geographic area and shared level terrain, asphalt-paved roads in the residential area, similar spacing between homes, and similar distance from the landfill. The phases of the route included the start of the route at the garage/yard, travel to and from the collection area, collection, travel with a full load to the landfill, idling in queue to dump, dumping, travel return unloaded back to the collection area, and break time. Trips to the landfill for dumping were repeated up to three times during the workday.

## 2.2. Vibration Measurement and Analysis

Two Quest Technologies (Oconomowoc, WI) HAVPro Human Vibration Meters were used to obtain WBV measurements. The meters comply with ISO 8041:1990(E) (including Amendment 1:1999(E)) and ISO 2631-1:1997 requirements [19] and were factory calibrated within one year prior to use. The meters were set to run in WBV mode, automatically setting parameters in accordance with ISO requirements. A standardized axes orientation was as follows: Y—vertical, X—fore and aft, and Y—transverse, shown in **Figure 2**. Settings included:

- $w_k$ weighting for the vertical (Z) axis;
- $w_d$ weightings for the horizontal (X and Y) axes;
- a 1.4 multiplier for the horizontal axes.

The accelerometers were Quest Technologies Model 5313AQ tri-axial LIVM seat pad accelerometers connected via cable to the HAVPro instruments.

The HAVPro time history buffer is limited to 240 samples in a set; therefore it can hold up to a 20-minute total of continuous 5-second interval data in its time history. If data collection continues past the full-point of the buffer, the initial 5-second datum is eliminated to enter the newest incoming 5-second datum while still maintaining an on-going time weighted average. The HAVPro instrument will hold up to 100 data time history sets. To maintain datum, upon reaching the 20-minute storage limit of the time history buffer, the data set is saved as a data time history set and a new buffer data set is begun. In this manner, a continuous record of recordings of approximately eight hours can be achieved.

WBV data from the HAVPro meters was downloaded via an RS232 port and USB adaptor to a computer. This data was entered into the Quest Technologies Quest Suite Professional II software, Version 3.7.2269, for storage and analysis.

## 2.3. Experimental Procedure

The experiments were carried out over a two day period, July 31 and August 1, 2007, with three different trucks and drivers. As previously stated, one driver and truck pairing was repeated on the second day. Each trial run recording time limit was approximately eight hours.

For Day 1, a series of 24 individual 20-minute sets were recorded during the workday to allow viewing of 5-second interval data for the different phases of the collection task to avoid the overwriting of data on the HAVPro memory buffer. On Day 2, this setting was maintained for one truck and changed for the other. The setting change entailed a reduction of the interval time to 1-second, with the intent of increasing resolution of vibration data during the collection tasks. With 1-second intervals, each time history set was four minutes long; a total of 99 4-minute time history sets were recorded for a total of 6.6 hours.

During daylight hours on the day prior to the experiments, an initial survey of the truck cabs was conducted to determine optimal equipment layout to avoid interference with the driver during normal duties. As shown in **Figure 2**, the seat pad accelerometer was placed on the seat and aligned to the manufacturer's markings of X, Y, and X axes. Also, a small piece of tape was affixed to the front of the pad to assist with proper and easy directional alignment. On Day 1, the HAVPros and seat pad accelerometers were placed in two trucks just prior to the daily shift, at approximately 4:00 a.m. Tape was used to secure the meters from movement while stowed behind and below the driver seats in **Figure 3**. The drivers were instructed about the basic nature of the experiment and asked to not deviate from their routine route tasks. As each driver prepared to leave the truck yard, the HAVPro instruments were activated. On Day 1, one of the trucks was followed to observe operational practices for several

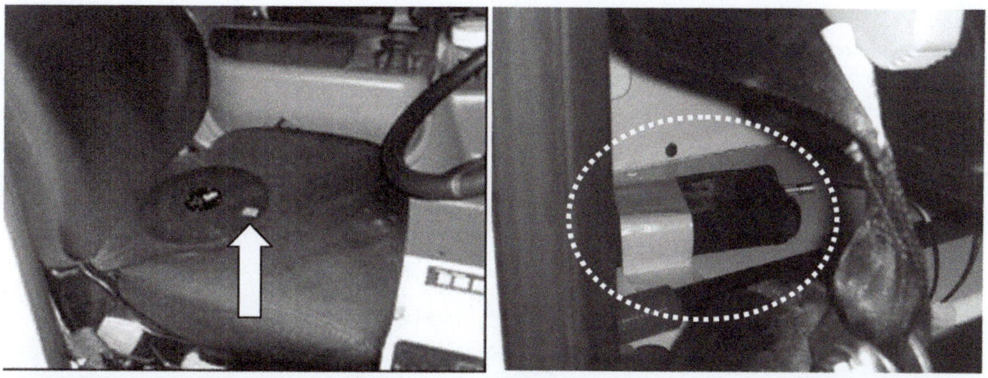

**Figure 3.** Seat pad accelerometer on driver seat and. HAVPro instrument secured behind and below the driver seat.

hours during the driving to the collection route and actual solid waste collection times. Truck logs were obtained to provide driver-recorded times of landfill delivery and unloading. The logs were compared with the graphic WBV results and verified the modes of operation during specific time periods. The instruments were retrieved from the trucks at the end of the workday. Data was downloaded into two separate computers, to assure backup in the event of one computer's failure.

To create a composite of the full-day individual time history sets, each data set for a given truck and day was combined to give a more accurate assessment of the total workday exposure. This was accomplished by using the data-combining feature of the Quest Technologies Quest Suite Professional II software. The individual time history sets were retained in original form to allow further detailed analysis of specific phases of the waste collection task such as during collection, unloading, etc.

# 3. Results

## 3.1. Composite Findings

A summary of the frequency weighted r.m.s. Aeq values in the X, Y, Z, and XYZ-sum axes is presented in **Table 1** for the three different trucks over the 2-day period. The respective averages are presented at the bottom of the table. The four values for the XYZ-sum were all within ±4% of the 0.99 $m \cdot s^{-2}$ average. The individual X, Y, and Z axis values for the trucks ranged from 0.36 - 0.40, 0.35 - 0.46, and 0.56 - 0.65 respectively for the three axes; the widest distribution was noted in the Y-axis data. **Table 1** also contains XYZ-sum axes Vibration Dose Value (VDV) ranging 3.31 to 9.77 and the Crest Factors (CF) 2.84 to 8.04.

## 3.2. Waste Collection Phase Analysis

**Table 2** presents data for WBV during three phases of the collection operation. The driving-only phase started with a full load from the collection area to the landfill and appears to have included travel on some unpaved roads near the landfill. The unloading of the hopper at the landfill starts with the truck waiting in a queue to approach the active cell for dumping. This part is characterized by idling, moving, and dumping. The XYZ-sum vibration appears to be about the same for driving-only and collection (1.03 to 1.13 r.m.s.), and are higher than unloading segment (0.85 r.m.s.).

During waste collection which is approximately 75% of the workday, the X, Y and Z axis values are relatively close to each other ranging from 0.36 to 0.58 r.m.s., with WBV only slightly higher in the Z axis. Also during collection, three of four Y and Z axes data sets were equal to or greater than the 0.50 $m \cdot s^{-2}$ action level. The meter settings of 5-second and 1-second averaging yielded similar results. **Figures 4(a)-(b)** graphically present a representative two-minute segment showing vibration during of the container collection process, measured on Truck #1, Day 1. The heavy bold graph lines on top in **Figure 4(a)** and **Figure 3(b)** are the XYZ sum data, with the lighter lines representing the individual X, Y, and Z components. The peaks in the Y axis correspond to the dumping of the residential container into the truck hopper. X axis peaks occur from the lurching of the truck as it starts and stops in short distances from house-to-house.

**Table 1.** Frequency Weighted Acceleration $m \cdot s^{-2}$ (r.m.s.). Aeq and Vibration Dose Value ($m \cdot s^{-1.75}$) for Full-day Combined Data Sets (5-second interval).

| Truck and day | Measured time (hr:min) | X | Y | Z | XYZ sum | VDV XYZ sum | CF XYZ sum |
|---|---|---|---|---|---|---|---|
| 1 - 1 | 7:43 | 0.36 | 0.41 | 0.61 | 0.98 | 3.31 | 2.84 |
| 2 - 1 | 8:23 | 0.36 | 0.35 | 0.65 | 0.95 | 8.53 | 3.04 |
| 1 - 2[a] | 6:36 | 0.38[a] | 0.44[a] | 0.59[a] | 1.01[a] | 5.92[a] | 8.04[a] |
| 3 - 2 | 8:13 | 0.40 | 0.46 | 0.56 | 1.02 | 9.77 | 7.33 |
| Average | | 0.38 | 0.42 | 0.60 | 0.99 | | |
| SD | | 0.02 | 0.05 | 0.04 | 0.03 | | |

[a] 1-second interval averaging.

**Table 2.** Frequency Weighted Acceleration $m \cdot s^{-2}$ (r.m.s.). Aeq for three 20-minute Operation Phase Segments (5-second interval), Truck #1.

| Segment | Day/Time | X | Y | Z | XYZ-sum |
|---|---|---|---|---|---|
| Collection | 7 - 31/2:20 a.m. | 0.36 | 0.50 | 0.57 | 1.03 |
| Collection | 8 - 1/7:17 a.m. | 0.44[a] | 0.54[a] | 0.58[a] | 1.13[a] |
| Driving | 7 - 31/9:00 a.m. | 0.20 | 0.32 | 0.89 | 1.03 |
| Unloading | 7 - 31/9:20 a.m. | 0.27 | 0.31 | 0.65 | 0.85 |

[a] 1-second interval averaging.

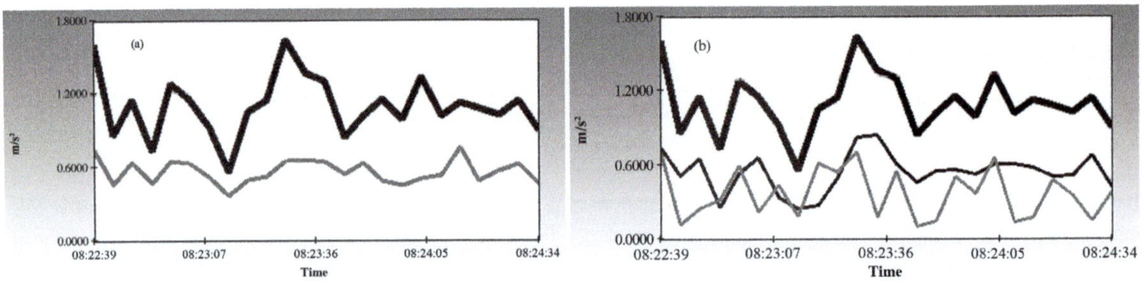


**Figure 4.** Measured vibration while collecting solid waste during a representative two minute interval on Day 1: (a) XYZ sum total and Z axis vibration; (b) XYZ sum total and Y and X axes. XYZ sum; Z axis; Y axis; X axis.

## 4. Discussion

### 4.1. General Results

It would be expected that the WBV levels should be relatively similar for the four experiments as trucks were operated on or with identical seats, similar terrain, equivalent spacing of containers, and same distance from the landfill. Truck #2 is of a different manufacturer than Trucks #1 and #3; however, all trucks were equipped with the identical McNeilus 31-yard side loader compactor. Truck #3 has the highest X and Y axis value, the lowest Z axis value, and the highest XYZ-sum value. The very slight differences in Y and Z axis values for Truck #1 on Days 1 and 2 are explained by minor differences in collection conditions. Truck #1—Day 2 data was recorded using 1-second averaging, resulting in a total data collection time of only 6.6 hours versus 8 hours for the three other experiments; however, results appear to be similar to the 5-second averages.

### 4.2. Comparison with Prior Studies

**Table 3** shows a comparison of prior studies with the current findings. Bovenzi [3] reported a fairly low WBV

**Table 3.** Comparison of findings with prior studies according to axis, $m \cdot s^{-2}$ (r.m.s.).

| Study author | X | Y | Z | XYZ sum | Waste Collection type |
|---|---|---|---|---|---|
| Bovenzi | 0.08 | 0.12 | 0.21 | 0.31 | Semi-automated |
| Maeda | 0.76 | 0.79 | 1.1 | | Manual |
| Melo | 0.17 | 0.19 | 0.43 | 0.60 | Semi-automated |
| Paschold | 0.38 | 0.42 | 0.60 | 0.99 | Automated |

exposure and may reflect a waste collection method in which the driver frequently exits the truck to move and position the container at a hydraulic lifter. The Melo [18] study reported findings of higher WBV levels, based on a waste collection method identical to Bovenzi. Maeda [17] presented higher WBV levels and it appears that this study involved drivers spending more time driving the vehicle between the collection site and the landfill. An explanation for the difference in WBV levels between this experiment and the Bovenzi and Melo studies is found in the amount of time that the driver is actually in the driver's seat; for automated collection drivers, it is almost 100% of the work day. These reports did not specify the types of waste collection operations, seat types, CFs, or vibration measurement duration; limited operational detail was supplemented by direct contact with the authors. These missing details could provide further insight into the differences in measurement outcomes.

### 4.3. Comparison with Health Standards

In the absence of a U.S. regulatory standard, a comparison of findings is made with European minimum requirements for the protection of workers from exposure to mechanical vibration. European Directive 2002/44/EC presents a daily exposure action value of 0.5 $m \cdot s^{-2}$ and a limit value of 1.15 $m \cdot s^{-2}$. All four measurements showed exposure to be above the action value and below the limit value. According to ISO 2631-1:1997(E), values in the action level zone present potential health risks that require attention. Precise assessment of risk within the action level zone is currently not possible; however, it should be noted the average XYZ-sum value, 0.99 $m \cdot s^{-2}$, was only 15% below the limit value that suggests health risks are likely.

The ISO standard states that the WBV assessment is made independently for each axis and the health risk evaluated according to the highest single axis value. However, section 7.2.2 explains when two or more of the axes vibration magnitudes are *similar* (not defined in the standard), the vector sum can be used for health risk assessment.

The four CF values are below the 9.0 threshold contained in ISO 2631-1:1997(E). If the 9.0 threshold were exceeded, the r.m.s. values could be considered invalid and exposure would have to be determined by the VDV method.

### 4.4. Limitations

This study measured four vehicles, a small sample size relative to the number of trucks. However, the results were fairly consistent. The 1997 ISO standard requires an adequate monitoring duration to assure precision and representation of typical exposure. This study differs from the prior solid waste collection truck and many other WBV studies, as an entire work day of monitoring was completed and has many combined sample sets of WBV data (up to eight hours).

Many other variables in a full scale study should be recorded and included such as driver anthropometry, seat type and adjustment, vehicle conditions (tire pressure, shock system, differing models, etc.), roads, driving speeds, variations in residential waste loads, and so on.

## 5. Conclusion

This study finds that automated residential waste collection drivers are exposed to WBV levels associated with an increased health risk while working in a seated position [3] [7]. When this study's higher values are compared to previously reported WBV values for solid waste truck operation those of Bovenzi [2] and Melo [18] in semi-automated operations, they are most likely explained by the robotic arm activity and more frequent starts and stops attributed to the nature of automated collection as compared to semi-automated or manual collection.

Also, the driver of the automated truck is in the vehicle virtually 100% of the work day. The X and Y axes levels are higher during collection than while driving; Z axis values are lower in collection than driving; and XYZ-sum values are equivalent for collection and driving. This study identifies a need to obtain improved operation mode detail from the operator or an assigned observer. Additional variables to consider for further research include the effects of operator posture, collection route characteristics (terrain, roads, home spacing), equipment type, and seat transmission (seat type, condition, adjustment).

## References


[1] European Parliament and the Council of the European Union (2002) Directive 2002/44/EC of the European Parliament and of the Council of 25 June 2002. http://www.vibrationdirective.com/media/pdfs/translation-english-ec-vibration-directive.pdf

[2] Bovenzi, M. and Hulshof, C. (1998) An Updated Review of Epidemiologic Studies on the Relationship between Exposure to Whole-Body Vibration and Low Back Pain. *Journal of Sound and Vibration*, **215**, 595-611. http://dx.doi.org/10.1006/jsvi.1998.1598

[3] Bovenzi, M., Rui, F., Negro, C., D'Agostin, F., Angotzi, G., Bianchi, S., Bramanti, L., Festa, G., Gatti, S., Pinto, I., Rondina, L. and Stacchini, N. (2006) An Epidemiological Study of Low Back Pain in Professional Drivers. *Journal of Sound and Vibration*, **298**, 514-539. http://dx.doi.org/10.1016/j.jsv.2006.06.001

[4] Gallais, L. and Griffin, M. (2006) Low Back Pain in Car Drivers: A Review of Studies Published 1975 to 2005. *Journal of Sound and Vibration*, **298**, 499-513. http://dx.doi.org/10.1016/j.jsv.2006.06.012

[5] Schwarze, S., Notbohm, G., Dupuis, H. and Hartung, E. (1998) Dose-Response Relationships between Whole-Body Vibration and Lumbar Disk Disease—A Field Study of 388 Drivers of Different Vehicles. *Journal of Sound and Vibration*, **215**, 613-218. http://dx.doi.org/10.1006/jsvi.1998.1602

[6] National Institute of Occupational Safety and Health (NIOSH) (1997) Musculoskeletal Disorders and Workplace Factors. DHHS (NIOSH) Publication, Cincinnati, 97-141.

[7] International Organization for Standardization (ISO) (1997) 2631-1, Mechanical Vibration and Shock—Evaluation of Human Exposure to Whole-Body Vibration—Part 1: General Requirements.

[8] American National Standard Institute (ANSI) (1997) S3.18-2002 ISO 2631-1-1997, Mechanical Vibration and Shock—Evaluation of Human Exposure to Whole-Body Vibration—Part 1: General Requirements.

[9] Mansfield, N. (2005) Human Responses to Vibration. CRC Press LLC, Boca Raton.

[10] American Conference of Governmental Industrial Hygienists (ACGIH) (2001) Documentation of the Threshold Limit Values for Physical Agents. ACGIH, Cincinnati.

[11] Occupational Safety and Health Administration (OSHA). General Industry Standards, Construction, and Maritime. Accessed 16 April 2014. www.ohsa.gov

[12] Bureau of Labor and Statistics. BLS Incidence Rates of Nonfatal Occupational Injuries and Illnesses by Industry and Case Types. Searched 9 April 2014. www.bls.gov

[13] Kuijer, P. and Frings-Dresen, M. (2004) World at Work: Refuse Collectors. *Occupational and Environmental Medicine*, **61**, 282-286. http://dx.doi.org/10.1136/oem.2002.001172

[14] Theisen, H. (2002) Collection of Solid Waste. In: Tchobanoglous, G. and Kreith, F., Eds., *Handbook of Solid Waste Management*, Second Edition, McGraw-Hill, New York, 7.1-7.27.

[15] Paschold, H. and Seergeev, A. (2009) Whole-Body Vibration Knowledge Survey of U.S. Occupational Safety and Health Professionals. *Journal of Safety Research*, **40**, 171-176. http://dx.doi.org/10.1016/j.jsr.2009.02.008

[16] Bovenzi, M. (2013) A Prospective Cohort Study of Neck and Shoulder Pain in Professional Drivers. *Proceedings of the 5th International Conference on Whole Body Vibration Injuries*, Amsterdam, 5-7 June 2013.

[17] Maeda, S. and Morioka, M. (1998) Measurement of Whole-Body Vibration Exposure from Garbage Trucks. *Journal of Sound and Vibration*, **215**, 959-964. http://dx.doi.org/10.1006/jsvi.1998.1676

[18] Melo, R. and Carvalhinho, F. (2006) Whole-Body Vibration Exposure of Waste Truck Drivers. *Proceedings of the IEA 2006 16th World Congress on Ergonomics*, Maastricht, 10-14 July 2006.

[19] Quest Technologies (2005) HAVPro Personal Vibration Monitor Owner's Manual. Quest Technologies, Inc., Oconomowoc.

# 18

# Developing a Risk-Based Approach for Optimizing Human Reliability Assessment in an Offshore Operation

**Ahmad Bahoo Toroody[1], Mohammad Mahdi Abaiee[2], Reza Gholamnia[3*], Mohammad Bahoo Torody[4], Nastaran Hekmat Nejad[5]**

[1]Faculty of Engineering, Kar Higher Education Institute of Qazvin, Qazvin, Iran
[2]Faculty of Marine Technology, Amirkabir University of Technology, Tehran, Iran
[3]School of Health Safety and Environment, Shahid Beheshti University of Medical Science, Tehran, Iran
[4]Faculty of Engineering, Azad University of Zanjan, Zanjan, Iran
[5]Faculty of Engineering, Kar Higher Education Institute of Qazvin, Qazvin, Iran
Email: [*]gholamnia@sbmu.ac.ir

## Abstract

**Human error plays a pivotal rule in all aspects of engineering activities such as operation, maintenance, design, inspection and installation. Industries are faced up to various significant human errors and consequently irrecoverable loss each year, but still there is a lack of heeds to qualify as well as quantify such errors. This paper tries to estimate the probability of failure in lifting of light structures in sea by considering human errors. To do this, a strong qualifying tool such as Functional Resonance Analysis Method (FRAM) is applied to develop high risk accident scenario by considering non-linear socio-technical interaction in system. Afterwards, human error probability is calculated for each activity using the Success Likelihood Index Method (SLIM) based on resonance that is carried out in FRAM network. Then Event Tree (ET) is conducted to assess consequences. The present study is aimed to interpret the importance of attentions to qualitative methods in implementing quantitative risk analyses to consider human error in calculation. The final outcome depicts that considering human error in the process of risk assessment will result in more accuracy and reliability in final Risk Probability Number (RPN). The developed methodology has been applied to a case study of an offshore installation.**

## Keywords

**Risk-Based Assessment, Probability of Failure, Accident Scenario, FRAM, SLIM, ET**

[*]Corresponding author.

## 1. Introduction

Risk assessment is a systematic process for analyzing and evaluating events that could affect the achievement of objectives, positively or negatively. Such events can be identified in the external environment and within an organization's internal environment. It is necessary to define both risk analyzing and risk evaluating to represent the risk assessment clearly. Risk analyzing is therefore defined as systematic use of available information to identify hazards and to estimate the risk to individuals, property, and the environment [1]. Besides, risk evaluating is means that process in which judgments are made on the tolerability of the risk on the basis of a risk analysis and taking into account factors such as socioeconomic and environmental aspects [1]. The risk evaluation will sometimes include a comparison of the results from the risk analysis with some risk acceptance criteria. Risk assessment is intended to provide management with a view of events that could impact the achievement of objectives.

The risk analysis may be qualitative or quantitative, depending on the objective of the analysis. Qualitative risk analysis is a risk analysis where probabilities and consequences are determined purely qualitatively. Qualitative assessments are the most basic form of risk assessment, categorizing potential risks based on either nominal or ordinal scales. Whereas quantitative risk analysis (QRA) is determined as risk analysis that provides numerical estimates for probabilities and/or consequences—sometimes along with associated uncertainties. A QRA is best suited for quantifying risk associated with low-probability and high-consequence events, and may range from specialized probabilistic assessment to large-scale analysis. The term semi-quantitative risk analysis is sometimes used to denote risk analyses that quantify probabilities and consequences approximately within ranges.

As part of these assessments, risk and reliability analysts are required to perform evaluations of human reliability in addition to the analyses of hardware systems which are the primary focus of a typical risk assessment. Increasing emphasis is being placed on a comprehensive assessment of the human role in system safety following the occurrence of major disasters in the petrochemical industry such as Piper Alpha and other industries like Chernobyl where human errors are seen as direct or indirect causes. A better estimate of human reliability would help to design more effective safety systems and evaluate more accurate risk assessments. Human reliability is the probability that a person correctly performs system-required activities in a required time period (if time is a limiting factor) [2]. Human reliability is related to the field of human factors engineering and involves the study of human Performance Shaping Factors (PSF) [3] The safety of offshore installations often requires claims on human action. Where safety of substantial human actions and administrative controls are required and their need is justified, the feasibility and reliability of the actions are demonstrated qualitatively using task analysis. This qualitative modelling is used to substantiate any human-based safety claims and the quantitative modelling of the probability of the associated human errors. It is necessary to carry out task decomposition and analysis of sufficient depth in order to understand what is being assessed, the demands and influencing factors on personnel and to assist with the identification of reasonably practicable design options or improvements to support human reliability. Task analysis provides the necessary support to the HRA process for this demonstration of adequacy. The first part of the paper will discuss the implications of qualitative analysis for quantitative risk assessment. A network of Functional Resonance Analysis Method (FRAM) [4] is conducted to represent a qualitative risk assessment which consists of task analysis as function definitions.

The Functional Resonance Accident Model and Functional Resonance Analysis Method [4] are being developed to enable Resilience Engineers to model complex socio-technical systems for the purposes of risk assessment [5] And accident analysis [6] [7]. In this approach Risk assessment goes beyond component failures to look at system dynamics and must consider issues of non-linear risk assessment, furthermore the emergence of faults from usual performance variability can be construed by the concept of functional resonance, derived from the theory of stochastic resonance. The non-linear socio-technical interaction in system was taken into account to consider the impact of human activities on system's safety level.

It should be noted that since all operation are assumed to be performed at same time, it is not possible to consider the variability and resonance of all functions in an entire accident scenario. Hence, the risk assessment is conducted for an operation as a specific resonance in the FRAM network with considering the human errors. This resonance is based on variabilities of the functions in the case study of the paper.

Quantifying the error potential of the actions is provided by an analytical approach: Success Likelihood Index Methodology (SLIM) [8]. Such an approach is also critical in order to ensure accurate quantification of risk. If significant human contributors to the likelihood of major accidents occurring are omitted, then the probability of the event occurring may be seriously underestimated. Conversely, the role of the human in enhancing the reliability of a system needs to be taken into account. Though dozens of Quantitative Risk Assessment techniques are employed today, most suffer from lack of calculation of Human Error likelihood. The main focus of the paper is improving Human Error Probabilities (HEP) to have better Quantitative Risk Assessment (QRA). For handling data scarcity, multi-expert judgment is used. Performance Shaping Factors (PSFs) are identified and evaluated. In the last step, an Event Tree (ET) for provided resonance based on FRAM network is presented to evaluate the Quantitative Risk Assessment (QRA).

## 2. Risk-Based Approach

A risk-based methodology is developed to assess the risk of studied operation as illustrated in **Figure 1** Methodology applied to estimate HEP and Consequences failures, including Task Analysis, Qualitative Risk Assessment and Quantitative Risk Assessment. Each of these levels consists of several steps. Qualitative Risk Assessment and Quantitative Risk Assessment are presented by applying Functional Resonance Analysis Method (FRAM), Success Likelihood Index Methodology (SLIM) and Event Tree (ET) respectively. The HEPs are estimated by applying the SLIM process. The probability of other hazard-promoting events is calculated by using Homogeneous Poisson Process (HPP). After obtaining the HEPs based on a specific scenario, the final value of the risk is calculated by integrating the HEPs and consequence analysis results (Risk Assessment).

## 3. Task Analysis

Task Analysis (TA) is a fundamental methodology in the assessment and reduction of human error and their relative risk. TA methods can be used to eliminate the preconditions that give rise to errors before they occur. They can be used as an aid in the design stage of a new system, or the modification of an existing system. Task

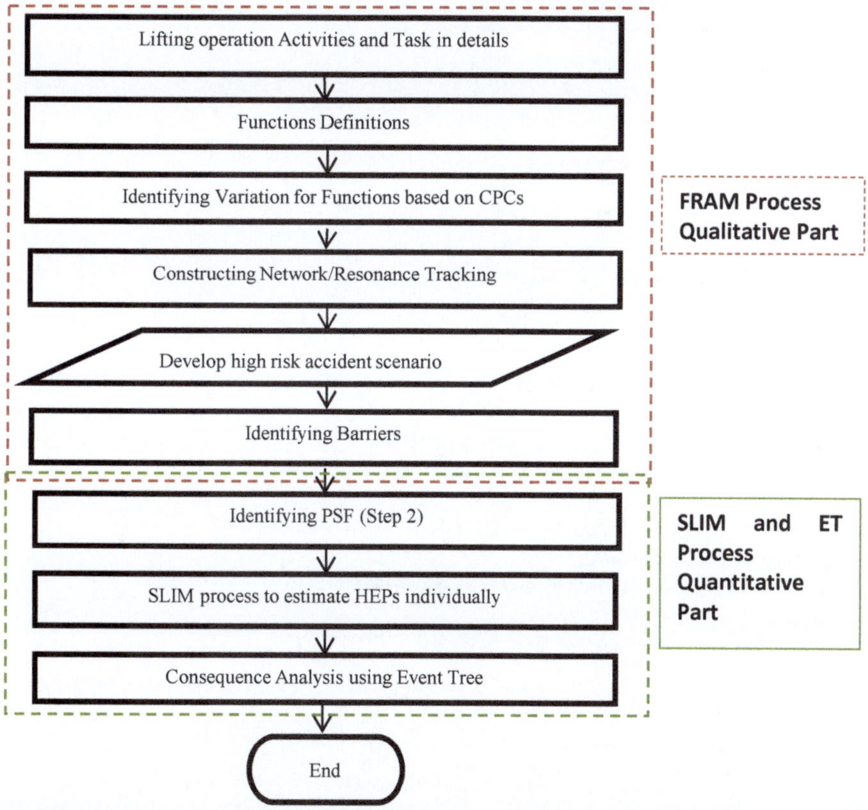


**Figure 1.** Methodology applied to estimate HEP and Consequences failures.

analysis can also be used in a retrospective mode during the detailed investigation of major incidents. The starting point of such an investigation must be the systematic description of the way in which the task was actually carried out when the incident occurred [9].

When the scenario is developed, the human related activities and the probability of error for each activity is identified. A summarized representation for different activities and tasks for lifting operation procedure are presented in **Table 1**. The major activity involves the mobilization tasks which require safely located and securely fitted and easily accessible and operated. However, all of possible interaction for lifting sub-activities in its equivalent FRAM network (see **Figure 2**).

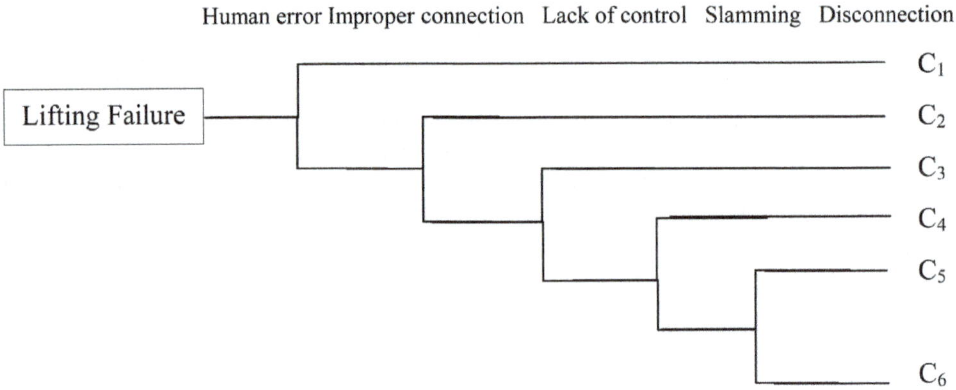


**Figure 2.** Event tree developed for light lifting in the sea.

**Table 1.** Human related Activities during lifting operation.

1. Draw up work
1.1 Planning work order
1.2 Conducting work permit system
1.3 Equipment diagnostics
1.4 Risk assessment of each planned activity
1.5 Tool box meeting before startup of each work
1.6 Documenting the permit to work system

2. Mobilization
2.1 Crane sea-fastening
2.2 Survey equipment installation
2.3 ROV system installation
2.4 Providing side effecting equipment

3. Positioning the vessel in the site
3.1 Apply GPS and thrusters to being stable in the site

4. Startup survey system
4.1 Check beacon, multi beam, USBL to ensure that they work
4.2 Property connect the beacon and underwater gyro cable to the load

5. Prepare lifting equipment
5.1 Connect properly wire and belts to the load
5.2 Check if wires/belts are out of order or not
5.3 Check the safety factors and breaking load for wires and working load for belts

6. Lower support down
6.1 Check the speed of lowering process specially near the surface

7. Check the position of support by survey team
7.1 Monitor the load transitional and rotational position in the sea

8. Check the position of support by ROV team
8.1 ROV takes the fix point for validation of the load position

9. Release the support in the seabed
9.1 Check the position, if it is ok release the load

## 4. Functional Resonance Analysis Method (FRAM)

The Functional Resonance Accident Model and its associated Functional Resonance Analysis Method (FRAM; [4]) are systemic models and methods that consider safety as an emergent property of the socio-technical system as a whole. In this approach functions and function performance are the units of analysis Rather than physical components and sequences of events. It is a risk model reviewing non-linear interactions and it is reviewing everyday activity when things are working as they should do. By describing operations when they are functioning you can find out how and why something goes wrong in the system. A function may be defined as "a set of actions that a system performs or is used for, which are valuable for the achievement of a set of goals" [10].

FRAM is based on four principles [7] Compatible with the postulates that form the basis of Resilience Engineering [11]. First, both successes and failures result from the adaptations that organizations, groups and individuals perform in order to cope with complexity. Success is measured in the ability of an organization, group or individual to anticipate the risk and critical situations, to recognize them in time and to make appropriate action. Failure is a result of temporary or permanent absence of this ability. Second, due to the fact that nothing could be as complete as predicted and planned, adjustments must always be made. The third principle states that how variability from multiple functions may combine in unexpected ways, leading to disproportionately large consequences. Fourth the variability of a number of functions may at some occasion reinforce each other and thereby cause the variability of one function to exceed its normal limits. The consequence may spread through tight couplings rather than via identifiable and enumerable cause-effect links.

The FRAM network of studied operation is presented (**Figure 2**). This network is a basis of quantitative risk assessment of this paper. In the FRAM network of present case study functions with green color are background function which provide a support for foreground function. The functions with blue color are foreground function. Foreground function directly can lead to a failure in lifting process.

As it is obvious from FRAM network the process has 14 functions, 7 background and 7 foreground. The functions are coupled with each other via their common aspects. There are some functions with barrier goal such as quality control, winch control and Connecting Wire/Belts and inspection of connection. Lack of functional barriers make some functions of the operation such as Under Water Gyro/Beacon, USBL system, Lifting support by Crane, vulnerable against unpredictable variabilities.

It should be noted that since all operation are assumed to be performed at same time, it is not possible to consider the variability and resonance of all functions in an entire accident scenario. Hence, the risk assessment is conducted for an operation as a specific resonance in the FRAM network with considering the human errors. This resonance is based on variabilities of the functions (it is specified by numbers). The resonance is a detectable signal that emerges from the unintended interaction of the variabilities of many functions that together may combine in unexpected ways, leading to consequence that are disproportionally large [12].

## 5. Human Error Probability Estimation by SLIM Process

After developing a high-risk accident scenario qualitatively based on FRAM, it is needed to find the probability of error for each human related lifting activity which is one of the main parts of consequences in event tree for risk assessment. Without estimating human error probability (HEP) the final failure estimation for top event in event tree will not be plausible. Since the majority of errors during the operation are due to human errors. Different activities and tasks are identified for this purpose in **Table 1**.

The SLIM integrates various Performance Shaping Factors (PSFs) relevant to a task into a single number called a success likelihood index (SLI). The SLI is calculated by the following formula (see Equation (1)). For numerous sub-activities for each task then SLI should be calculated for each sub-activity separately and consequently the related HEP should be calculated by Equation (2) in which, "$n$" is the number of sub-activity and "$m$" is the number of PSFs to find related SLI for task $j^{th}$, besides, R and W are the Rate and Weight of each PSF respectively.

$$SLI = \sum_{i=1}^{m} R_i W_i \tag{1}$$

$$SLI_j = \sum_{j=1}^{n} \sum_{i=1}^{m} R_{ij} W_i \ . \tag{2}$$

For a given SLI, the human-error probability (HEP) for a task is estimated by using the Equation (3):

$$Log\left(HEP\right) = a \times SLI + b \tag{3}$$

$$HEP = 10^{a \times SLI + b} \tag{4}$$

where $a$ and $b$ are constants determined from two or more tasks for which HEPs are known. In this study $a$ and $b$ are considered as −1.95 and 10 E−04, respectively.

Identifying PSFs is a substantial step of presenting the SLIM. The first step of Human Reliability Assessment is to focus on human behavior and identify a set of human factors believed to be related to performance. These PSFs are then employed to estimate the probability of human error in a given situation [13].

## 5.1. Identification PSFs

Performance shaping factor is provided basis for considering potential influences on human performance and systematically considering them in quantification of Human Error Probabilities (HEPs). PSFs often characterized as internal and external. Internal PSFs are influences that the individual brings to the situation such as mood, fitness, stress level, etc. External PSFs are influences in the situation or environment that affect the individual such as temperature, noise, work practices, etc. Currently there is no standard set of PSFs used in HRA methods, but most sets use PSFs identified in human performance literature [14]. Personal factors include, attention [8] attitude [15], personality [8], fatigue [8], knowledge [8], [16], experience [16], [18], motivation [17]. Additional factors include communication [16] [18], teams [19] [20], leadership [21], safety culture [22], ergonomics [23], training [8], [16], environment [17], management [19], time [24] and workload [19]. PSFs are used to meet multiple goals in HRA and the study of human performance. PSFs are used to pin-point positive or negative influences on human performance and to predict conditions that lead to human errors. Several HRA methods use the state (level of influence) of the PSFs to estimate HEPs or to gain qualitative insight about the scenario. PSF states are defined on different scales depending on the selected method, but they generally range from low to high influence.

## 5.2. PSFs Assessment

Determining the weight of PSFs to estimate the SLIs is one of the most pivotal steps. Human performance data with greater detail is difficult to find in real world situations, which requires the use of expert judgment techniques [13]. In this assessment, the PSFs with highest ranks are taken into account as the related PSFs, listed in **Table 2**. The number in the second column denotes the normalized importance (weight Wi) of a particular PSF for the task under consideration, as determined by experts.

Rating the PSFs is another important step in the SLIM procedure. Participant experts such as technical engineers select rating R from 0 to 1 for each of PSFs. Each PSF rating has an ideal value of 1 at which human performance is judged to be optimal. These ratings are based on six PSFs demonstrated in **Table 2** as the most important ones in lifting of light structures. It is necessary to mentioned that, Human error assessment is faced up with uncertainty, especially in any modern and novel industry like offshore industry. In these technologies, the problem of Minimum field data in regard to major component is inevitable. The main reason of uncertainty in

**Table 2.** Rank and weight of PSFs.

| PSF | Rank | Weight |
|---|---|---|
| Experience | 10 | 0.21 |
| Skill | 9 | 0.19 |
| Motivation | 8 | 0.17 |
| Stress level | 7 | 0.15 |
| Work Memory | 7 | 0.15 |
| Time pressure | 6 | 0.13 |

such technologies is the lack of knowledge with regard to inappropriate or missing experimental and operational data. As a result, a combination of qualitative and quantitative risk assessment with expert judgment could result in a better interpretation of system based on epistemic knowledge and subsequently a better ability to cope with scarce in operational experience and uncertainty.

By applying Equation (2) SLI were obtained for each activity. Afterward, Equation (3) and Equation (4) are used to calculate the HEP of each task. Human Error Probability of activities is presented in **Table 3**.

It is obvious that the first top event for related ET is related to human error. So, it is necessary to compute the cumulative probability of the first top event, Human error, using the probabilities of the sub-activities based on Equation (4). Assuming that these activities and sub-activities are independent to find the worst-case scenario (The reason is that the independency indicates that the output of following event occurs if any of the input sub-events occur), then, the probability of human error, $HEP_T$, for light structure's lifting in the offshore industry can be calculated using Equation (5).

$$HEP_T = 1 - \prod_{j=1}^{n} \left(1 - HEP_j\right). \quad (5)$$

**Table 3.** Human error probability.

| Activity | HEP | Uncertainty | |
|---|---|---|---|
| | | Lower bound | Upper bound |
| 1. Draw up work<br>1.1 Planning work order | 9.73E−02 | 6.21E−02 | 1.53E−01 |
| 1.2 Conducting work permit system | 6.56E−02 | 3.81E−02 | 1.23E−01 |
| 1.3 Equipment diagnostics | 7.24E−02 | 4.62E−02 | 1.23E−01 |
| 1.4 Risk assessment of each planned activity | 7.21E−02 | 5.51E−02 | 1.13E−01 |
| 1.5 Tool box meeting before startup of each work | 9.88E−02 | 6.3E−02 | 1.32E−01 |
| 1.6 Documenting the permit to work system | 8.4E−02 | 5.36E−02 | 1.44E−01 |
| 2. Mobilization<br>2.1 Crane sea-fastening | 9.03E−02 | 5.34E−02 | 1.55E−01 |
| 2.2 Survey Equipment Installation | 9.48E−02 | 6.05E−02 | 1.48E−01 |
| 2.3 ROV system Installation | 8.87E−02 | 5.15E−02 | 1.52E−01 |
| 2.4 Providing side effecting equipment | 1.08E−01 | 6.12E−02 | 1.79E−01 |
| 3. Positioning the vessel in the site<br>3.1 Apply GPS and thrusters to being stable in the site | 1.05E−01 | 5.71E−02 | 2.11E−01 |
| 4. Start-up survey system<br>4.1 Check beacon, multi beam, USBL to ensure that they work property | 9.71E−02 | 5.23E−02 | 1.92E−01 |
| 4.2 Connect the beacon and underwater gyro cable to the load | 9.92E−02 | 6.77E−02 | 2.11E−01 |
| 5. Prepare Lifting Equipment<br>5.1 Connect properly wire and belts to the load | 8.18E−02 | 4.75E−02 | 1.46E−01 |
| 5.2 Check if wires/belts are out of order or not | 1.01E−01 | 6.08E−02 | 1.58E−01 |
| 5.3 Check the safety factors and breaking load for wires and working load for belts | 8.00E−02 | 5.24E−02 | 1.82E−01 |
| 6. Lower support down<br>6.1 Check the speed of lowering process specially near the surface | 1.27E−01 | 6.11E−02 | 2.11E−01 |
| 7. Check position of support by Survey Team<br>7.1 Monitor the load transitional and rotational position in the sea | 1.09E−01 | 5.15E−02 | 1.78E−01 |
| 8. Check position of support by ROV team<br>8.1 ROV takes the fix point for validation of the load position | 1.09E−01 | 5.15E−02 | 1.78E−01 |
| 9. Release the support on seabed<br>9.1 Check the position, if it is ok release the load | 1.36E−01 | 6.99E−02 | 2.18E−01 |

## 6. Quantitative Risk Assessment by Using Event Tree

Applying a continuous improvement in the system Safety Management in any kinds of process in the industries needs different approach such as Risk Assessment. The expected achievements of risk Management is that the hazards and risks of the system were identified, analyzed, assessed, evaluated, controlled and finally reduced. In the assessment phase, different approaches are used such as Quantitative Risk Assessment (QRA) and probabilistic safety assessment (PSA). Among several techniques available as a Quantitative Risk Assessment (QRA), Event Tree (ETs) have widely been used to explore the probability of consequences resulted from an initiating event. It has been used successfully in the nuclear industry, the chemical process industry, and in several other application areas. Event tree analysis is also commonly used for human reliability assessment. The method is inductive and follows a forward logic [1].

### 6.1. Homogeneous Poisson Process (HPP)

In probability theory and statistics, the Poisson distribution is a discrete probability distribution that expresses the probability of a given number of events. The occurrences of the hazardous event are often modeled by a homogeneous Poisson process with frequency of failure rate $\lambda$, which is the expected number of occurrences per year (or some other time unit).

We use Poisson distribution to find probabilities of occurrences of hazard-promoting factors sequences. The failure rate estimates are based on Recorded failure events, Expert judgment and Laboratory testing and in some cases a combination of these. The HPP of hazardous events of the event tree in **Figure 3** are calculated by using Equation (6) [25] and indicated in **Table 4**. However, the probability of the first top event, Human error, has been derived using Equation (5)

$$\Pr\left(N(t)=n\right)=\frac{(\lambda t)^n}{n!}e^{-\lambda t} \quad \text{for } n=0,1,2,\cdots \tag{6}$$

### 6.2. Event Tree Approach Developed for Light Lifting

It should be noted that since all operation are assumed to be performed at same time, it is not possible to consider the variability and resonance of all functions in an entire accident scenario. Hence, the risk assessment is conducted for an operation as a specific resonance in the FRAM network with considering the human errors. Based on accident reports, related research and expert opinions, the most probable accident scenario is the provided resonance in the FRAM network in which a human error may lead to inappropriate checking of the connection between belt and trunnion. Besides inadequate control in the speed of the load in lowering process, specially near the surface may increase the probability of failure that is slamming of the load and consequently disconnection of load and derrick. The resulting diagram displays the mentioned accident scenarios (see **Figure 2**). The occurrences of the hazard-promoting event are often modeled by a homogeneous Poisson process with frequency $\lambda$ [1].

Applying the probabilities of hazard-promoting factors, the probabilities of consequences can be calculated for light lifting in the sea (**Table 5**).

## 7. Frequency Histogram for Light Lifting Accident

By working through the entire event tree, we produce a spectrum of light lifting failure and their probability for the various accident sequences (**Figure 3**). As it is obvious from histogram, the probability of hazard-promoting

**Table 4.** Probability of each hazard-promoting factors of event tree.

| Hazard-promoting factors | Probability |
|---|---|
| Human error (HEPT) | 0.86 |
| Improper connection between belt and trunnion | 0.08 |
| Lack of speed control in lowering process | 0.11 |
| Slamming | 0.10 |
| Disconnection | 0.12 |

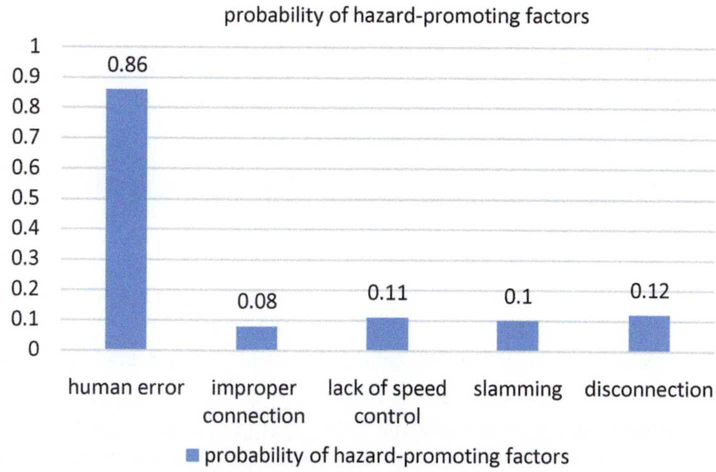


**Figure 3.** Frequency histogram for lifting failure.

**Table 5.** Risk assessment of lifting failure accidents.

| Index | End state description | Probability |
|---|---|---|
| $C_1$ | *Safe condition* | 1.4E−01 |
| $C_2$ | Mishap | 7.91E−01 |
| $C_3$ | Near miss | 6.1E−02 |
| $C_4$ | Improper fastening of connection points like shackles, belts or slings; successful installation; without damage | 6.81E−03 |
| $C_5$ | Slamming; successful installation; minor property damage | 6.65E−04 |
| $C_6$ | Disconnection; unsuccessful installation; major property damage; major injury | 9.08E−05 |

factors reached a pick in human error. It depicts two matters, firstly the importance of human error rule and secondly is that without considering human error in the processing of light lifting risk assessment there is a striking difference in the final outcomes.

## 8. Conclusion

A risk-based approach has been conducted to assess the risks of light lifting in offshore industry. This methodology consists of Task Analysis, Qualitative Risk Assessment and Quantitative Risk Assessment. The aim is to illustrate the importance of the role of human error in evaluating and estimating the risks of system with pinpoint accuracy. To do so, risk assessment in this study is divided into two parts: Qualitative Risk Assessment and Quantitative Risk Assessment. The HEPs are estimated by applying the SLIM process. The probability of other hazard-promoting events is calculated by using Homogeneous Poisson Process (HPP). Although quantitative risk assessment and probabilistic safety analyzing are vital and necessary for system safety management, the study showed that by eliminating the role of human reliability in the assessment, the effectiveness and accuracy of evaluation will decline enormously. After obtaining the HEPs based on a specific scenario, the final value of the risk is calculated by integrating the HEPs and consequence analysis results (Risk Assessment). The main challenges experienced for this study are data scarcity and uncertainty in expert opinion. By applying multi-expert knowledge, this paper presents a methodology to overcome these two major limitations.

## References

[1] Rausand, M. (2011) Risk Assessment. John Wiley & Sons, New Jersey. http://dx.doi.org/10.1002/9781118281116

[2] Swain, A.G. (1983) Handbook of Human Reliability Analysis with Emphasis on Nuclear Power. US Nuclear Regula-

tory, Washington.

[3] Blackman, H.G. (2008) Human Error Quantification Using Performance Shaping Factors in the SPAR-H Method. 52*nd Annual Meeting of the Human Factors and Ergonomics Society*, New York, 22-26 September 2008, 1733-1738. http://dx.doi.org/10.1177/154193120805202109

[4] Hollnagel, E. (2004) Barriers and Accident Prevention. Ashgate, Aldershot.

[5] Lundblad, K.S. (2008) FRAM as a Risk Assessment Method for Nuclear Fuel Transportation. *Proceedings of the* 4*th International Conference Working on Safety*, Crete, 30 September-4 October 2008.

[6] Woltjer, R. (2007) The Alaska Airlines Flight 261 Accident: A Systemic Analysis of Functional Resonance. *Proceedings of the* 2007 (14*th*) *International Symposium on Aviation Psychology*, Dayton, Ohio, 24 April 2007, 763-768.

[7] Hollnagel, E.P. (2008) A Functional Resonance Accident Analysis of Comair Flight 5191. 8*th International Symposium of the Australian Aviation Psychology Association*, Sydney, 8-11 April 2008.

[8] Kirwan, B. (1994) A Guide to Practical Human Reliability Assessment. Taylor & Francis, London.

[9] Embrey, D. (2000) Task Analysis Techniques. Human Reliability Associates Ltd.

[10] Woltjer, R. (2009) Functional Modeling of Constraint Management in Aviation Safety and Command and Control. Linkoping Studies in Science and Technology, Linkoping University, Sweden, Dissertation No. 1249.

[11] Hollnagel, E.W. (2006) Resilience Engineering: Concepts and Precepts. Ashgate, Aldershot.

[12] Herrera, I. (2010) Comparing a Multi-Linear (STEP) and Systemic (FRAM) Method. *Reliability Engineering and System Safety*, **95**, 1269-1275. http://dx.doi.org/10.1016/j.ress.2010.06.003

[13] Mashrura Musharraf, J.H. (2013) Human Reliability Assessment during Offshore Emergency Conditions. *Safety Science*, **59**, 19-27. http://dx.doi.org/10.1016/j.ssci.2013.04.001

[14] Groth, K. (2009) A Data-Informed Model of Performance Shaping Factors for Use in Human Reliability Analysis. Doctoral Dissertation, University of Maryland, College Park, MD.

[15] Triandis, H. (1991) Attitude and Attitude Change. In: *Encyclopedia of Human Biology*, Academic Press, San Diego, CA, 485-496.

[16] Gertman, D.I., Hallbert, D.I. (2001) Review of Findings for Human Error Contribution to Risk in Operating Events. Idaho National Engineering and Environmental Laboratory, Idaho, Idaho Falls.

[17] Barnes, V.E. and Haagensen, B.C. (2001) The Human Performance Evaluation Process: A Resource for Reviewing the Identification and Resolution of Human Performance Problems. US Nuclear Regulatory Commission, Washington DC, Tech. Rep. NUREG/CR-6751.

[18] Gemuenden, M.H. (2001) Teamwork Quality and the Success of Innovative Projects: A Theoretical Concept and Empirical Evidence. *Organization Science*, **12**, 435-449. http://dx.doi.org/10.1287/orsc.12.4.435.10635

[19] Reason, J. (1997) Managing the Risks of Organizational Accidents. Ashgate, London.

[20] Copper, B.M. (1994) The Relation between Group Cohesiveness and Performance: An Integration. *Psychological Bulletin*, **115**, 210-227. http://dx.doi.org/10.1037/0033-2909.115.2.210

[21] Sasou, K., Takano, K. and Yoshimura, S. (1996) Modeling of a Team's Decisionmaking Process. *Safety Science*, **24**, 13-33. http://dx.doi.org/10.1016/S0925-7535(96)00029-X

[22] Burack, R.C. (1987) Measuring Norms and Expectations with the OCI. Human Synergistics, Organizational Culture Inventory Level V Manual, Chicago, Illinois.

[23] Kecklund, L.J. and Svenson, O. (1997) Human Errors and Work Performance in a Nuclear Power Plant Control Room: Associations with Work-Related Factors and Behavioral Coping. *Reliability Engineering and System Safety*, **56**, 5-15. http://dx.doi.org/10.1016/S0951-8320(96)00137-8

[24] Edwards, J.F. (1998) Time Pressure, Task Performance, and Enjoyment. In: McGrath, J., Ed., *The Social Psychology of Time*, Sage Publications, Newbury Park.

[25] Ross, S. (2010) A First Course in Probability. 8th Edition. Pearson, NJ.

# 19

# Review on Rockburst Theory and Types of Rock Support in Rockburst Prone Mines

**Eugie Kabwe, Yiming Wang**

Department of Mining and Mineral Resources Engineering, University of Science and Technology Beijing, Beijing, China

Email: kabweeugie@yahoo.com

## Abstract

**As the depth of mining increases, the occurrence of seismic events is becoming a more common phenomenon causing serious problems regarding to the stability and safety of mines. However, seismicity and the accompanying rockburst phenomena are not a well-defined and well-understood in underground excavations in these times. Efforts to clearly explain the mechanisms are underway. This paper is an overview on the mechanism of rockburst and supports applicably in rockburst prone excavations, predicted classification scales for damage on rock surfaces and rock supports. Current design methods for support systems are reviewed, which are mostly dependent on practical approaches and are geared towards static support design. Based on this, the current review focuses on ground supports under dynamic conditions.**

## Keywords

**Cone Bolts, Jacketted Rock Bolt, Rockburst, Rock Support, Yield-Lok**

## 1. Introduction

The overall stress field attains high values when mining reaches greater depths. This stress is further redistributed due to the activity resulting in excessive stress concentrations in certain sections of the rock mass. Accumulation of stress in the rock leads to destructive stress that causes fracture which can be a seismic source. The passing dynamic stress wave as a result of fracturing in the rock mass is defined as a seismic event. A seismic event can also be a sudden inelastic deformation within a given size of rock [1] [2]. Effectively, a seismic event is the vibration of rock breaking. Seismic events are a normal response of a rockmass to the stress changes caused by the creation of mining excavations. The term seismicity is given various definitions with different authors. One definition can be stated seismicity as the rock mass response to deformation and failure [2].

## 2. Rockburst

The rockburst phenomena, came into sight in the early 1900s in the mines of South Africa [3]. Today several deep underground mines in all around the world face the problem of rockbursts. A rockburst is a mining-induced seismic event that affects and cause destruction to excavations in the rock [2]. It is vital to differentiate between seismic event and rockburst. A seismic event does not necessarily cause damage in openings, while a rockburst will, with varying severity of damage. A very good definition of rockburst given by the Canadian Rockburst Research Program is "damage to an excavation that occurs in a sudden or violent manner and is associated with a seismic event" [3] [4]. The role of seismic events in a rockburst phenomenon can be examined as to the actual course of burst; if it's the seismic event or the stored stress energy in the rock around an excavation released during breaking. A seismic event can be located at a certain distance away from the rockburst location, or can also overlap with the rockburst phenomenon as in the case of strain bursts. Hence the cause of rockbursts can generally be classified as self-initiated and remotely triggered [5].

Self-initiated rockbursts arise from the stresses close to the boundary of an excavation which exceed the strength of a rock mass and failure can be in unstable manner. The stress redistribution that is the outcome of a nearby mining increases the concentration of the stresses at the spot. Degradation of strength with time and loss of confinement can also deteriorate the capacity of the rock mass. One or all of these conditions lead the strength of the rock mass to be exceeded by the stress and consequently lead to failure. Loss of structural stability, a factor independent of the strength of the rock mass, can also cause a self-initiated rockburst as can be demonstrated in the sudden buckling of column or slab of rock [6].

Remotely triggered rockbursts are caused by large magnitude seismic events. Remotely triggered rockbursts commonly occur in some hard rock mines, usually after the mine has been significantly mined out and where faults intersect stopes or large mined-out areas and sill pillars. Large vibrations and the accompanying dynamic stresses as a result of incoming seismic waves can lead to both fracturing of the rock mass and structural instability. Furthermore the mining stage can influence the phenomena of rock bursting. In the initial stage of the mine's life highly localized stress concentrations near drifts that are relatively isolated from each other are the usual causes of the events. In the later life stage of the mine multiple opening and numerous stopes cause additional rockburst mechanisms [6].

Rock bursts are generally divided into three classifications:

- Strain bursts: These are caused by high-stress concentrations at the edge of mine openings that exceed the strength of the rock. Events can range from small slivers of rock being ejected from the walls to the collapse of a complete wall as it tries to achieve a more stable shape.
- Pillar bursts: Severe rock bursts, involving thousands of tons, have been caused by the complete collapse of support pillars. These tend to occur in extensively mined-out areas, and the resulting damage can be severe.
- Fault slip bursts: Recognized in the 1980s, this type of rock burst occurs when slippage suddenly occurs along a geological weakness plane. This is the same mechanism as for an earthquake.

Damage may occur as rock bulking by fracturing, ejection of rock due to seismic energy, or rockfalls by seismic quaking (**Figure 1**) [7]. Each mechanism may result in different levels of damage to an excavation and its support system. The damage severity depends on many factors, including:

- Failure potential near the opening;
- Support effectiveness;
- Local rock stiffness;
- Magnitude of seismically induced stress, rock accelerations, or velocities;
- Opening geometry, size, and orientation;
- Geological structure.

Three levels of rock burst damage severity are defined in **Figure 1**. Damage level determination is usually based on observations of previous damage, where such observations are available, plus analytical methods [8]. Severity levels are summarized in **Table 1**. Only a limited number of ground support elements are suitable for dynamic loading conditions, and under the most severe conditions even these are restricted by maximum practical support limitations [8].

## 3. Rock Reinforcement Systems

The main purpose of rock reinforcement in underground mining is to ensure stability of the rock mass around

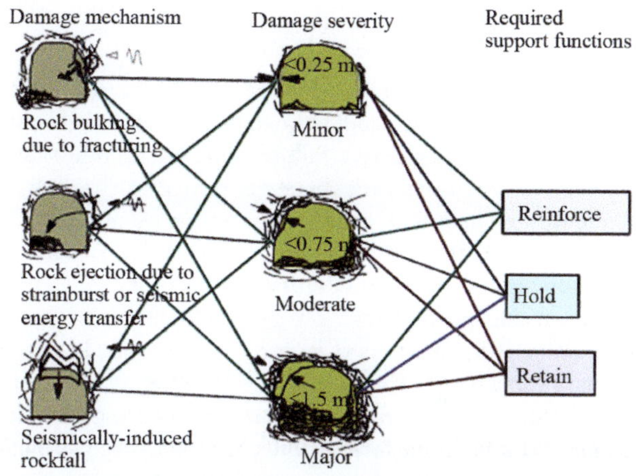


**Figure 1.** Type of damage mechanism, damage intensity and applicable support functions (After Kaiser *et al.* (1996)).

**Table 1.** Rock-burst damage mechanisms and the nature of anticipated damage (After Kaiser *et al.* (1996)).

| ***Damage Mechanism*** | ***Cause of Rock*—Burst** Damage | ***Damage*** Severity | ***Thickness*** m | ***Weight*** kN/m$^{2*}$ | ***Closure***[†] mm | ***Ve,***[‡] m/s | ***Energy*** kJ/m$^2$ |
|---|---|---|---|---|---|---|---|
| Bulking without ejection | Rock with high stress and small amount of stored stress energy | Negligible<br>Reasonable<br>Foremost | <0.25<br><0.75<br><1.5 | <7<br><20<br><50 | 15<br>30<br>60 | <1.5<br><1.5<br><1.5 | Not critical<br>Not critical<br>Not critical |
| Bulking causing ejection | Rock with high stress with substantialextrastress energy | Negligible<br>Reasonable<br>Foremost | <0.25<br><0.75<br><1.5 | <7<br><20<br><50 | 50<br>150<br>300 | 1.5 - 3<br>1.5 - 3<br>1.5 - 3 | Not critical<br>2 - 10<br>5 - 25 |
| Ejection by remote seismic event | Seismic energy transfer to jointed or broken rock | Negligible<br>Reasonable<br>Foremost | <0.25<br><0.75<br><1.5 | <7<br><20<br><50 | <150<br><300<br>>300 | >3<br>>3<br>>3 | 3 - 10<br>10 - 20<br>20 - 50 |
| Rockfall | Inadequate strength, forces increased by seismic acceleration | Negligible<br>Reasonable<br>Foremost | <0.25<br><0.75<br><1.5 | $<7\ g/(a+g)$<br>$<20\ (a+g)$<br>$<50\ (a+g)$ | -<br>-<br>- | -<br>-<br>- | -<br>-<br>- |

excavation openings while maintaining access to the mine, allowing the rock mass to support itself with its natural strength, while keeping the intended functionalities.

Support systems may be classified into two broad categories: internal and external, which can be either active or passive. A support becomes active when stresses are induced in it at the time of installation. Therefore, such supports reinforce the rock mass structure by exerting an "induced" stress on the ground immediately after its installation [8]. The common examples are pre tensioned rock bolts, hydraulic props and power-driven supports for long wall faces. Active supports are applied in situations where an excavation is believed to cause excessive deformation in the ground [9]. An example of this situation is separation of rock wedges from the rock mass. Passive supports do not reinforce the rock mass immediately after installation, but their effect is seen as subsequent mining activities take place. Examples of passive supports are steel arches, timber sets, composite packs and un-tensioned bolts. External supports are generally of the passive type, they are placed around the boundary of the excavation to help restrain the movement of the rock walls and avoid the failure of rock mass. Steel arches, wooden cribs and fiber reinforced shotcrete are some types of external supports. Backfill is another common type of external, passive support used in hard rock mining. Internal supports continue to see technological developments in the mining industry. The basic mechanism of internal support is to bound rocks together to maintain the overall stability of the rock mass around an excavation. Internal supports which are pre-tensioned at the time of installation are considered to be of the active type. Swellex, Split Sets, grouted bars and mechanical anchors are some of the common examples of internal supports. Rock bolts are most preferred during support installations in underground excavations than other types of supports because they provide unbarred excavation openings and have minimum repairs and set up, they require less storage space, they are very easy to install and require less installation time.

Instability of the rock mass is usually triggered by the mining activities, alteration of the in-situ stresses, causing either extreme high stresses that are way above the rock strength, this result in shear failure or relatively extensive spalling or low stresses causing progressive failure along planes of low strength. In cases of increasing depth or at later age of the mine, seismicity and rockbursts arise due to slippage on natural or mining-induced planes of weakness and fracture of the intact rock itself (strain burst), usually close to excavation boundaries [5]. For a successful support and reinforcement systems, the maximum effect is achieved by the interaction between the bolt and shotcrete, thus retaining and holding. (**Figure 2**) shows the concept between rock mass, and the reinforcement components: bolts, mesh and shotcrete.

The five most important characteristics of support in burst-prone grounds are: 1) Peak load capacity of the support; 2) Displacement at peak capacity; 3) Ultimate load bearing capacity or yield support; 4) Ultimate displacement capacity or stretch limit; 5) Energy dissipation capacity of support. The choice of a support system will depend on the geo-mechanical conditions, stress field, geology, the expected loading and nature of the damage.

### 3.1. Dynamic Supports for Burst Prone Ground

Reinforcing mechanisms generally restrict and control the bulking of the rock mass. Typically, reinforcing elements like grouted rebars act as stiff support elements, however split set bolts, yielding Swellex or Cone bolts [10], act as ductile or yielding elements under high stress conditions. The holding function of supports is aimed to hold the retaining elements of the system and loose rock to stable ground.

Dynamic support design effectively implies rock-burst-resistant support design and is a function of the size of the seismic event and the location of the event hypocenter relative to the mine infrastructure in question. Local site factors such as the orientation of the infrastructure can also play a significant role in support and ground performance. Support ductility is required to achieve displacement capacity and energy absorption capacity [8].

The main issue with underground excavation is the stabilization of the rock around excavations at depth. The main difference between deep rock and surface rock is the increment of the in situ stresses. Significantly this increase in the rock stresses, result in rockburst occurrence in hard rocks, large squeezing deformations in soft and weak rocks. From observation in most underground mines, rockburst phenomena begins at depths from 600 m to 800 m and more pronounced below 1000 m. Most metal mines in Sweden, Canada, Australia and South Africa, at the moment mining operations are at depths between 1000 m and 3000 m. At such depths, conventional support devices [11] [12] fail to cope with the severe rock conditions.

Conventional rock bolts are categorised in three types looking at their anchoring mechanisms: 1) Two-point fixed mechanical bolts; 2) Fully encapsulated rebar bolts; 3) Frictional bolts. Fully encapsulated rebar bolts are fused to the grout/rock lengthwise with a link between the bolt ribs and the grout. Rebar has a high load-bearing capacity, though, rebar cannot accommodate large rock dilations, would only endure a deformation of 2 - 3 cm when exposed to a fracture this indicates that it's a tough but rigid rock bolt (**Figure 3**). Interaction of frictional rock bolts with the rock is through friction amid the hole wall and the cylinder-shaped surface of the bolt.

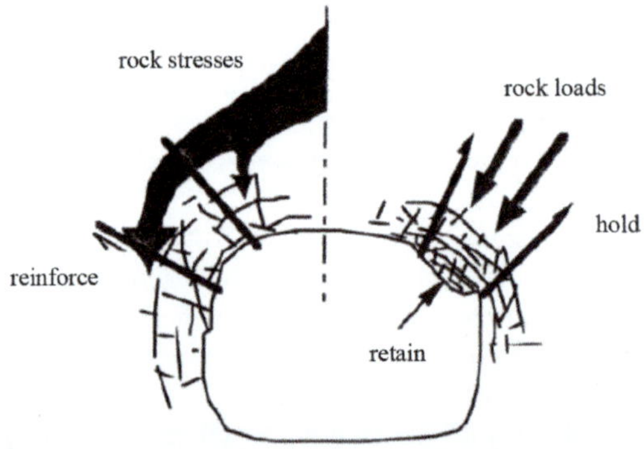


**Figure 2.** Interaction of reinforcement components (McCreath and Kaiser, 1992).

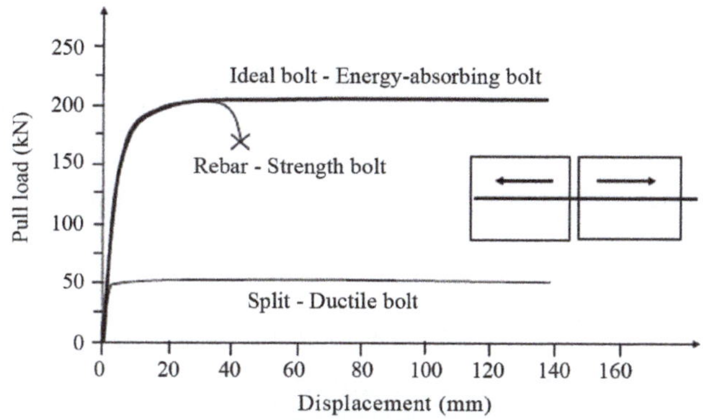


**Figure 3.** Ideal bolt properties in relative to other bolts (After Li CC (2013)).

The bolts can endure large rock deformations with a low load-bearing capacity. A standard Split Set bolt cannot carry a load greater than 50 kN (**Figure 3**). A perfect support scheme needs to be tough but rigid like are bar and endure deformation like a Split Set bolt (**Figure 3**), with the ability to handle load similar to the strength of the rock bolt material [12].

It is not suitable to use rebar bolts as rock reinforcement devices at depths with high in situ rock stresses. When used with weak and soft rocks, it is often observed that either the face plate of the rebar is greatly loaded, or the thread of the bolt is pulled to failure [12] [13]. In the case of hard rocks, the rock spalls and peels behind the bolt plate, leaving a short part of an extruding bolt out of the rock surface. In Kiirunavaara mine in Sweden, old rebar bolts are occasionally visible on the work faces of successive cut slices. It is seen that some of those bolts fail all because of minor shear and initial movement at rock fractures [14].

The untimely failure of the rebar bolts (**Figure 3**), indicates that a rebar is too stiff to withstand rock dilations in high stress rock masses [15]. To withstand large rock dilations, Split Sets are used as reinforcement devices in many deep mines. They are capable of enduring large rock deformations, but their load-bearing capacity is very low. Both rebar and Split Sets are low energy-absorbing devices. Ortlepp pointed out that support devices applicable in deep mines should be able to handle high loads and also withstand large deformations, in that they should be able to absorb a large amount of energy prior to failure [12] [16]. The first energy-absorbing device designed to counteract rock burst problems is the cone bolt [12] [17].

## 3.2. Ideal Bolt

Rock bolts can be strength bolt, ductile bolt, or an energy-absorbing bolt. Strength bolts are those that provide a support load equal or close to the intrinsic strength of the bolt material. Ductile bolts are able to accommodate large rock deformations. Energy-absorbing bolts are characterized by their high load capacity and also their large deformation capacity [12]. The installed rock support system must be able to absorb dynamic energy while at the same time be able accommodate large rock deformation due to rock failure.

Strategically, installed support system is focused on the control of rock performance after failure, not on the prevention of rock failure occurrence [18]. In fact, it is actually un-economical to prevent rock damage from occurring by the increase in rock support capacity. In rock damage control, the support behaviour must be essentially altered to allow for yielding. This section reviews the current support applicable in burst prone ground that have energy absorption or yielding capacity. Enforcement of stability for large rock blocks, and the holding together of weak lamented rock are achieved by rock bolts of the correct dimension. Widespread research and advance work on yielding rock support has been conducted, energy-absorbing bolts have been successively developed and applied. Currently, dozens of energy-absorbing bolt types and their yielding mechanism can be illustrated as structural components sliding and steel deformation (**Figure 4**) [18].

### 3.2.1. Swellex Rock Bolt

Swellex is a friction bolt; the dowel varies up to a length of 12 m (**Figure 5**). The Swellex Mn2 4 is a 3 m long and a 25 mm diameter tube before folded during manufacturing. Pm24C Swellex bolts can be used to provide

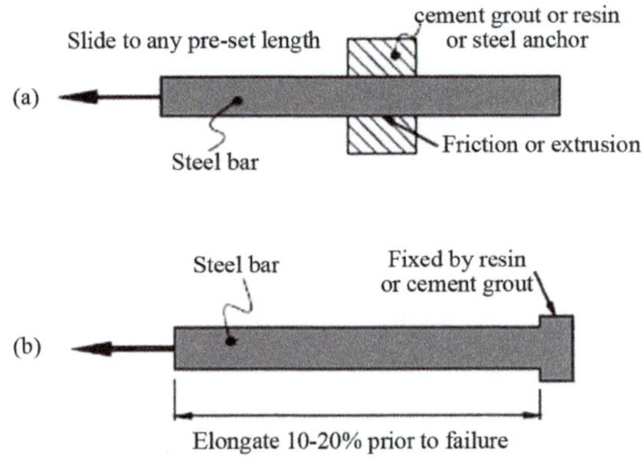


**Figure 4.** Illustration of Yielding mechanism of a rock bolt (After Wang *et al.* (2013)).

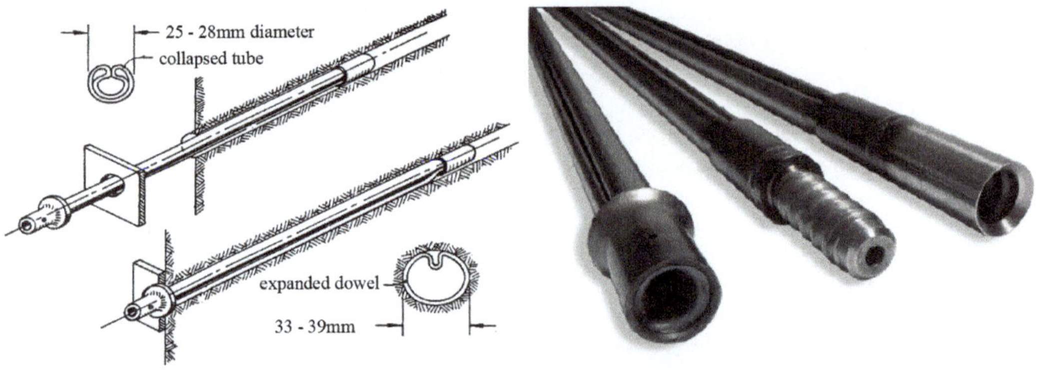


**Figure 5.** Friction anchored rockbolt—Swellex.

deep tendon support under dynamic loading (**Table 2**). When inserted into a drill hole (32 - 39 mm) no extensive driving force is required. It has expansion capability of 42 mm when activated by injection of high water pressure, causing a tight contact with the drillhole walls after the inflation. The concept of the bolt entails that the rock is secured by immediate and full support action. The moment the bolt is expanded in the hole, it interacts with the rock to maintain its integrity. The excellence of the bolt installation is confirmed when the pump stops, and is independent of rock mass conditions or operator experience [19]. The Swellex rockbolts are intended to enhance the efficiency of each bolt, so that the bolting operation ties with the required safety levels as intended [20].

### 3.2.2. Roofex

Roofex is made up of a steel bar encased in a smooth plastic cover fixed inside the borehole with cement or resin grout (**Figure 6**). The bolt has an energy absorber, it works like a sliding element over the steel bar. Enabling the bolt to extend outwards during sudden movements from rock bursts while at the same time providing constant load capacity (**Figure 7**). This capability makes the Roofex rock bolt especially appropriate for developing new underground excavations in rock burst prone areas [20]. The bolt can be made in different lengths usually used in mining, and capacity of displacement can be designated at the time of bolt manufacturing.

### 3.2.3. Cone Bolts

The cone bolt is made up of a smooth steel bar coated with a thin layer of waxy lubricating material, so that it easily detaches from the grout under a tensile load, the bar has a flattened conical bell-shaped forged section at one end (**Figure 8**). The bolt is entirely encapsulated with cement grout or resin in a borehole. Installed in a specifically formulated Coneloc or Fasloc resin, clearly mixed by the resin mixing blade. A plastic sleeve is installed over the shaft of the bolt to debond the resin from the bar, thus allowing the cone to yield or plough

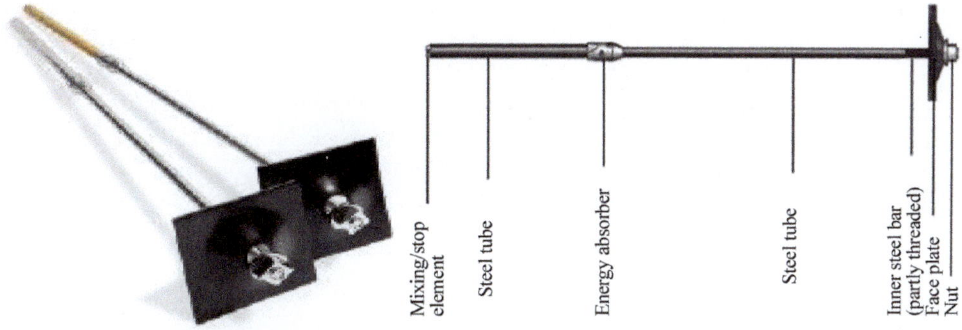


**Figure 6.** Roofex rock bolt.

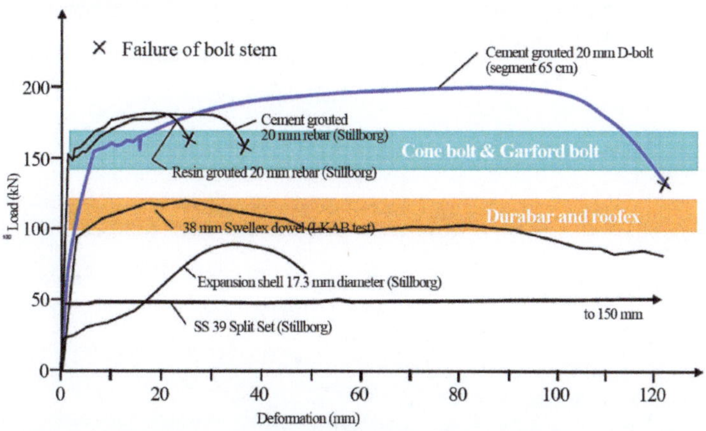


**Figure 7.** Displacement characteristics of rock bolts (after Li, 2013).

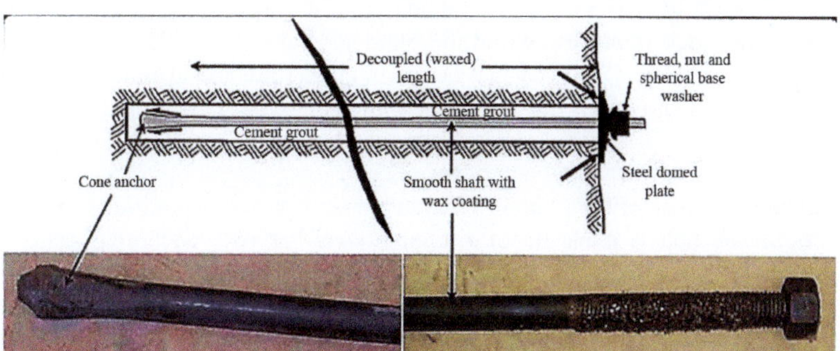


**Figure 8.** Cone bolt.

through the resin. Cone bolts show high-ductility and energy absorption characteristics. Cone bolts should provide high-shear resistance [8].

A pull load is induced in the bolt shank when the rock mass dilates in between the cone and the bolt plate. The bolt has been designed in that the conical end section works through the grout material when exposed to a pull load exceeding a predefined value. This enables the bolt to absorb energy from the rock, initially it was designed for use with cement grout later adapted to resin grout [12].

### 3.2.4. Modified Cone Bolt

The Modified Cone Bolt cone bolt is basically a smooth bar with a threaded section at one end, forged cone and mixing blade at the other end which works as a resin mixer (**Figure 9**). These bolts are usually applied in many

**Table 2.** Swellex properties.

| Type | Swellex load capacity | | |
|---|---|---|---|
| | ***Minimum breaking load***, expanded profile (kN) | ***Minimum yielding load***, expanded profile (kN) | Minimum elongation (%) |
| Pm16 | 160 | 140 | 10 |
| Pm24 | 240 | 200 | 10 |
| Pm24C | 240 | 200 | 10 |
| Pm24H | 200 | 190 | 10 |
| Mn24 | 220 | 180 | 20 |
| Sp24 | 240 | 200 | 6 |

**Figure 9.** Modified cone bolt.

burst-prone mines especially in Canada as the main dynamic rock support systems. Ground observations carried out in mines revealed susceptibilities in the exterior anchoring point of the bolt especially at the bolt plate. The anchoring may be lost if the rock fractures behind the plate and falls down [21], this causes the bolt to completely lose its reinforcement ability. Observed that the resin takes time and at times fails to harden, indicating that the cone bolt mixing quality of resin is not always definite and to the standard [22].

### 3.2.5. D-Bolts

The D-bolt was developed to stabilize both burst-prone and squeezing rock conditions that are a major factor in underground excavation. The bolt is made up of a smooth steel bar with anchored sections along its length. These can firmly be fixed within a borehole with the aid of cement grout or resin, the smooth segments between the anchors may easily bend in reaction to rock stretching (**Figure 10**). Reinforcement performance of the other segments can never be affected if one of the other segment fails. The bolt is designed to completely utilize the strength and the deformation capability of the bolt material lengthwise. The bolt has large load-bearing and deformation capacities. Through static pull trials and dynamic drop trials it was seen that the bolt extent lengthens from about 14% - 20% at a load equal to the strength its material, thus absorbing a larger quantity of energy (**Figure 11**). The impact normal load of a 20 mm D bolt was about 200 - 230 kN, with some percentage of the load transmitted to the bolt plate [12] [23].

### 3.2.6. Garford Bolt

The Gar ford dynamic bolt is comprised of a compact steel bar, an anchor and a coarse-threaded steel cover at one end (**Figure 12**). Its characterised by its anchor this permits the bolt to expanse by a great extent when the rock dilates. Varden *et al.* presented some tests and selection of the gar ford dynamic bolt (**Figure 13**) for use under dynamic conditions and explained that the bolt comprises of a 20 mm mild steel solid bar and a resin mixing device of 350 mm long, 43 mm diameter rough threaded steel cover folded at the end. The dynamic segment is an original sliding anchor mechanism which is constrained on to the bolt beneath the mixing device.

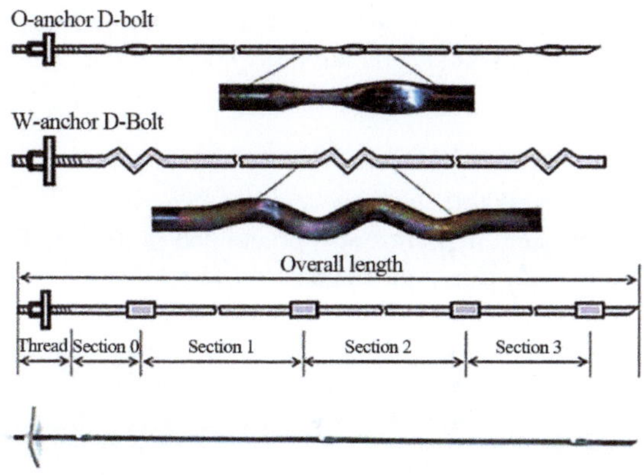


**Figure 10.** D-bolt.

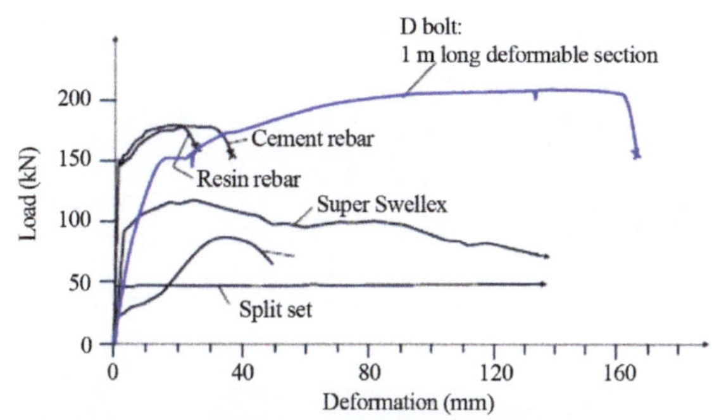


**Figure 11.** Load-displacement of D bolt under pull loading test.

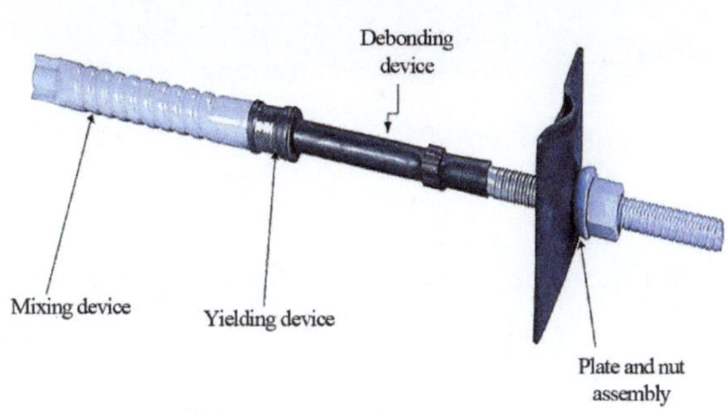


**Figure 12.** Garford bolt.

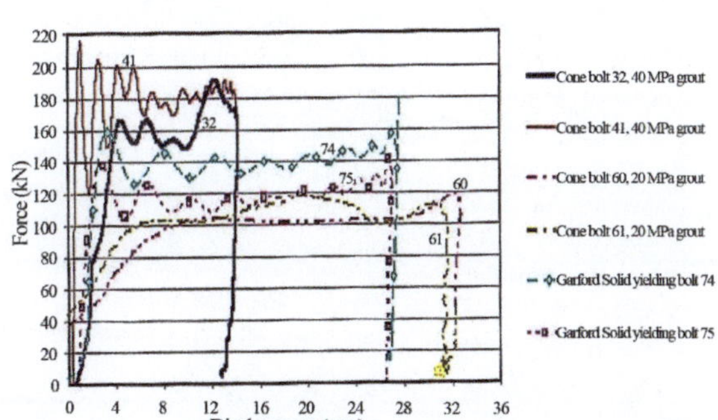


**Figure 13.** Dynamic force displacement curves (After Varden *et al.* 2008).

The rest of the bolt is enclosed in a polyethylene sheath, this provides a debonding action, which debonds the bolt below the dynamic section. Under a dynamic condition the bolt is pushed through the narrow opening and it stretches [24].

### 3.2.7. Dura Bar

This is an energy-absorbing yielding rockbolt. It is anchored at two-points. The inner anchor of the bolt is a crinkled segment of the smooth bolt stem that slips at a predefined pull load. The external anchor is the bolt plate, the rod is bent to form a wave and act as a ductile anchor. The collar end is completed by an eye for cable lacing or a threaded segment for nut and washer (**Figure 14**). The entire length of the bar is coated with wax excluding the collar end for improved de-bonding. The hole is injected using cement grout the anchor of Dura bar is a creased segment of the bar. During face plate loading, the anchor slips alongside the creased outline in the cement grout at a definite pull load [25]. (**Figure 15**) shows displacement of the bar under a dynamic test.

### 3.2.8. Dynatork Bolt

Its Dywidag systems international dynamic bolt, it's made with a coiled mixing blade to ensure ultimate mixing of DSI resin in the borehole (**Figure 16**).

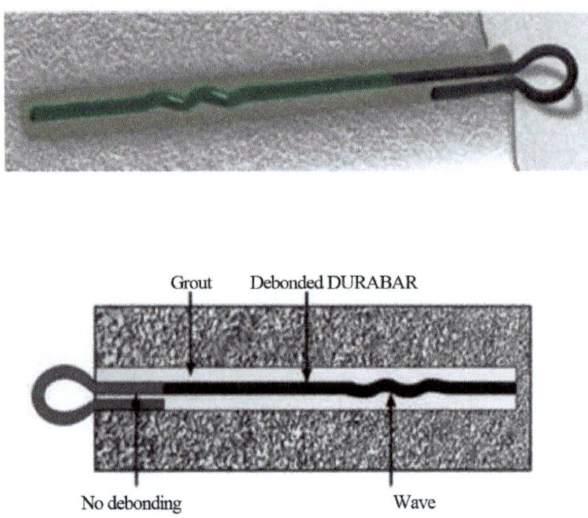


**Figure 14.** Dura bar.

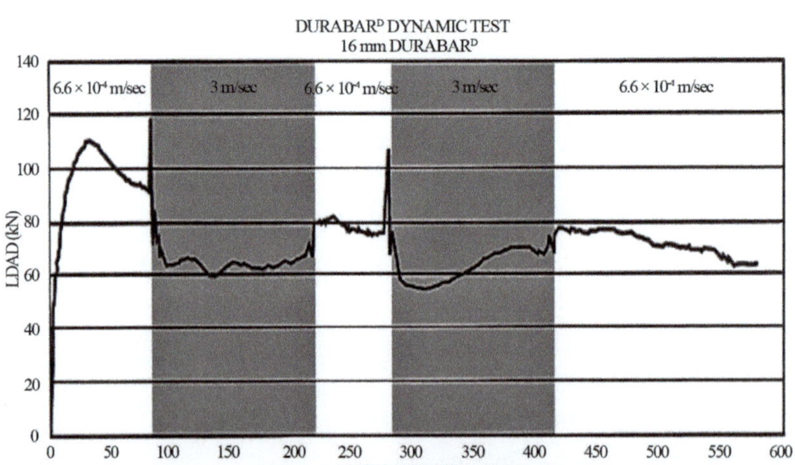


**Figure 15.** Dynamic test results of DURABAR yieldable bolt (CANMET-MMSL, 2012).

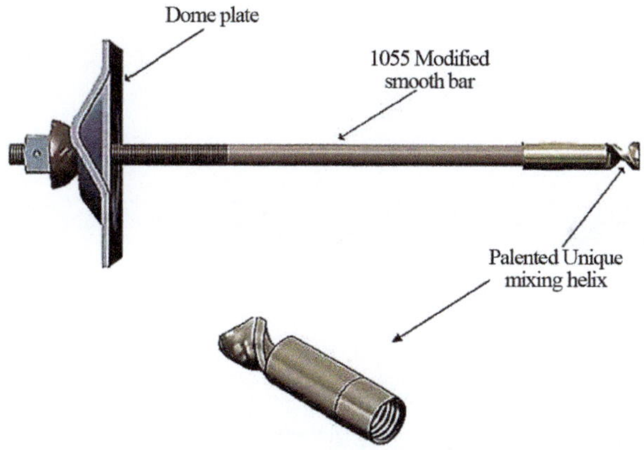


**Figure 16.** DSI dynamic bolt.

This blade design, according to DSI has a double function, mixing the resin correctly and providing perfect docking for support in static conditions. The bolt is engineered to yield and aid transmission of dynamic load into the resin during rock dilation in rock burst conditions, absorbing the energy through controlled deformation (**Figure 17**). The energy is expected to be absorbed during the ploughing effect of the cone through the resin [25].

### 3.2.9. Split Set Bolts

Comprise of a slotted, hollow highly-strengthened steel tube and faceplate. The bolts are fitted by pushing them into an undersized hole. The circular coil force created by compression of a C-shaped tube delivers frictional anchorage along the complete length of the tube (**Figure 18**). The toe end of the tube is slanted inward to facilitate quick and easy installation. It can be particularly useful in mild rock-burst conditions where the bolt will slip and not break, when used with wire mesh it will hold shattered rock caused by a mild burst.

These bolts can also be used to pin up mesh in highly fractured and yielded ground prior to the application of shotcrete. Provided the load demand does not exceed capacity, this system can work well for short-term support. However, susceptibility to corrosion rules out this system for most long-term support requirements. The bolts are installed as dowels. Split Set bolt pull-out capacity is rated in metric tons per meter (t/m) and is derived from in-situ pull tests. Nominal pull-out strengths vary from 1 to 2 t/m [8].

### 3.2.10. The Yield-Lok

This bolt can be fully or partially grouted with resin or cement mortar. The yielding performance is based on the interactions between the polymer coating, the head and the grouting media (**Figure 19**). The angled head of the polymer coating helps to scrap resin cartridge packing when the bolt is inserted into the resin and this improves anchorage. Mixing of the resin is enabled by deformations on polymer coating similar to rebar, thus the bolt can be fully inserted without rotation, and then spun afterwards to mix the resin. The Bolt can be constructed in dynamic condition a case of YL-Dynamic Bolt or for specified loads in static condition a case of YL-Static Bolt (**Table 3**) [25] [26].

When the rock bolt is tensioned, it immediately provides primary support upon being installed. During static loading conditions, the bolt performance is actually similar to a rebar bolt (**Figure 20**). The head transmits the effects on the polymer coating during dynamic loading, this brings about restricted compression, thermal softening and flow of the polymer around the upset head, creating a pulling effect. The energy from dynamic loading is absorbed by pulling the upset through the polymer, some of the energy is used up during friction between the smooth bar and the polymer coating [25].

The feature of yielding elements can be well controlled, ensure the steady performance of the bolt in the full length of polymer coating. The purpose of grouting material is to provide confinement of the polymer coating, instead of working as a yielding element as in the case of cone bolts. It means that if the bolt is completely grouted and confined, the performance of the bolt is not dependent on the grouting material. Since the mechanism of displacement is confined within the polymer, de-bonding agents like grease cannot help achieve the

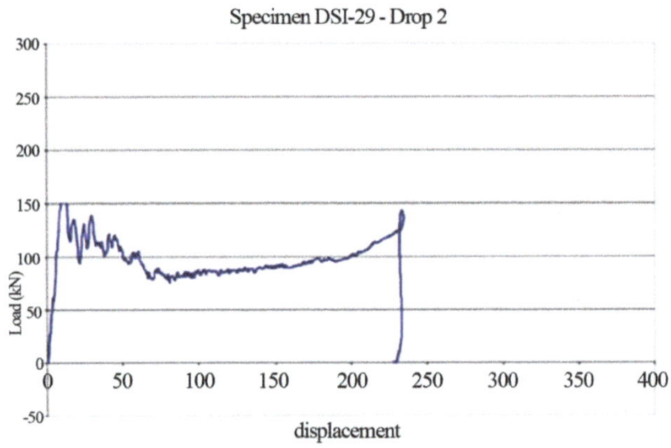


**Figure 17.** dynamic laboratory test results of DSI-Dynatork bolt (CANMET-MMSL, 2012).

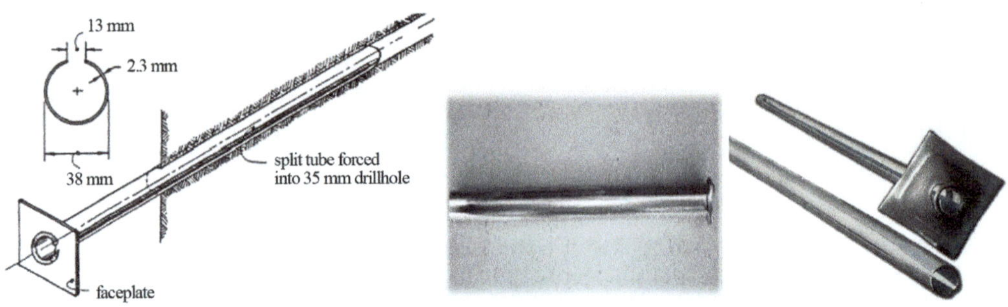


**Figure 18.** Friction anchor or Split Set.

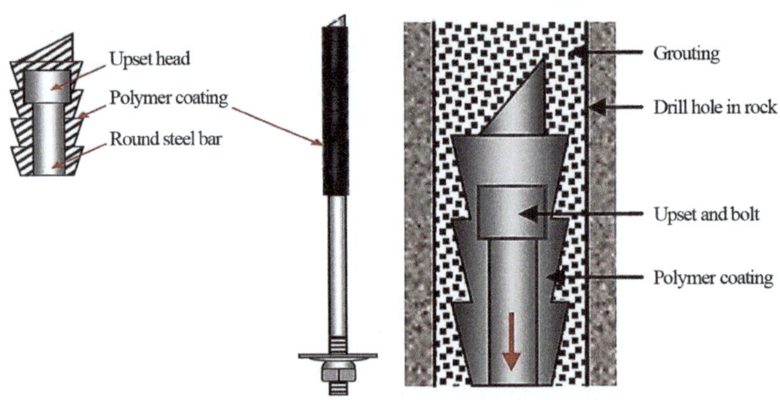


**Figure 19.** Yield-Lok.

definite effect and steady performance of the bolt [26].

### 3.2.11. Jacketted Rock Bolt

The function of jacketted rock bolts proposed by Chileshe (2015) was to balance the stress and strain, imposed by surrounding rock on grout and/or grouted rock bolts. If the high strain energy in the excavation periphery were to be accommodated over the whole length of a grouted rock bolt to the toe by a load balancing rock bolt, localised strain and stress excess would be mitigated. The low shear stress induced in the minimal strain zone in the deepest toe part of the grouted drill-hole would be increased, while that at the collar of the hole would be reduced. The result would be equitably distributed average shear stress on the rock bolt. The likelihood of grout or grout-rock or grout-bolt or rock bolt failure would be significantly reduced thus increasing the effectiveness

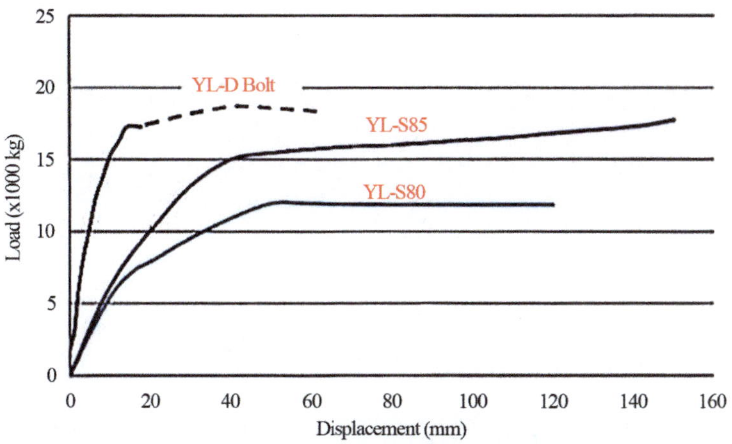


**Figure 20.** Static test results of Yield-Lok Dynamic and static bolts (CANMET-MMSL, 2012).

**Table 3.** Yield-Lok Bolt properties.

| Type | YIELD-LOK Bolt | | |
|---|---|---|---|
| | ***Minimum breaking load***, expanded profile (kN) | ***Minimum yielding load***, expanded profile (kN) | Minimum elongation (%) |
| | 178 | 133 | 8 |
| | ***Dynamic yield load*** (kN) | ***Dynamic yield displacement*** | ***Displacement capacity, max*** (mm) |
| YL—dynamic bolt | 71 - 89 | 200 mm per effect with energy input 16.4 kJ | 762 |
| YL—static bolt | ***Static yield load*** (kN) 116 - 133 | - | (500 mm) or more up to length of polymer coating |

of the grouted rock bolt. The overall effect of the load-balancing rock bolt would be to increase the effective anchorage length, especially with longer rock bolts [27].

The bolt is assembled in concentrically tubular steel jackets (**Figure 21**). Each jacket is 0.6 m to 1.1 m in length, and fitted within each other. The jackets overlap by, say, 0.2 - 0.4 m, thus forming a total length and diameter compatible with a typical rock bolt hole. The tightly or spirally or groove-fitted or lined or indented overlaps would provide friction between jackets and thus transmit energy from one jacket to the other in a frictional balancing system. The capacity to slip is essential. The entire jacketted assembly is grouted into the drill hole or simply in force-fitted direct contact with the rock [27].

Alternatively, to accommodate extreme cases of gross deformation, such as rock-bursts, the innermost tube can be replaced with a similarly sized solid rod (**Figure 22**) of a length that allows extension, even protrusion out of the hole [27].

### 3.2.12. New Yielding Bolt Proposed by (Wang *et al.*, 2012)

This bolt has similar appearance with Roofex, but a different structure of energy-absorber, its more simpler and stable, this increases its resistance. The bolt uses the elastic properties of steel through its operations. The bolt is intended to disperse and control large amount of energy liberated from the rock mass during the process of deformation. The new bolt is made up of a drawing die and drawing rod (**Figure 23(a)**). The rod is a metal bar with different diameters coated with a thin layer of wax, for easy detachment from the grout under a pull load, the die which also acts as an anchoring point is fixed inside the borehole with cement grout or resin. Operations of the bolt involve the metal being pushed through a die by tension applied at the end of the die. The elastic movement is sorely due to compression, from the metal with the die [15] [23].

The deformation of the rock induces a tension in the bolt, this causes sliding between the drawing die and rod. When the drawing die orifice diameter gets smaller than the thick end of drawing rod, pressure is induced. When

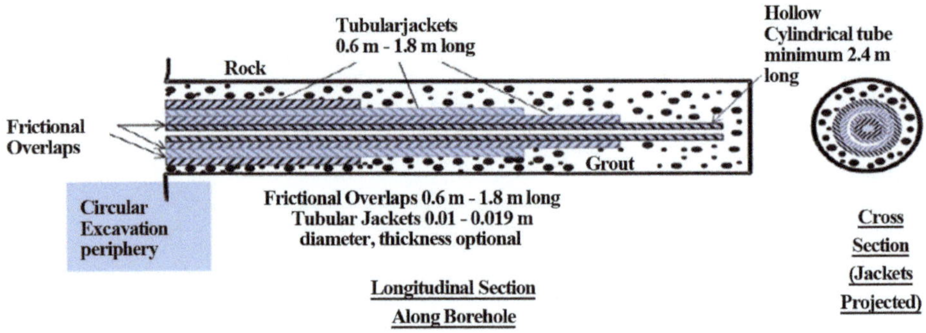


**Figure 21.** Jacketted Rock Bolt.

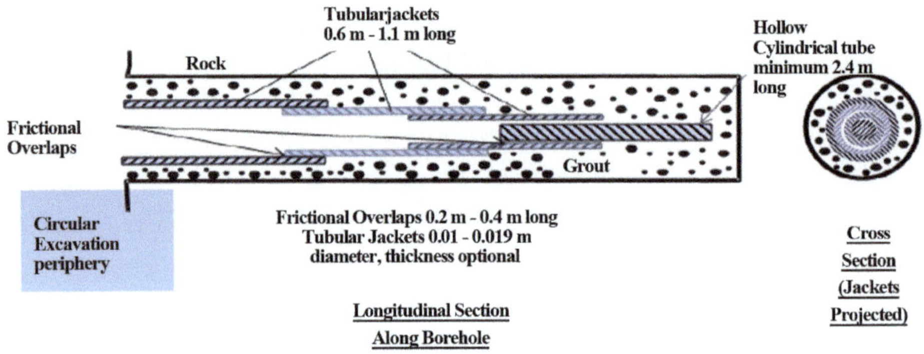


**Figure 22.** Jacketted Rock Bolt.

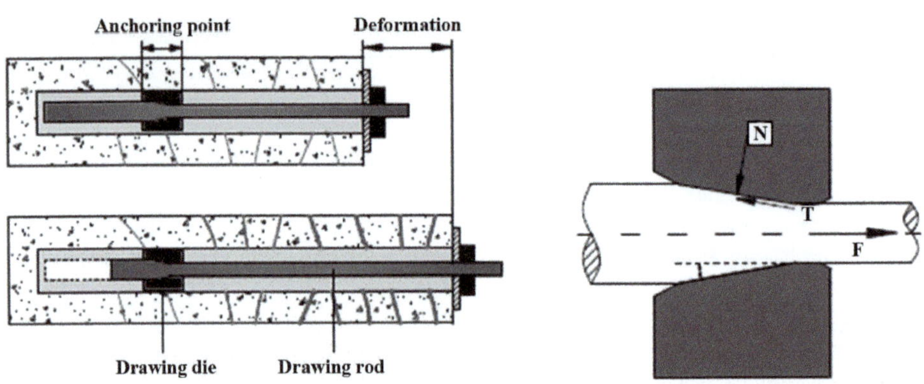


(a) (b)

**Figure 23.** (a) & (b) Mechanism of the new bolt (After Wang *et al.* (2012)).

the tension exceeds the defined drawing force, the rod is slowly pulled out (**Figure 23(b)**). The new bolt is designed to act like a stiff rock reinforcement element until the defined drawing load is exceeded. In the process of drawing, the bolt is capable of absorbing a large amount of energy (**Figure 24**). In addition to the continuous resistance and the large deformation, the effect of bolt operation is an increase in the rod's tensile strength. Upon installation, suitable pre-stress is enforced to support the rock mass this enhances the supporting effect considerably [15] [23].

## 4. Integrated System Support

To form an integrated system support to be employed in high stress ground conditions rockbolts can be combined

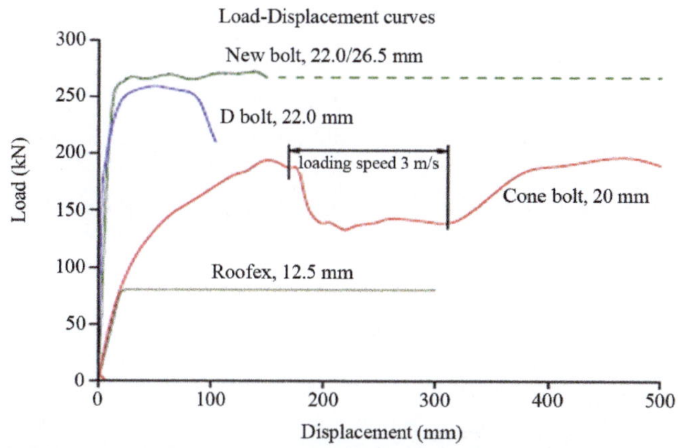


**Figure 24.** Load versus displacement behaviors of various boltunder pull loading test.

with surface support elements. Installing rock bolts as the only support system reduces the anticipate defficiency of rockburst support. This was observed underground in real rockburst conditions and was also established by a simulated rockburst trial conducted in a South African mine [28]. In the simulated rockburst trial, the support comprised of rebars, mesh, and lacing. Before blasting, mesh and lacing was removed and only there bars were left. Rebars were affected by the dynamic loading but not a single bolt failed due to dynamic loading of maximum peak particle velocity of 3.3 m/s and an ejection velocity of 2.5 m/s which was verified. Milev and Spottiswood (2005) [28] clarified that the cause was due to rapid weakening of peak particle velocity from the surface to the more competent rock mass [18].

Support displacement capacity is critical in areas of generated dynamic loading [29]. Cablebolts can be effective where fractured ground has relaxed and become vulnerable to vibration. Support loads generated over larger displacements absorb more of the kinetic energy of such a groundmass. Stiff supports will rupture before the mass can be decelerated and stabilized. Cable bolts are rock support elements which can reinforce and hold the rock mass. Cablebolts can survive higher values of ultimate displacement before failure; even though the energy that they could absorb is moderate [8]. It may be advisable to combine long partially bonded cables, with modified geometry or plated end lengths, with closely spaced primary support such as grouted rebar [30]. Under dynamic loading the rebar will maintain a reinforced film at the excavation surface. The large displacement of this film can be accommodated by the cables, maintaining ultimate holding capacity after the disturbance has passed. Without the rebar, the surface skin may loosen and disintegrate, rendering the cable bolts ineffective [31].

Tannant and Kaiser (1995) [32] describe an innovative support element for support of blocky ground under dynamic loading. A cable is anchored in a borehole using a Swellex friction bolt. A length of cable extends beyond the down hole extent of the Swellex and has an optional button or swage at the end. Even if the Swellex breaks in several places under loading, the cable remains frictionally anchored within the Swellex segments. The load-displacement characteristics are ideal for dynamic conditions; a consistent yield load and a large displacement capacity before rupture [33].

Wire mesh is used to contain ejected rock and hold back protruding pieces on the excavation surface; it's usually used in combination with shotcrete. Steel fibre reinforced shotcrete and mesh reinforced shotcrete is extensively used and preferred in underground mining. In poor quality and loose rock masses, where shotcrete to the rock surface is poor due to adhesion, the mesh delivers a significant reinforcement. Thus when stabilizing and building walls for underground fill, weld mesh is applied in the stabilization of the surface [34].

Steel arches were generally used to support haulages especially in coal mines where they are often essential to withstand large deformations. Deformations can be accommodated by employing yielding arches comprising of elements intended to slip at predetermined loads [35]. Hydraulic props are set with a pre-load to provide active support, The load bearing capacities of individual props may vary from as little as 5 t for a very light prop to more than 100 t for a 0.3-m-diameter prop. Rapid-yielding hydraulic props are widely used to provide concentrated, active support of the hanging wall close to the face. Their rapid-yielding capability allows the energy released by rock bursts to be absorbed rapidly and safely, thereby minimizing the damage caused [35].

## 5. Discussion

Charette and Hill (2011) [36] analysed basic static and dynamic loading data of six grouted rock bolt types capable of long term support. They concluded that the rebar and D-Bolt had a higher static load performance, while the MCB cone, Yield-Lok, D-Bolt and Roofex provided large deformation capacities at various loading levels. The MCB33 cone is best coupled with rebar reinforcement, while the D-Bolt and Yield Lok are designed to be used as single pass reinforcement bolts. Laboratory testing appeared to show that rebar and MCB cone bolt worked best together, at bolt displacements of less than 25 mm, while a dynamic event exceeding 7 kJ occurring would fail fully-grouted rebars (displacement larger than 25 mm). Thereafter, only the MCB cone bolt is able to provide static support [27].

Yielding bolts application such as cone bolts or friction sets is a priority in rockburst prone mines, Rebars are considered to be too stiff and that they can't be used for rockburst support. Rebar bolts can provide suitable support to uneven blocks to prevent rock falls, considered suitable in shallow excavations with low in situ rock stresses, and the main risk is from rock falls aided by gravity. The key task of the bolt in a shallow tunnel is to equilibrate the weight of loosened blocks that are reinforced by the bolt. Thus, the bolt strength is more significant than its deformation capacity in low in situ stress conditions [12] [13].

Combination of mesh or shotcrete with rebar bolts, can control the fracturing and rock bulking perfectly in hard rock mines, if the stress is somewhat low to moderate. When a rockburst strikes, rebars usually break at the threaded section and lose their function of holding but still maintain their reinforcing function to some point. However, if yielding bolts with straps are added to the rock support system, then a two tiered mechanism is formed. Rebars streng then the rock mass to ensure that it does not go through pre-mature fracturing. During rock mass failure, yielding bolts ensure the broken rocks are correctly retained. When developing a standard bolt applicable for rock burst prone ground, significant factor is its high initial load capability and stiffness for static loading as well as its high yielding and energy absorbing capacity for dynamic loading [18].

Application of shotcrete in burst-prone grounds is considered unlikely because it has been observed that shotcrete became part of the ejected rock when a rockburst occurs. Shotcrete improves the installed rockbolt-mesh support system by enhancing its link between the bolt and mesh. Prevent main blocks in blocks from moving and thus enhances the overall integrity of the rock mass and it needs to be used effectively in the support system. In preventing shotcrete being ejected off with rocks, mesh-over-shotcrete is applied. The problem of ejected rock can be solved by mounting yielding bolts and mesh/mesh straps over shotcrete. Thus, it is very vital to appreciate the three vital support function of for each support element [18]. As stated above its best to have a combined reinforcement to withstand ground deformations from seismic activity and rockburst, but cone bolts proves to be the best yielding bolt in line with its deformation capacity, but however can also work best in an integrated system of supports.

## 6. Conclusion

Yielding rock bolts are most preferred during support installations in rockburst prone ground because they provide obstructions in excavation openings, they are easy to repair, install and set up, they occupy small space during storage and they also require less installation time. However, with these advantages of rock bolts over other supports, rock bolts are not very effective installed by themselves, and they are best in an integrated system. Support components have many functions in terms of reinforcement, holding and retaining in that they can be strong on one side and weak on the other. It's very important that support components are combined in order to increase their capabilities for support in rockburst-prone mines. A very good support system is one which has many and different elements to form a combined system.

## Acknowledgements

The authors would like to thank the department of mining and mineral resources engineering of the University of Science and Technology Beijing.

## References

[1] Mendecki, M.A. (1999) A Guide to Routine Seismic Monitoring in Mines. Creda Communications, Cape Town.

[2] Larsson, K. (2004) Mining Induced Seismicity in Sweden. Master's Thesis, Luleå University of Technology, Luleå.

[3] Blake, W. and Hedley, D.G. (2003) Rockbursts, Case Studies from North American Hard-Rock Mines. Society for Mining, Metallurgy and Exploration Inc., Littleton.

[4] Per John, L. (1983) Hard Rock Pillar Strength Estimation an Applied Empirical Approach. Masters's Thesis, The University of British Columbia, Vancouver.

[5] Woldemedhin, Y.B. and Mwagalanyi, H. (2011) Investigation of Rock-Fall and Support Damage Induced by Seismic Motion at Kiirunavaara Mine. Master's Thesis, Luleå University, Luleå.

[6] Malmgren, L. (2005) Interaction between Shotcrete and Rock: Experimental and Numerical Study. PhD Thesis, Luleå University of Technology, Luleå.

[7] Cai, M. and Kaiser, P.K. (2012) Design of Rock Support System under Rockburst Condition. *Journal of Rock Mechanics and Geotechnical Engineering*, **4**, 215-227.

[8] Bawden, F.W. (2011) Ground Control Using Cable and Rock Bolting. In: Darling, P., Ed., *SME Mining Engineering Handbook*, Society for Mining, Metallurgy and Exploration Inc., Littleton, 616-617.

[9] Stillborg, B. (1994) Professional Users Handbook for Rock Bolting. Trans Tech Publications, Clausthal-Zellerfeld.

[10] Rajua, D., Mitria, H. and Thibodeaub, D. (2011) Mine Safety Enhancement by Designing Dynamic Rock Supports. *1st International Symposium on Mine Safety Science and Engineering*, Sudbury, November 2011, 1591-1602.

[11] Hoek, E., Kaiser, P.K. and Bawden, W.F. (1995) Support of Underground Excavation in Hard Rock. Balkema, Rotterdam.

[12] Li, C.C. (2010) A New Energy-Absorbing Bolt for Rock Support in High Stress Rock Masses. *International Journal of Rock Mechanics & Mining Sciences*, **47**, 396-404. http://dx.doi.org/10.1016/j.ijrmms.2010.01.005

[13] Li, CC. (2007) A Practical Problem with Threaded Rebar Bolts in Reinforcing Largely Deformed Rock Masses. *Rock Mechanics and Rock Engineering*, **40**, 519-524. http://dx.doi.org/10.1007/s00603-006-0094-7

[14] Li, C.C. (2010) Field Observations of Rock Bolts in High Stress Rock Masses. *Rock Mechanics and Rock Engineering*, **43**, 491-496. http://dx.doi.org/10.1007/s00603-009-0067-8

[15] Wang, G., Wu, X.Z. and Jiang, Y.J. (2012) A New Yielding Bolt for Rock Support in High Stress Rock Masses. *Applied Mechanics and Materials*, **204-208**, 366-369. http://dx.doi.org/10.4028/www.scientific.net/AMM.204-208.366

[16] Ortlepp, W.D. (1992) The Design of Support for the Containment of Rockburst Damage in Tunnels—An Engineering Approach. In: Kaiser, P.K. and McCreath, D.R., Eds., *Rock Support in Mining and Underground Construction*, Balkema, Rotterdam, 593-609.

[17] Jager, A.J. (1992) Two New Support Units for the Control of Rockburst Damage. In: Kaiser, P.K. and McCreath, D.R., Eds., *Rock Support in Mining and Underground Construction*, Balkema, Rotterdam, 621-631.

[18] Cai, M. (2013) Principles of Rock Support in Burst-Prone Ground. *Tunneling and Underground Space Technology*, **36**, 46-56. http://dx.doi.org/10.1016/j.tust.2013.02.003

[19] Smith, M. (2007) Rock & Soil Reinforcement. Ulf Linder, Örebro.

[20] Smith, M. (2007) Underground Mining Methods. Ulf Linder, Örebro.

[21] Li, C.C. and Marklund, P.I. (2004) Field Tests of the Cone Bolt in the Boliden Mines. In: Hamre, L., Rohde, J.K.G., Berg, K.R. and Nilsen, B., Eds., *Fjellsprengninsteknikk/Bergmekanikk/Geoteknikk*, Norsk Jord og Fjellteknisk Forbund, Oslo, 12.

[22] Simser, B., Andrieux, P., Mercier-Langevin, F., Parrott, T. and Turcotte, P. (2007) Field Behaviour and Failure Modes of Modified Conebolts at the Craig, LaRonde and Brunswick Mines in Canada. In: Hadjigeorgiou, J., Stacey, D. and Potvin, Y., Eds., *Challenges in Deep and High Stress Mining*, Australian Centre for Geomechanics, Perth, 347-354.

[23] Wang, G., Wu, X.Z., Jiang, Y.J., Huang, N. and Wang, S.G. (2013) Quasi-Static Laboratory Testing of a New Rock Bolt for Energy-Absorbing Applications. *Tunnelling and Underground Space Technology*, **38**, 122-128. http://dx.doi.org/10.1016/j.tust.2013.05.010

[24] Varden, R., Lachenicht, R., Player, J., Thompson, A. and Villaescusa, E. (2008) Development and Implementation of the Garford Dynamic Bolt at the Kanowna Belle Mine. *10th Underground Operators' Conference*, Launceston, 14-16 April 2008, 95-102.

[25] Guntumadugu, D.R. (2013) Methodology for the Design of Dynamic Rock Supports in Burst Prone Ground. PhD Thesis, Department of Mining and Materials Engineering, McGill University, Montreal.

[26] Wu, R., Oldsen, J. and Campoli, A. (2011) Application of Yield-Lok Bolt for Bursting and Convergence Grounds in Mines. *30th International Conference on Ground Control in Mining*, Morgantown, 26-28 July 2011, 1-6.

[27] Chileshe, P.R.K. (2015) Review and Design Concept for Grouted Multi-Jacketted Friction Balancing Rock Bolt. *International Journal of Scientific & Technology Research*, **4**, 92-100.

[28] Milev, A.M. and Spottiswood, S.M. (2005) Strong Ground Motion and Site Response in Deep South African Mines.

*The Journal of the South African Institute of Mining and Metallurgy*, **105**, 515-524.

[29] Kaiser, P.K., McCreath, D. and Tannant, D. (1996) Rockburst Research Handbook. CD-ROM Version, Ontario.

[30] Kaiser, P.K. and Maloney, S.M. (1997) Scaling Laws for the Design of Rock Support. *Pure and Applied Geophysics*, **150**, 415-434. http://dx.doi.org/10.1007/s000240050085

[31] Hutchinson, D.J. and Diederichs, M.S. (1996) Cablebolting in Underground Mines. Bi Tech Publishers, Richmond.

[32] Kaiser, P.K., McCreath, D.R. and Tannant, D.D. (1995) Rockburst Support Handbook. Geomechanics Research Centre, Sudbury.

[33] Thompson, A.G., Villaescusa, E. and Windsor, C.R. (2012) Ground Support Terminology and Classification: An Update. *Geotechnical and Geological Engineering*, **30**, 553-580. http://dx.doi.org/10.1007/s10706-012-9495-4

[34] Wood, D.F. (1992) Specification and Application of Fiber Reinforced Shotcrete. In: Kaiser, P.K. and McCreath, D.R., Eds., *Rock Support in Mining and Underground Construction*, Balkema, Rotterdam, 149-156.

[35] Brady, B.H.G. and Brown, E.T. (2004) Rock Mechanics for Underground Mining. 3rd Edition, Springer Science and Business Media, Dordrecht.

[36] Charette, F. and Hill, M. (2011) Single Pass Bolting Approach for the Stabilisation of Rock Masses in High Stress Conditions. *Metsco Proceeding CIM/ICM Conferences & Exhibitions*, Montreal, 2011.

# 20

# Underlying Factors of Occupational Accidents: The Case of Turkey

**Fatih Yilmaz, Selcuk Alp**

Department of Industrial Engineering, Yildiz Technical University, Istanbul, Turkey
Email: ftyilmaz@yildiz.edu.tr, alp@yildiz.edu.tr

## Abstract

**An average of 80 thousand occupational accidents occurs each year in Turkey and 1500 workers are lost their lives. In accordance to "policy of preventive approach", many legal arrangements have been made in Turkey. In this context, weight to issues such as risk assessment and preventive OHS services is given; however, employee training, employee participation, labour inspect, and periodic checks in working areas such as maintenance and audit have not been achieved. The level of implementation of legal arrangements was inadequate. Research was carried out to determine the significance rate of the factors that cause accidents in 5 sectors with the highest number of occupational accidents in Turkey. Questionnaires administered to experienced safety specialists were reviewed by analyzing with the fuzzy TOPSIS method. According to the results, especially in the construction and coal mining sectors and all sectors in general, employee training, employee participation and periodic maintenance in the workplace are more important accident prevention factors than risk assessment and preventive OH & S services. It is seen that for the prevention of accidents, all measures should be implemented by systematic and by a multidisciplinary approach.**

## Keywords

**Safety, Fuzzy TOPSIS, Occupational Accident, Risk Assessment**

## 1. Introduction

Occupational accidents cause important social and economic problems by the loss of life and physical injuries. In every year, 2 million fatal accidents occur worldwide and 2 workers die because of occupational accident in every minute. Many sectors such as construction, coal mining, textile etc. involve high risk due to its production processes and labour intensive characteristic and occupational accidents bring up financial loss in large scale in these sectors. In developing countries, these risky sectors are the most important sectors that have a great contribution to economic development with its employment capacity and an added-value to the economy. On the other hand, due to the lack of preventive measures, occupational accidents occur frequently [1]-[3].

According to the official sources, 65 - 80 thousand occupational accidents occur in Turkey every year and as a result of these accidents, 700 - 1700 workers die and approximately 2000 workers become permanently incapable. Occupational accidents cause approximately $3 billion income loss in social security system. Majority of occupational accidents are not recorded and because 99% of occupational diseases are not identified, they are not considered in statistics. Estimations of Turkish Statistical Institute and Turkish Social Security Institution indicate that the expected number of occupational accidents and diseases can be 4 times more than the recorded data [4] [5].

**Table 1** shows the numbers of accidents at work and the fatalities of 5 sectors with the highest incidence rate of occupational accidents in Turkey. Occupational accidents, fatalities and disabilities in Turkey are at the alarming levels. Metal products sector ranks first in the number of accidents in Turkey. Also, construction industry is ranked first in the number of fatalities. 30% of the fatal accidents occur in construction sector every year. Transportation sector at which many of the fatal accidents occur as a result of vehicle accidents, takes the second place after construction sector. Fabric sector is one of the highest employment sectors in Turkey. Accordingly, the number of accidents in this sector is high.

Turkey has ratified the ILO Convention No. 155 and 161 and in order to ensure compliance with European Union directives, it has made a number of legislations in occupational health and safety (OH & S). After the "Labour Act" enacted in 2003 and the related directives came into force, the "preventive approach based on risk assessment" is adopted and all directives revised in accordance with these principle. New legal obligations, such as making risk assessment in workplaces, preventive OH & S services, safety representatives, committees, and providing training to employees, were introduced. In particular, great importance is given to issues such as risk assessment, safety specialist assignments and workplace OH & S units. All enterprises were classified according to the hazard class and obliged to provide these preventive services via physician, technical staff and engineers according to workplace hazard class. In addition, qualifications, duties, working hours and responsibilities of "External OH & S Service Units" were identified and implemented.

However, despite more than 10 years have passed, the expected benefits of the regulations have not been achieved. Occupational accidents began to rise again. Especially in high risky sectors and the overall work-related accidents showed a slight decrease trend until 2008, but after 2008 it was seen that started to rise again (also see **Table 1**).

**Table 1.** Accidents and fatalities in the sectors that most occupational accidents occur in Turkey.

| Years | Coal Mining | | Metal Products | | Construction | | Transportation | | Textile | |
|---|---|---|---|---|---|---|---|---|---|---|
| | Accidents | Fatal | Accidents | Fatal | Accidents | Fatal | Accidents | Fatal | Accidents | Fatal |
| 2003 | 5647 | 54 | 9682 | 23 | 8198 | 274 | 3762 | 88 | 7382 | 21 |
| 2004 | 5481 | 38 | 11,584 | 37 | 8106 | 263 | 4098 | 109 | 6839 | 14 |
| 2005 | 6011 | 82 | 10,283 | 25 | 6480 | 290 | 3928 | 163 | 5869 | 29 |
| 2006 | 6722 | 35 | 11,039 | 31 | 7143 | 397 | 4478 | 165 | 5155 | 26 |
| 2007 | 6293 | 38 | 11,224 | 45 | 7615 | 359 | 4483 | 146 | 5639 | 20 |
| 2008 | 5728 | 30 | 6971 | 34 | 5574 | 297 | 1974 | 111 | 3641 | 9 |
| 2009 | 8193 | 3 | 7314 | 11 | 6877 | 156 | 2329 | 36 | 3771 | 12 |
| 2010 | 8150 | 92 | 6918 | 43 | 6437 | 475 | 2206 | 133 | 3474 | 16 |
| 2011 | 9217 | 58 | 7268 | 72 | 7749 | 570 | 2363 | 194 | 3239 | 22 |
| 2012 | 8828 | 20 | 7045 | 25 | 9230 | 256 | 2549 | 73 | 5127 | 18 |
| 2013 | 11,289 | 36 | 12,061 | 34 | 26,967 | 521 | 7597 | 183 | 10,996 | 20 |
| 2014 | 10,026 | 335 | 12,357 | 14 | 29,699 | 501 | 7287 | 172 | 12,128 | 17 |

Significant improvements have been provided related to "risk assessment" and "preventive services" in the last 10 year, however, labour inspection, training, employee participation, maintenance and control (supervision) services were inadequate. Arrangements are not given enough attention to these issues. After the directives, "preventive services" can be obtained from outside of workplace from "external" health and safety units. A "market" emerged consisting of these enterprises; however, due to competition of these firms, quality of the services has decreased rapidly. Although enacted "Occupational Health and Safety Act" in 2012, the arrangements and practices of participation of employees were inadequate.

The obligations imposed by the Act are not sufficiently inspected. Therefore, work-related accidents have been increased in last 3 years. Defective policies were also caused great accidents which resulted deaths of hundreds of workers in coal mining and the construction industries. The purpose of the article; to identify the significance of the risk factors that cause accidents at work in Turkey, both in general and also the most work-related accidents sectors with fuzzy TOPSIS method. Based on the obtained results, the study aims to make a critical evaluation of occupational health and safety policy in Turkey, and to offer suggestions.

## 2. Material and Method

The data used in this article was obtained from the results of a survey applied to 6 safety consultants experienced in 5 industries. Participants are requested to evaluate the causes of accidents in the 5 sectors with the highest number of accidents in Turkey according to the Social Security Institute's statistics. The data were analysed with Fuzzy TOPSIS method and interpreted.

Decision-making is a process for finding the best option out of suitable alternatives. Multi-criteria decision-making (MCDM) is used in the ranking of the alternatives in terms of multiple criteria and has a very wide range of applications [6] [7]. The methods developed for solving MCDM problems, are often used in cases where a large number of decision-making process and with conflicting criteria [8] [9]. TOPSIS (Technique for Order Preference by Similarity to Ideal Solutions) is one of the MCDM methods which don't takes relations between criteria into account but the benefits or costs of the features will impose to the results to be obtained of the criteria discussed. Method analyses the results in terms of proximity to positive ideal solution and negative ideal solution results. These two alternatives for the proximity measure obtained by using the proximity index, is obtained an accurate ranking results from large to small, the best alternative is designated with this ranking [7] [10] [11].

MCDM methods, is a method used in research related to health and safety in the workplace. For example; to examine the safety of workers in hot and humid environment conditions [12], to measure the safety factor and performance in road construction [13], in determining the increase in chemical tanker safety requirements [14], in determination of faulty behavioural risks in behavioural based safety management in the workplace [15], in determining the risk of occupational safety in the manufacturing industry [16], in working in dusty environments for determining the risk of workplace safety [17], the excavation of construction safety risks and the determination of losses arising from occupational accidents [18], the evaluation of health and safety risks of working in hazardous waste recycling services [19] various MCDM methods are used. This method is widely used for occupational safety, occupational health, issues related to the efficiency and performance measurement [20]-[23].

In this study, the responses are analysed obtained from safety specialists who has practical experience over than 30 years in 5 industries. They were requested to identify deficiencies, important factors of preventing work-related accidents and determine their significance rates in 5 sectors based on the individual experiences. The resulting data were analysed by using the Chen *et al.* (2006) proposed fuzzy TOPSIS model algorithm.

The Fuzzy TOPSIS model's which developed by Chen *et al.* (2006), algorithm will be explained in this part of the study. In the implementation of this model, decision-makers consisting of experts in the field assess criteria and available alternatives according to these criteria, primarily. The assessments made by linguistic variables are converted to trapezoidal fuzzy numbers by using the values in **Table 2** and **Table 3**.

Decision-makers have made the decision criteria relating to weight the importance of some of the alternatives and the evaluation criteria, respectively: $i = 1, 2, \cdots, m$ and $j = 1, 2, \cdots, n$, including $\tilde{x}_{ijk} = (a_{ijk}, b_{ijk}, c_{ijk}, d_{ijk})$ and $\tilde{w}_{ijk} = (w_{jk1}, w_{jk2}, w_{jk3}, w_{jk4})$ as the accepted decision makers of fuzzy criterion values obtained with the alternative assessment of the criteria $\tilde{x}_{ij} = (a_{ij}, b_{ij}, c_{ij}, d_{ij})$ and decision criteria importance value $\tilde{w}_{jk} = (w_{j1}, w_{j2}, w_{j3}, w_{j4})$ will be displayed. These values in this step,

**Table 2.** Linguistic variable for decision criteria and Trapezoidal fuzzy numbers.

| Linguistic variable | Trapezoidal fuzzy numbers |
|---|---|
| Very high (VH) | (0.8, 0.9, 0.9, 1.0) |
| High (H) | (0.7, 0.8, 0.8, 0.9) |
| Fairly high (FH) | (0.5, 0.6, 0.7, 0.8) |
| Medium (M) | (0.4, 0.5, 0.5, 0.6) |
| Fairly low (FL) | (0.2, 0.3, 0.4, 0.5) |
| Low (L) | (0.1, 0.2, 0.2, 0.3) |
| Very low(VL) | (0.0, 0.1, 0.1, 0.2) |

**Table 3.** Linguistic variable for decision criteria and Trapezoidal fuzzy numbers.

| Linguistic variable | Trapezoidal fuzzy numbers |
|---|---|
| Very high (VH) | (8, 9, 9, 10) |
| High (H) | (7, 8, 8, 9) |
| Fairly high (FH) | (5, 6, 7, 8) |
| Medium (M) | (4, 5, 5, 6) |
| Fairly low (FL) | (2, 3, 4, 5) |
| Low (L) | (1, 2, 2, 3) |
| Very low(VL) | (0, 1, 1, 2) |

$$\left.\begin{aligned} a_{ij} &= \min_k \left\{ a_{ijk} \right\} \\ b_{ij} &= \frac{1}{K}\sum_{k=1}^{K} b_{ijk} \\ c_{ij} &= \frac{1}{K}\sum_{k=1}^{K} c_{ijk} \\ d_{ij} &= \max_k \left\{ d_{ijk} \right\} \end{aligned}\right\} \tag{1}$$

$$\left.\begin{aligned} w_{j1} &= \min_k \left\{ a_{jk1} \right\} \\ w_{j2} &= \frac{1}{K}\sum_{k=1}^{K} w_{jk2} \\ w_{j3} &= \frac{1}{K}\sum_{k=1}^{K} w_{jk3} \\ w_{j4} &= \max_k \left\{ w_{jk4} \right\} \end{aligned}\right\} \tag{2}$$

are calculated using equations above.

Decision problem, in the form of the matrix is shown as follows,

$$\tilde{D} = \begin{bmatrix} \tilde{x}_{11} & \tilde{x}_{12} & \cdots & \tilde{x}_{1n} \\ \tilde{x}_{21} & \tilde{x}_{22} & \cdots & \tilde{x}_{2n} \\ \vdots & \vdots & \ddots & \vdots \\ \tilde{x}_{m1} & \tilde{x}_{m2} & \cdots & \tilde{x}_{mn} \end{bmatrix} \tag{3}$$

$$\tilde{W} = \left[\tilde{w}_1 \ \tilde{w}_2 \ \tilde{w}_3 \ \tilde{w}_4\right]. \tag{4}$$

Here $\tilde{x}_{ij}$ and $\tilde{w}_j$ is trapezoidal fuzzy numbers, $\tilde{D}$ fuzzy decision matrix, $\tilde{W}$ shows the vector of fuzzy weights. Fuzzy decision matrix, in case of utility maximization (5), or in the case of cost minimization are normalized using the equation number (6). Normalized fuzzy decision matrix are shown as $\tilde{R} = \left[\tilde{r}_{ij}\right]_{m \times n}$. $B$ defined as the utility criteria and $C$ is the cost criteria for the values of $\tilde{r}_{ij}$ in equations (5) and (6) calculated as,

$$\tilde{r}_{ij} = \left(\frac{a_{ij}}{d_j^*}, \frac{b_{ij}}{d_j^*}, \frac{c_{ij}}{d_j^*}, \frac{d_{ij}}{d_j^*}\right), \ d_j^* = \max_i d_{ij}, j \in B \tag{5}$$

$$\tilde{r}_{ij} = \left(\frac{a_j^-}{d_{ij}}, \frac{a_j^-}{c_{ij}}, \frac{a_j^-}{b_{ij}}, \frac{a_j^-}{a_{ij}}\right), \ a_j^- = \min_i a_{ij}, j \in C. \tag{6}$$

The weighted normalized fuzzy decision matrix $\left(\tilde{V}\right)$ calculation is required because of each decision criteria may have different importance weights. Fuzzy decision matrix are expressed as $\tilde{V} = \left[v_{ij}\right]_{m \times n}, i = 1, 2, \cdots, m; j = 1, 2, \cdots, n$. And calculated as,

$$\tilde{v}_{ij} = \tilde{r}_{ij} \times \tilde{w}_j. \tag{7}$$

After calculating the weighted normalized fuzzy decision matrix, fuzzy positive ideal solution $\left(A^*\right)$ and fuzzy negative ideal solution $\left(A^-\right)$ is determined as follows,

$$A^* = \left(\tilde{v}_1^*, \tilde{v}_2^*, \cdots, \tilde{v}_n^*\right) \tag{8}$$

$$A^- = \left(\tilde{v}_1^-, \tilde{v}_2^-, \cdots, \tilde{v}_n^-\right). \tag{9}$$

Here $i = 1, 2, \cdots, m$ and $j = 1, 2, \cdots, n$ including, $\tilde{v}_j^* = \max\left\{v_{ij4}\right\}$ and $\tilde{v}_j^- = \min\left\{v_{ij1}\right\}$. The distance of each alternative in the problem to $\left(A^*\right)$ and $\left(A^-\right)$ can be calculated as follows, respectively.

$$d_i^* = \sum_{j=1}^{n} d_v\left(\tilde{v}_{ij}, \tilde{v}_j^*\right), i = 1, 2, \cdots, m \tag{10}$$

$$d_i^- = \sum_{j=1}^{n} d_v\left(\tilde{v}_{ij}, \tilde{v}_j^-\right), \ i = 1, 2, \cdots, m \tag{11}$$

where $i = 1, 2, \cdots, m$ and $j = 1, 2, \cdots, n$, including $d_v(.,.)$, $m = \left(m_1, m_2, m_3, m_4\right)$ and $n = \left(n_1, n_2, n_3, n_4\right)$ shows the distance between two trapezoidal fuzzy numbers and is calculated by the following equation.

$$d_v\left(\tilde{m}, \tilde{n}\right) = \sqrt{\frac{1}{4}\left[\left(m_1 - n_1\right)^2 + \left(m_2 - n_2\right)^2 + \left(m_3 - n_3\right)^2 + \left(m_4 - n_4\right)^2\right]}. \tag{12}$$

Relative closeness is calculated to determine the ranking of the alternatives after the calculation of the distance. Relative closeness,

$$CC_i = \frac{d_i^-}{d_i^* + d_i^*}, \ i = 1, 2, \cdots, m \tag{13}$$

is calculated by the equation above. According to relative closeness, the alternatives are ranked from highest value to lowest value [9] [24] [25].

## 3. Results

The data obtained from questionnaires filled in by the experts were analysed by fuzzy TOPSIS model. According to results of the analysis impact of accident criteria's that cause accidents at work in 5 sectors is determined generally. Weighted value of importance of the criteria was found as shown in **Table 4**. The results are also provided in **Figure 1**.

**Table 4** and **Figure 1**, shows that "Inadequacy of Education" and "Inadequacy of Employee Participation" criteria have the highest impact on occupational accidents with values closed to each other. "Inadequacy of Supervision", "Inadequacy of OH & S Preventive Services", "Inadequacy of Maintenance" criteria have the highest values after those two highest criteria. The third part criteria which have relatively lower weights are obtained as; "Inadequacy of Risk Assessment".

Following the estimation of weights, the importance of alternatives was determined according to the criteria. The inadequacy determined is presented in **Table 5**.

As examined in **Table 5**, the most important criteria for causing occupational accidents in the construction industry is seen "Inadequacy of Supervision" criterion that engineers, technical staff or field control personnel to be done. The important causes of the accidents for the construction industry are, "Inadequacy of Maintenance" and "Inadequacy of Employee Participation" criteria, respectively.

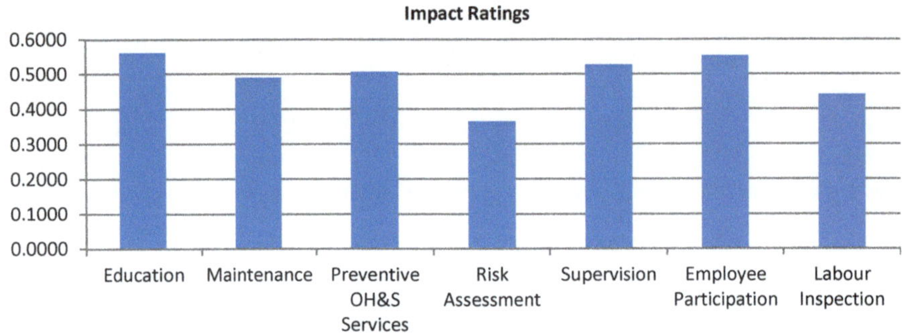


**Figure 1.** Distribution of impact ratings of occupational accidents factors in Turkey.

**Table 4.** Impact ratings of occupational accidents factors in Turkey.

| Occupational accidents factors | Impact ratings |
|---|---|
| Inadequacy of education | 0.5612 |
| Inadequacy of maintenance | 0.4897 |
| Inadequacy of preventive OH & S services | 0.5059 |
| Inadequacy of risk assessment | 0.3637 |
| Inadequacy of supervision | 0.5175 |
| Inadequacy of employee participation | 0.5525 |
| Inadequacy of labour inspection | 0.4414 |

**Table 5.** Occupational accident factors impact ratings by sector in Turkey.

| | Inadequacy of education | Inadequacy of maintenance | Inadequacy of preventive OH & S services | Inadequacy of risk assessments | Inadequacy of supervision | Inadequacy of employee participation | Inadequacy of labour inspection |
|---|---|---|---|---|---|---|---|
| **Construction** | 0.5173 | 0.5504 | 0.4682 | 0.4662 | 0.5686 | 0.5326 | 0.4406 |
| **Coal mining** | 0.5310 | 0.5024 | 0.4356 | 0.2171 | 0.4586 | 0.5326 | 0.4136 |
| **Metal products** | 0.5850 | 0.4188 | 0.5479 | 0.4038 | 0.5015 | 0.5511 | 0.4433 |
| **Textile** | 0.6044 | 0.5366 | 0.5060 | 0.4314 | 0.5322 | 0.5725 | 0.4647 |
| **Transportation/Logistics** | 0.5703 | 0.4397 | 0.5776 | 0.2776 | 0.5248 | 0.5750 | 0.4460 |

Construction sector is the most fatal occupational accidents occur in Turkey. Construction is high risky sector which has labour-intensive features, conducted in outdoors and different climatic conditions, requiring the use of a large number of equipment and materials. Poor maintenance and improper use of equipments, improper and careless work operations cause serious accidents at construction sites. Therefore; maintenance and periodic controls of equipment, site audits and safety trainings plays a decisive role in construction accidents.

"Inadequacy of Employee Participation" criterion is the forefront criterion for Coal Mining Industry. Then the "Inadequacy of Education" and "Inadequacy of Maintenance" comes. Coal mines is operated in rural areas, workers with low levels of education and wages and in poor conditions in Turkey. Some coal mines are old and incapable of technologically. Coal enterprises, increase the production and increase the workload accordingly to resolve the growing energy demand. Some of the coal mine operators has no experience of mining and has not sufficient information on issues related to safety, experience and sensitivity. After the privatization of coal mines in Turkey, private sector enterprises did not also take the necessary measures after privatization. There is no adequate labour inspection in mines. Therefore, the coal mining industry has become the sector the most occupational accidents occurring in Turkey. Thus, especially Soma and Ermenek, as a result of major accidents occurring in 2014, 354 workers lost their lives [26]-[30].

In the metal products industry, "Inadequacy of Education", "Inadequacy of Employee Participation" and "Inadequacy of Preventive OH & S Services" criteria were considered important. "Inadequacy of Preventive OH & S Services", "Inadequacy of Employee Participation" and "Inadequacy of Education" criterias proportions are calculated very close together in Transportation/Logistics industry. "Inadequacy of Education", "Inadequacy of Employee Participation" criteria came to the forefront in textile industry. "Inadequacy of Education" is proportionally higher degree of importance in textile industry. Textile industry is labour-intensive, creating high levels of employment, however, using the unskilled labour force. Textile sector has low level of fatalities, but the injury rate is more than the other sectors. In this case, the sector causes much more workday losses and costs. Metal Products, Textile and Transport sectors, has better conditions in terms of safety relative to construction and the coal mining. These sectors are composed of more favourable enterprises to improvements with effective inspection and low cost investments.

In Transportation/Logistics industry "Inadequacy of Employee Participation", "Inadequacy of Preventive OH & S Services" and "Inadequacy of Education" criteria were calculated close. Without exception, the lowest impact criteria in all sectors "Inadequacy of Risk Assessments" are seen as the criteria.

Before the new regulations brought in preventive OH & S services, many large enterprises had been organized "preventive services" within the workplace itself. After providing the opportunity to external services, in order to reduce costs, many enterprises have begun to supply these services from the outside of workplaces. Time and the quality of the services has dropped. 98% of workplaces in Turkey are small and medium-sized enterprise. 70% of work-related accidents occur each year in these enterprises. Insufficient external services weaken the workplace control processes particularly in medium-sized and hazardous industrial and construction enterprises. This situation leads to increase in work-related accidents.

Regulations cannot be applied to practice in Turkey. Despite the legislative work in safety in recent years, applications and needs with those written on formally is quite different. Today OH & S policy is based on the preventive principles and risk assessment. In this context, to provide preventive OHS services in the workplace is one of the most important applications in this regard. However, to prevent occupational accidents and improve the safety of the conditions can be possible with the determination and implementation of an OH & S policy which covers all the necessary criteria. Occupational safety requires a multi-criteria, interdisciplinary and systematic approach. Importance should be given to all preventive measures and all legal arrangements should be inspected. While the policies and measures are identified, industrial conditions and their influence levels of risk in the workplace must be taken into consideration. Applications focusing on specific issues on overall quality are not sufficient to prevent occupational accidents.

## 4. Conclusions

Considering the statistics in Turkey, safety and health practices cannot achieve the expected benefits. Except for legislation, assignment of safety specialists and risk analysis, and other preventive activities were insufficient. Prevention of occupational accidents are impossible if establishments do not take necessary precautions such as allocating necessary financial funds for the measures, establishing a permanent monitoring system to audit for all works, creating a work discipline, appointing enough technical staff and engineers, doing enough labour inspect, and keeping records of accidents in the workplace.

In Turkey, occupational health and safety is approached with a general overview. Uniform regulations covering all sectors are made. Production conditions and tools in different sectors, status of employees, regional differences, the risks in different sectors, impact levels and results are not considered to be affected. These deficiencies also reduce the effectiveness of regulations and labour inspection. According to statistics, it is understood that separate perspective, policy and planning are needed for risky sectors such as coal mining, construction and metal product.

Risk analysis should be carried out in mines and construction sites, the enterprises in very poor condition must be stopped. Particularly in mining and construction sector, enterprises should not be licensed until fully provided all necessary safety conditions.

Preventive health and safety services are insufficient in terms of time and quality. It reduces the effectiveness of the control mechanism and increases the risk of accidents. In addition, many "External OH & S Units" are insufficient about staff, technical facilities and expertise. The service prices of the external OH & S units are determined by the competition in the market. In this case, external OH & S units have no other task than to prepare necessary documents such as risk assessment report, emergency action plans, safety expert contracts and safety training for a fee and sending those to workplaces.

Due to insufficiency of labour inspector and inspection times, hundred thousands of enterprises cannot be inspected at all. In this case the obligations and penalties are not applied, occupational accidents cannot be prevented.

Risk assessment becomes legally obliged to all workplaces in Turkey. However, there is no uniformity in practice. Which risk assessment method will be used according to various parameters such as sector, type of enterprise, number of employees, raw materials and equipment used, and at what intervals it will be done are not clearly defined. The knowledge and experience of employers, safety specialists and physicians in this field are insufficient. In addition, regulations on this subject have been prepared poorly and is not guiding.

Occupational health and safety trainings done in the workplaces in the corporate firm emphasis on safety culture but most of the medium and small businesses are insufficient on this subject. Education is seen as a legal obligation or does not given in many workplaces. Starting with the external OH & S units offering these services, especially in SME's educations has turned out to be "participation certificates of education" prepared and archived.

Regulations have been made on "employee participation" which is a significant deficiency in Turkey in recent years. According to the Occupational Health and Safety Act, "OH & S Representatives" have an important function in Turkey. Representatives have rights to inform the hazards associated with occupational safety to employer and government, and to demand preventive actions. But employers and managers are not giving enough importance to this issue. This topic is not inspected enough by labour inspectors.

According to the results obtained in our research, especially in construction industry, "Inadequacy of Maintenance" is also has a significant importance. In September, the elevator accident occurred in construction site which resulted in the death of 10 people is demonstrated that the importance of "maintenance services". Employers, site supervisors and engineers are not care enough on this issue. There are many regulations and hesitations about what will be implemented. The qualification of periodic inspection firms and duration of the periodic inspection of equipments is not clear. In many construction sites, periodic inspection and maintenance is neglected. In some cases it is still used even though it is known that the equipment is defective.

According to our research results, preventive actions that weight for both general and sectoral differ from each other. In Turkey, significant efforts have been made regarding the preventive OHS services and risk assessments in recent years. However, other necessary measures are not applied adequately such as labour inspection, maintenance, education, employee participation and workplace supervision. Occupational health and safety is a system of both enterprises and national level. If any stage of this system remains inadequate, it decreases the level of prevention of other measures.

## References

[1] International Labour Office (ILO) (2004) Labour Statistics. ILO Publication, Geneva.

[2] Rubio, M.C., Martinez, G. Rubio, J. and Ordoñez. J. (2008) Role of the Civil Engineer as a Coordinator of Safety and Health Matters within the Construction Sector. *Journal of Professional Issues in Engineering Education and Practice*, **134**, 152-157. http://dx.doi.org/10.1061/(ASCE)1052-3928(2008)134:2(152)

[3] Takala, J. (1999) Global Estimates of Fatal Occupational Accidents. *Epidemiology*, **10**, 640-646. http://dx.doi.org/10.1097/00001648-199909000-00034

[4] Turkey Social Security Institute (2000-2012) Statistical Yearbooks. Social Security Institute Publication, Turkey.

[5] Turkey Statistical Institute (2013) Occupational Accidents and Work Related Health Issues Research Report. Turkey Statistical Institute Publication, Turkey.

[6] Tilehnoei, M.H. and Aref, M.A. (2013) Temporal Dimension Evaluation by Fuzzy TOPSIS Method. *International Journal of Architecture and Urban Development*, **3**, 55-60.

[7] Özdemir, A.I. and Seçme, N. (2009) Analysis of Two-Stage Strategic Supplier Selection with Fuzzy TOPSIS Method. *Afyon Kocatepe Üniversitesi İ.İ.B.F. Dergisi*, **10**, 79-112.

[8] Başkaya, Z., and Öztürk, B.A. (2012) Group Decision Making with Fuzzy TOPSIS in Supplier Evaluation Problem and Relationships between Decision Maker's Individual Decisions. *Uludag Journal of Economy and Society*, **21**, 153-178.

[9] Santos, F.J.J. and Camargo, H.A. (2010) Fuzzy Systems for Multicriteria Decision Making. *CLEI Electronic Journal*, **13**, 1-8.

[10] Nang Idayu, N.Z. and Lazim, A. (2012) Evaluation of Sustainable Development Indicators with Fuzzy Topsis Based on Subjective and Objective Weights. *IIUM Engineering Journal*, **13**, 13-26.

[11] Mahdavi, I., Mahdavi-Amiri, N. Heidarzade, A. and Nourifar. R. (2008) Designing a Model of Fuzzy Topsis in Multiple Criteria Decision Making. *Applied Mathematics and Computation*, **206**, 607-617. http://dx.doi.org/10.1016/j.amc.2008.05.047

[12] Guozhong, Z., Neng, Z., Zhe, T., Ying, C. and Binhul, S. (2012) Application of a Trapezoidal Fuzzy AHP Method for Work Safety Evaluation and Early Warning Rating of Hot and Humid Environments. *Safety Science*, **50**, 228-239.

[13] Janackovic, G.L., Savic, S.M. and Stankovic, M.S. (2013) Selection and Ranking of Occupational Safety Indicators Based on Fuzzy-AHP: A Case Study in Road Construction Companies. *South African Journal of Industrial Engineering*, **24**, 175-189. http://dx.doi.org/10.7166/24-3-463

[14] Celik, M. (2010) Enhancement of Occupational Health and Safety Requirements in Chemical Tanker Operations: The Case of Cargo Explosion. *Safety Science*, **48**, 195-203. http://dx.doi.org/10.1016/j.ssci.2009.08.004

[15] Dağdeviren, M. and Yüksel, I. (2008) Developing a Fuzzy Analytic Hierarchy Process (AHP) Model for Behavior-Based Safety Management. *Information Sciences*, **178**, 1717-1733. http://dx.doi.org/10.1016/j.ins.2007.10.016

[16] Tadic, D., Djapan, M., Misita, M., Stefanovic, M. and Milanovic, D.D. (2012) A Fuzzy Model for Assessing Risk of Occupational Safety in the Processing Industry. *International Journal of Occupational Safety and Ergonomics*, **18**, 115-126. http://dx.doi.org/10.1080/10803548.2012.11076922

[17] Jiang, Z.-A. and Wang, P. (2012) Evaluation System and Method for Analyzing Occupational Hazards in Dust Workplace. *Proceedings of the International Conference on Computer Distributed Control and Intelligent Environmental Monitoring* (*CDCIEM*), Zhangjiajie, Hunan China, 5-6 March 2012, 301-304.

[18] Kim, D.I., Yoo, W.S. Cho, H. and Kang, K.I. (2014) A Fuzzy AHP-Based Decision Support Model for Quantifying Failure Risk of Excavation Work. *KSCE Journal of Civil Engineering*, **18**, 1966-1976. http://dx.doi.org/10.1007/s12205-014-0538-7

[19] Hatami-Marbini, A., Tavana, M., Moradi, M. and Kangi, F. (2013) A Fuzzy Group Electre Method for Safety and Health Assessment in Hazardous Waste Recycling Facilities. *Safety Science*, **51**, 414-426. http://dx.doi.org/10.1016/j.ssci.2012.08.015

[20] Huang, Y.F., Hsu, K.H. Chen, P.S. and Dong, S.H. (2011) Discussing Performance Index of Human Resource Valuation with AHP-Occupational Safety Section in T Company in Taiwan as the Case Study. *Information Technology Journal*, **10**, 549-556. http://dx.doi.org/10.3923/itj.2011.549.556

[21] Janaćković, G.L. (2013) Delphi-Fuzzy AHP Ranking of the Occupational Safety Community of Practice Performance Indicators. *Journal of Management and Marketing*, **1**, 9-16.

[22] Novin, V., Givehchi, S. and Hoveidi, H. (2014) A Novel Fuzzy-Based Modelling for Route Safety Management of Hazardous Materials. *International Journal of Scientific & Engineering Research*, **5**, 277-282.

[23] Sadoughi, S., Yarahmadi, R., Taghdisi, M.H. and Mehrabi, Y. (2012) Evaluating and Prioritizing of Performance Indicators of Health, Safety, and Environment Using Fuzzy TOPSIS. *African Journal of Business Management*, **6**, 2026-2033.

[24] Ataei, E. (2013) Application of TOPSIS and Fuzzy TOPSIS Methods for Plant Layout Design. *World Applied Sciences*

*Journal*, **24**, 908-913.

[25] Liao, C.N. and Kao, H.P. (2011) An Integrated Fuzzy TOPSIS and MCGP Approach to Supplier Selection in Supply Chain Management. *Expert Systems with Applications*, **38**, 10803-10811. http://dx.doi.org/10.1016/j.eswa.2011.02.031

[26] Arıoğlu, E. and Arıoğlu, N. (1997) Statistical Evaluation and Minimizing of Occupational Accidents in the Turkish Construction Sector. *Concrete-Prefabrication Magazine*, **2**, 16-21.

[27] Ercan, A. (2010) Assessment of the Occupational Health and Safety in Construction Sector in Turkey. *Journal of Polytechnic*, **13**, 49-53.

[28] Güyagüler, T. (2002) Analyses of the Firedamp Explosions in Turkey and Suggested Preventive Measures. *Proceedings of the* 13*th Turkish Coal Congress*, Zonguldak , 29-31 May 2002, 45-51.

[29] Chamber of Mining Engineers (2003) Closure of Authority of Coal Enterprises regional offices and Redevance Applications. *Mining Bulletin*, **3**, 61-66,

[30] Sari, M., Düzgün, H.S.B., Karpuz, C. and Selçuk, A.S. (2004) Accident Analysis of Two Turkish Underground Coal Mines. *Safety Science*, **42**, 675-690. http://dx.doi.org/10.1016/j.ssci.2003.11.002

# 21

# On the Design of Energy Absorbing Crash Buffers for High Speed Roadways

**Mohammad Sharif Uddin[1*], Jacob Quintel[1,2], Grad Zivkovic[2]**

[1]School of Engineering, University of South Australia, Mawson Lakes, Australia
[2]Automotive Safety Engineering Pty Ltd., Lonsdale, Australia
Email: [*]Mohammad.Uddin@unisa.edu.au

## Abstract

**The aim of this paper is to design and analyze cost-effective and high energy absorbing buffer systems for high speed roadways. Unlike conventional crash cushions, the proposed buffer design is based on the assembly of a series of cylindrical hollow tubes (cells) with thorough slots around the cells. The idea is that during the collisions, the kinetic energy of the errant vehicles will be absorbed by the progressive deformation of the cells, hence minimizing damage to the vehicle and allowing a comfortable ride down deceleration of the vehicle's occupants. As the cell was the fundamental unit of the buffer design, three cells with different geometry were studied to understand the underlying deformation of the individual cells. Nonlinear quasi-static tests using three-dimensional (3D) finite element (FE) simulation and experimental techniques were performed to evaluate the deformation and energy absorption capacity of the cells. Simulation results matched closely with experimental ones with relatively small errors. Based on the experimental results of single cells, a number of potential buffer systems were designed for 80 and 100 km/h speed roadways. Results indicate that the buffers with larger diameter cells are favorable to be used in high speed zones as they reduce the overall size of buffers and contain less number of cells, while being able to absorb the required amount of impact energy. Consequently, they are found to result in a reduced cost associated with materials and fabrication. All the buffer designs were relatively shorter than commercially available buffers used in roadways. In addition, due to their reduced and compact size, the designed buffers can potentially be used in a space limited and hazardous road environment to reduce the vehicle crash with the fixed objects.**

## Keywords

**Collisions, Fixed Objects, Crash Buffers, Cylindrical Hollow Cells, Finite Element Method**

---

[*]Corresponding author.

## 1. Introduction

Road crashes resulting in severe injuries and deaths are a growing concern across the world. For instance, road crashes in Australia cost about $15 billion annually, representing approximately 1.4% of Australia's gross national product [1]. Among them, a large number of road fatalities occur as a result of collisions with fixed objects on roadways. These collisions, often involving trees and poles, are particularly severe due to the narrow shape and rigidness of such objects. This has been an ongoing problem for a number of years. It has been reported that about 36% of the fatal road crashes in South Australia from 2000 to 2011 involved collisions with the fixed objects [2]. The problem is similar in other parts of the world. For example, in US, approximately 20% of motor vehicle crash deaths result from a vehicle leaving the roadway and hitting a fixed object alongside the road, in which, collisions with trees accounting for about 50% and about 13% with utility poles [3]. In Germany, this was about 38% [4]. An example of a collision between a vehicle and a fixed object (Stobie pole) is shown in **Figure 1(a)**, indicating that such collision is capable of causing the lethal and devastating impact into the vehicle structure and the vehicle occupants [5].

While improved vehicle and road design has decreased the overall fatality rate, a similar percentage of people are still being killed or injured in collisions with fixed objects every year [2] [6]. As a result of this, the health care burden and long-term severe personal suffering are becoming a growing concern [7]. Due to stringent regulation on road safety, various safety measures have been taken in the design of vehicles. For instance, along with revolution of air bags as safety cushion, the design of the car frame has considered potential safety aspects including innovative shock absorbing structure and material, which can protect vehicle occupants from potential collisions [8]. In this regard, common considerations are light weight, composite material along with unique geometric shapes and structures and their locations on the car frames, which will be able to absorb maximum impact energy through a progressive deformation pattern at a short period of time [9]-[11]. In addition, energy absorbing structures for ensuring safety of railway vehicles are studied by many researchers who emphasized the importance of safety and the innovation design of material and structure to prevent potential fatalities and damages [12].

While in-car safety measures, often called active safety, are able to withstand the collision impact, hence reducing the risk of injury or death to vehicle occupants, they are not a complete solution for an errant car colliding with roadside narrow objects such as tree or pole. It has been reported that among all collisions, head to head or frontal collision is the most common incidence on roadways, in which, a vehicle collides with other vehicle head to head or a vehicle front hits the fixed objects located on road sides. In the case of head to head crash, the frontal contact of two cars is relatively large and hence the deformation and the scale of damage to the car structure are low. On the other hand, in the case of collision between the car front and the rigid object, the contact between the front of a car and the narrow rigid object is too small, the deformation into the car structure is significantly large; in other words, the total impact energy is absorbed by the car itself rather than the rigid object. It must be noted that the impact is even significantly higher when the car side hits the rigid objects as the side of a car is more vulnerable than the front of the car as the side of a car contains a simple door structure as opposed to the front which houses an engine and other energy absorbing structures [13].

In order to minimize the risk of injury or causality due to collision with fixed objects, passive road safety measures such as road side barriers are employed across the world. The main aim of such a road safety system is to protect vehicle occupants from fixed roadside objects that cannot be removed, relocated or made breakaway [14]. In other words, the objective is to absorb most of the kinetic energy of an errant vehicle in a controlled manner, thus greatly reducing the severity of a crash [15].

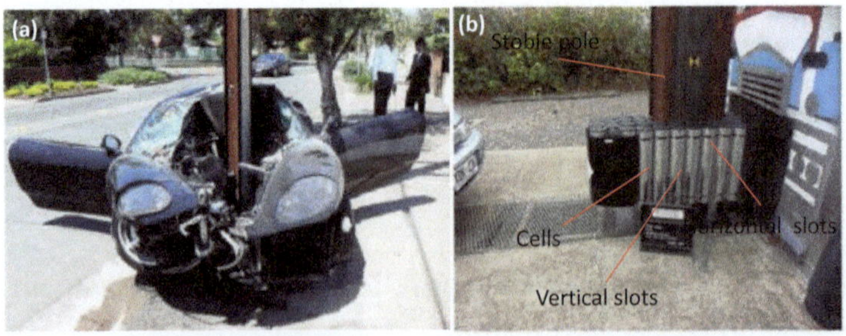


**Figure 1.** (a) Collision with fixed object (e.g. Stobie pole) [5]; (b) hollow tube based buffer system installed around Stobie pole [29].

In addition, as new materials and designs for energy absorbing structures are continually growing and evolving due to the strict regulation of safety aspects to be enforced in both car and road environment, their immediate assessment is crucial to understand the underlying effectiveness and for potential optimization. In this regard, experimental crash tests are often expensive and time consuming [16]. Therefore, finite element analysis, as a powerful computational tool, is a promising means and widely deployed in crash modelling to predict the performance of energy absorption of the safety cushions [17] [18]. In particular, the tool offers tremendous benefits in analyzing new candidate of material and design for extensive parametric studies [19].

This paper aims to design and analyze of a cost effective, compact and high energy absorbing safety buffers, which are capable of absorbing the impact energy of an out of control vehicle in high speed zones of 80 km/h and 100 km/h. The buffer is consisted of a series of metallic hollow tubes (often, called as cells) integrated together and installed around a fixed object. The key idea is most of the kinetic energy of the errant vehicle will be absorbed by the deformation of the individual cells, hence, minimizing impact onto the vehicle structure and protecting vehicle occupants from potential injuries or fatalities. Both finite element and experimental analyses were performed to determine the energy absorbing capability of an individual cell or tube, which led to design the geometric size of a buffer system for high speed zones (80 - 100 km/h). While the design of the cell based buffer system presented in this paper is shown for the frontal impact, the design can be modified into a shape which can be able to accommodate side impacts at any angles (e.g. 45°, 90°) with respect to the car axis.

## 2. Commercial Impact Attenuators

There are a variety of safety buffers, or often called, impact attenuators, available today for use on the roadside and each of which is designed to absorb the kinetic energy of an out of control vehicle, hence reducing casualties due to collisions [20]-[23]. They vary with the size, shape, and material and are deployed depending on the road environment, hazards, and road speed limit. An extensive review and assessment of a number of safety barriers as an acceptance to use in State of Queensland in Australia is described in [24]. To be noted are Rubber Speed Cushion [25], Universal TAU-II [26], QuadGuard [27], Absorb360 [28]. Rubber Speed Cushion made by the SafeRoads is shown to be an effective energy absorber to withstand the impact resulting from the car colliding with a speed up to 80 km and typically used in road break ways. However, they are relatively narrow and long and hence are not suitable for placement in front of tree or Stobie pole. Typical length of Rubber Speed Cushion is 5.44 m for 60 km/h and 8.29 m for 80 km/h. Universal TAU-II metallic crash cushion designed by ACP has a length of 3.41 m for 60 km/h, 5.15 m for 80 km/h, 7.75 m for 100 km/h and 9.48 m for 110 km/h. For the given speeds, the designs are significantly shorter in length than Rubber Speed Cushions. However, its complex design and need for a large concrete foundation for installation lead to higher costs. Quad Guard by Boylan is metallic and flexible structure, in which, the cells are compressed under the impact and hence absorbing the collision energy. It has a length of 3.08 m for 60 km/h and 4.91 m for 80 km/h zones. Similar to Universal TAU-II, this crash cushion is an effective energy absorber but still too long in length, costly, and complex for use for individual trees or poles. Absorb360 made by RMS (road management solutions) is a plastic segmented safety buffer and used for frontal impact. It is made with different dimensions and typically has a length of 3.7 m for 60 km/h areas, 6.7 m for 80 km/h, 9.7 m for 100 km/h and 11.7 m for 110 km/h areas. This crash cushion has a more simple design but it is far too long to be used in many locations and it is not suitable for side impact. Another type of crash cushion used as a road side barrier is water filled plastic barrier fitted with foam [17]. They are generally bulky, heavy and require significant effort to install and maintain depending on the locations of placement. In addition, they may not be environment friendly due to the unexpected leakage or spill of water from the cushion system in the event of collision.

While ideally each of the above safety buffers is, to some extent, effective energy absorbing devices for protection from fixed objects, they are continuous, indiscrete and non-flexible, making them only suitable for open roads. They may not be applied to the closed roads, in which, fixed objects such as trees, poles, commonly exist. Most of them are only designed to withstand the frontal collisions, which may not be the case always in real practice; in other words, they cannot withstand any side impact. Furthermore, due to their complex shape and large size, the fabrication, installation and maintenance cost of these devices is relatively high. Therefore, it is important to design and develop a compact, shorter, flexible, and cheaper safety buffer for road side fixed objects, thus reducing the hazards along the road network and creating a more forgiving roadside environment.

## 3. Energy Absorbing Pole/Tree Buffers

An important aspect when selecting possible sites for a buffer system is to ensure that the safety system itself doesn't pose a threat to road users, particularly, when the protecting system is located much closer to traffic than the fixed object. Therefore, a key principal in eliminating hazards posed to vehicles by fixed objects is, developing an effective process to identify objects in high-crash locations and an effective set of criteria to select the best option to deal with these objects [14]. Design of such buffer system must ensure that, in the event of a collision, no debris from the system or the vehicle causes a hazard to other vehicles. Energy absorbing pole/tree buffer is first introduced by [29], in which, the buffer contains a series of cylindrical hollow tubes assembled together by the bolt-nut fasteners. The configuration of a buffer fixed with a Stobie pole is shown in **Figure 1(b)**. Each cell contains vertical and horizontal slots at the top and bottom. Vertical slots allow more flexibility within the cells to deform. Horizontal cells are to accommodate the bolts and nuts that fix the cells together. The buffer systems are attached to the fixed objects on the streets. The cells can be arranged in a variety of ways depending on the fixed objects on the roadways.

The current buffer design [29] as described earlier is only suitable for low-to-medium speed zone, e.g. 60 km/h, and this has been tested successfully. However, they may not be applied in high speed zones, e.g. 80 km/h or 100 km/h. Under such high speeds, the impact of an errant vehicle against a fixed object is expected to be significantly high, and the buffer system needs to absorb a large amount of impact energy to eliminate potential injury to the vehicle's occupants. Hence, a design modification is urgently required to develop an improved, sustained, compact and cost effective buffer that will be suited to faster speed zones and this is of particular interest and the focus of the current paper.

## 4. Rationale for Buffer Design for High Speed Zones

Design of a buffer system involves a number of important geometric parameters. They include length, width, cell geometry (e.g. diameter, thickness). Generally, at the impact, the energy is absorbed through the elasto-plastic deformation of each individual cell within the buffer system. Given that the total amount of impact energy to be absorbed at a certain stopping distance is known, it is possible to determine the size of the buffer system by modifying important geometric parameters of the individual cells.

After the impact, the stopping distance of the vehicle is crucial to determine how quickly or smoothly the vehicle comes to rest position without posing any threat to the occupants. The larger the stopping distance, the more smooth and gradual absorption of energy and the less impact to the vehicle's occupants. In other words, the larger stopping distance allows more deformation into the buffer system until the vehicle stops. One way to allow for greater stopping distance is to increase the size of the buffer. The significance of this can be clearly observed by Equation (1), which shows the average deceleration ($a$) is inversely proportional to the stopping distance ($h$).

$$a = v^2/h \tag{1}$$

where, $v$ is the velocity of vehicle. The increase in buffer length can be achieved by the addition of more cells in the buffer. However, the larger length of the buffer means that many cells need to be used, which may increase the fabrication cost, setup and installation time. An alternative to adding more cells to increase the length is increasing the size of the individual cells, *i.e.* increasing the diameter of the cells. Increasing the size of the individual buffer cells could potentially lead to several significant benefits. These include more compact (shorter) buffer, greater ease of assembly of the cells and allow much greater accessibility. This is particularly true for the buffers with higher speed zones, where more cells are needed. The cost associated with material and fabrication will be dramatically reduced as less number of cells is required.

## 5. Materials and Method

### 5.1. Modified Cell Design

This paper focuses on design of modified cell geometry over the previous cell design by Zivkovic and Winfield [30], with an aim to (1) reduce the number of cells, (2) increase energy absorption capacity and (3) reduce manufacturing and handling cost. The previous cell design contains 114.3 mm outside diameter with a wall thickness of 4.5 mm. If four of these cells are attached in series to form a buffer, it will give a width of 457.2 mm. **Figure 2** shows the geometric definition of a cell. It would be preferable to replace these four cells with either two or

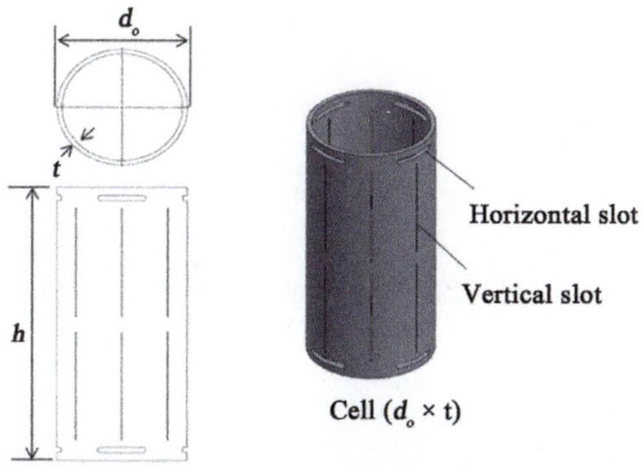


**Figure 2.** Geometric definition of a buffer cell.

three relatively larger cells, which would result in the overall width of the buffer remaining the same or similar. Note that the width of a buffer is defined as the product of the number of cells and the outer diameter of a cell.

With this aim, three different cell geometries with larger outer diameters (Cells A, B, and C) are considered in this study as shown in **Table 1**. Outer diameter of Cells B and C is the same except for thickness. For instance, three-cell assembly made with Cell A design would give a width of 495.3 mm, while two-cell assembly made with Cell B or C design would give a width of 438.2 mm. The width of both assemblies is close to that of the previous design (of foul-cell assembly) and would be considered acceptable size.

As can be seen in **Figure 2**, in the modified cell design, the location of horizontal slots and thus the length of vertical slots were modified to allow more stiffness at the fixation points between the cells. It must be noted that although it is likely that other geometry of cell would be preferred for the buffer, only geometries available from the local supplier were considered, for the sake of the justification of the proposed design modification. The performance of these three geometries was tested using both finite element simulations and experiments to find which cell geometry provides a cost effective and high energy absorbing buffer system for high speed zones.

## 5.2. Finite Element Modelling

A nonlinear quasi-static (*i.e.* implicit) simulation using finite element (FE) method was conducted in order to gain a greater understanding of how the individual cells deform and how much force would be needed to initiate the deformation. This, in turn, enabled to estimate the energy absorption or workdone capacity of the cells under the given impact load. For the sake of simplicity and being computationally effective, finite element analysis was initially applied on single cells. The simulation setup was designed in such a way that it could be replicated by a real experiment for future comparison. **Figure 3(a)** shows the finite element model of a cell.

The FE model consisted of a cell placed between a top and a bottom plate, the bottom plate was fixed while the top plate was allowed to move only in the vertical direction. The top plate was assigned a vertical displacement that would force the cell to be compressed laterally. This would replicate a push plate attached to a hydraulic cylinder during real life experimentation (see details in "Experiments" section). A reaction force probe was applied to the bottom surface of the top plate to estimate the force against the displacement of the cell.

ANSYS, a commercially available finite element code [31], was applied to solve the contact problem between the cell and the top and bottom plates. As the surface model of the cell was imported into the ANSYS platform, 4-node shell elements were used to mesh the simulation model. They were found to allow large strain deformation. Solid elements were also studied. However, both solid and shell elements had insignificant influence on the deformation. Interestingly, shell elements were shown to reduce the computational time, as opposed to 8-node solid elements. A mesh convergence study was performed to investigate the effect of mesh size. Mesh size was varied and after simulation, the change in deformation and von Mises stress of the cell were observed. Fine meshing was considered at the critical locations such as slots while course meshing was used at other areas. An optimum mesh size was found, for which, the percentage of error was 1% - 5%, which was considered an acceptable limit for engineering analysis. Mesh convergence study was repeated for all cell geometries as well.

The material of the cells studied was mild steel. The data for multi-linear isotropic hardening stress-strain curve for the material was obtained from experimental study [30]. **Figure 3(b)** shows the stress-strain data curve

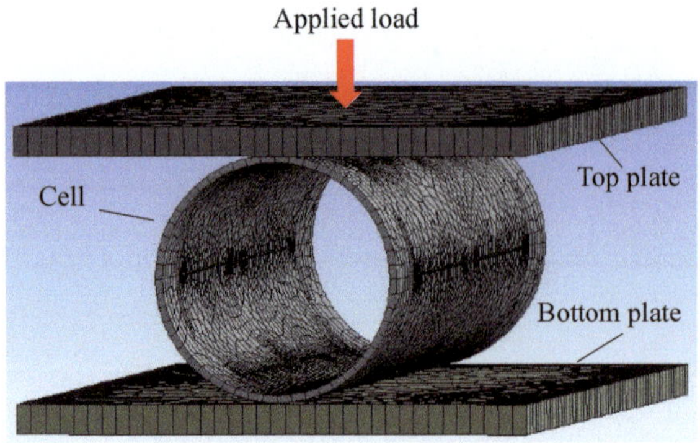


(a)

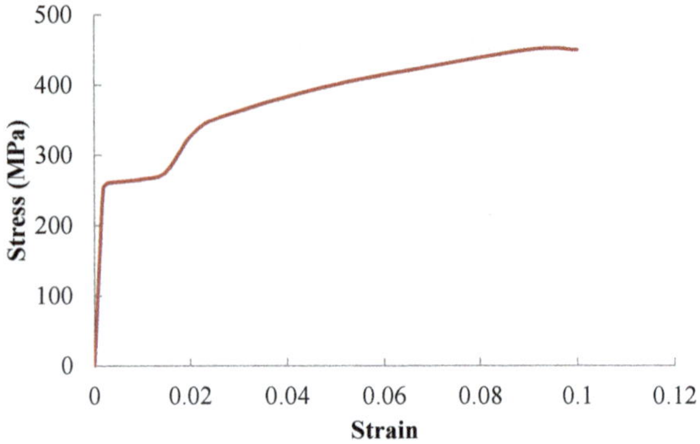


(b)

**Figure 3.** (a) FE model of a cell (b) stress-strain data of cell material (mild steel) used in FE analysis.

**Table 1.** Geometric dimension of the cells studied.

| Cell geometry | Outer diameter, $d_o$ (mm) | Wall thickness, $t$ (mm) | Cell height, $h$ (mm) |
|---|---|---|---|
| Cell A | 165.1 | 5.0 | 510 |
| Cell B | 219.1 | 6.4 | 510 |
| Cell C | 219.1 | 8.2 | 510 |

for mild steel, which is feed into the material library of the software during simulation. The effect of frictional contact between the cell and the top and bottom plates was studied. It was found that that friction had a negligible effect on the deformation pattern and magnitude of the cell. Hence, the contact was assigned as frictionless throughout simulation. This was also found to reduce computational time.

By following the same loading and boundary conditions used for the single cell geometry, the simulation for two-cell and four-cell assembly was performed to investigate the deformation behaviour of a buffer-like system comprising a series of cells. In this case, two cells or four cells were placed in between the top and the bottom plates in the simulation environment. The spot weld contact was assumed at connection between two cells to represent the actual bolt joint connection of the cells. Initial simulation results showed that the magnitude of deformation and energy absorption of single cell geometry had a linear relationship with that of two-cell and four-cell assemblies and for the limitation of space, the modelling details and the simulation results for two-cell and four-cell are not presented here. Therefore, the result and analysis presented in the paper were based on the deformation of a single cell. After each simulation, load and displacement data were estimated. This was used to calculate the amount of workdone and energy absorbed by the cells. The results were plotted and analysed to design an optimum buffer system.

## 5.3. Experiments

Experiments involving lateral compression were conducted in order to investigate the deformation of the individual cells. The aim was to validate the quasi-static FEA simulations. The tests involved the deformation of a cell located between two plates at top and bottom. The bottom plate was fixed and the top plate (push plate) was attached to a 20-tone hydraulic cylinder so that the cell was deformed in a controlled way that would enable the reaction force acting on the top plate to be recorded for each of the desired displacements. **Figure 4** shows the experimental setup for the cell deformation. At a certain interval of displacement, applied loads and displacements were measured by using a hydraulic pressure gauge and a linear encoder, respectively. During the tests, the deformation of the cell was allowed to 75% of its total deformation capacity. Note that the hydraulic pressure gauge was calibrated with a commercially available standard load cell to justify the accuracy of the load readings. Load-displacement and energy absorption plots were created for each test and used to compare with the corresponding plots from the FE simulations.

## 5.4. Buffer Designs

Based on the experimental and simulation results and data obtained for individual cell geometries, potential buffers were designed. The main focus is to determine the amount of energy that can be absorbed by the different cell. Energy absorption data for the individual cells obtained from the experiments were used to design the size of the buffer system as they were assumed to be the most reliable. Note that the design of the buffer presented in this study is assumed for the frontal collision; in other words, the front of an errant vehicle will hit front of the buffer.

The idea is that by knowing the amount of kinetic energy that a vehicle will have in any speed zone and the amount of the energy a cell can absorb, one can determine the number of cells that will be arranged in series to form the buffer system that will be able to absorb the desired amount of the energy due to the impact. It is reported that major of road crashes involve the front of the car hitting other objects such as the front of another cars or fixed roadside objects [32]. Therefore, in this study, the buffer system is designed assuming that the energy absorption or the deformation of the cells would be due to the frontal impact between a car and a fixed object.

It is assumed that during an impact, half of the total kinetic energy of the vehicle is absorbed by the vehicle itself. Hence, the aim of the buffer system is to be able to absorb the rest half of total kinetic energy through the deformation of the individual cells. Assume that a vehicle of mass, $m_{vehicle}$ travelling at a velocity of $v$ hits a fixed object. The total kinetic energy of the vehicle can be estimated as:

$$K.E._{total} = \frac{1}{2} m_{vehicle} * v^2 . \qquad (2)$$

Hence, the kinetic energy to be absorbed by the buffer system can be calculated as:

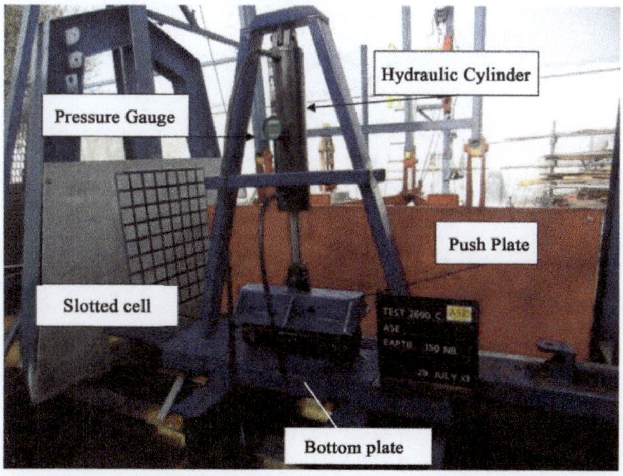


**Figure 4.** Experimental setup for the cell deformation.

$$K.E._{buffer} = \frac{1}{2} K.E._{total} . \tag{3}$$

Let's assume that the buffer system comprises a series of rows of the cells attached to together. The kinetic energy is absorbed through the continuous deformation of the cells. **Figure 5** shows an example arrangement of cells in a buffer design. If the energy absorption capacity of an individual cell at its 75% deformation, $K.E._{cell}$ and the number of cells in a row, *m*, are known, the number of rows, $N$ that will form the buffer system can be estimated as:

$$N = \frac{K.E._{buffer}}{K.E._{cell}} * m . \tag{4}$$

Accordingly, for the given cell geometry (*i.e.* outer diameter of a cell, $d_o$), the size of the buffer system can be estimated as:

$$\text{The width of the buffer, } W = (\text{number of cells in a row}) * (\text{outer diameter of a cell}) = m * d_o \tag{6}$$

$$\text{The length of the buffer, } L = (\text{number of rows in a buffer}) * (\text{outer diameter of a cell}) = N * d_o \tag{7}$$

This way, one can determine the size of the buffer system that will be able to absorb the impact energy of an out-of-control vehicle travelling at any speed.

# 6. Results and Discussion

## 6.1. Comparison between Simulation and Experiment

**Figure 6** shows a comparison of cell deformation patterns obtained from experiment and simulation. Both experiment and simulation display exactly the same trend of the deformation until the cell reaches its maximum deformation level. The cell undergoes both elastic and plastic deformations, except for a broken weld spot observed at the contact between the cell and the bottom plate. Damage at the weld can be due to improper welding or unexpected lateral movement of the cell under experimental loading. Overall, the results for a single cell justify the accuracy of the current FE model, thus indicating that the model would be able to reasonably estimate the accurate deformation of a buffer system consisting of a series of individual cells.

**Figure 7** illustrates a comparison of load-displacement and energy absorption patterns obtained from experiment and simulation for all three cells. The amount of energy absorption by the cell or work done by the top

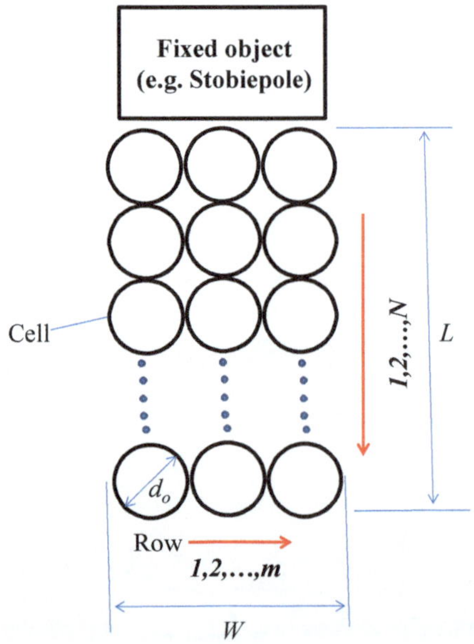


**Figure 5.** Illustration of parameters of a potential buffer design.

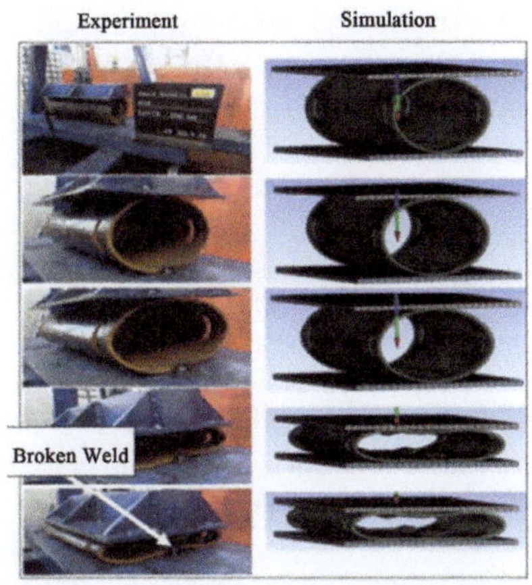


Figure 6. Comparison of simulated and experimental deformation of an individual cell.

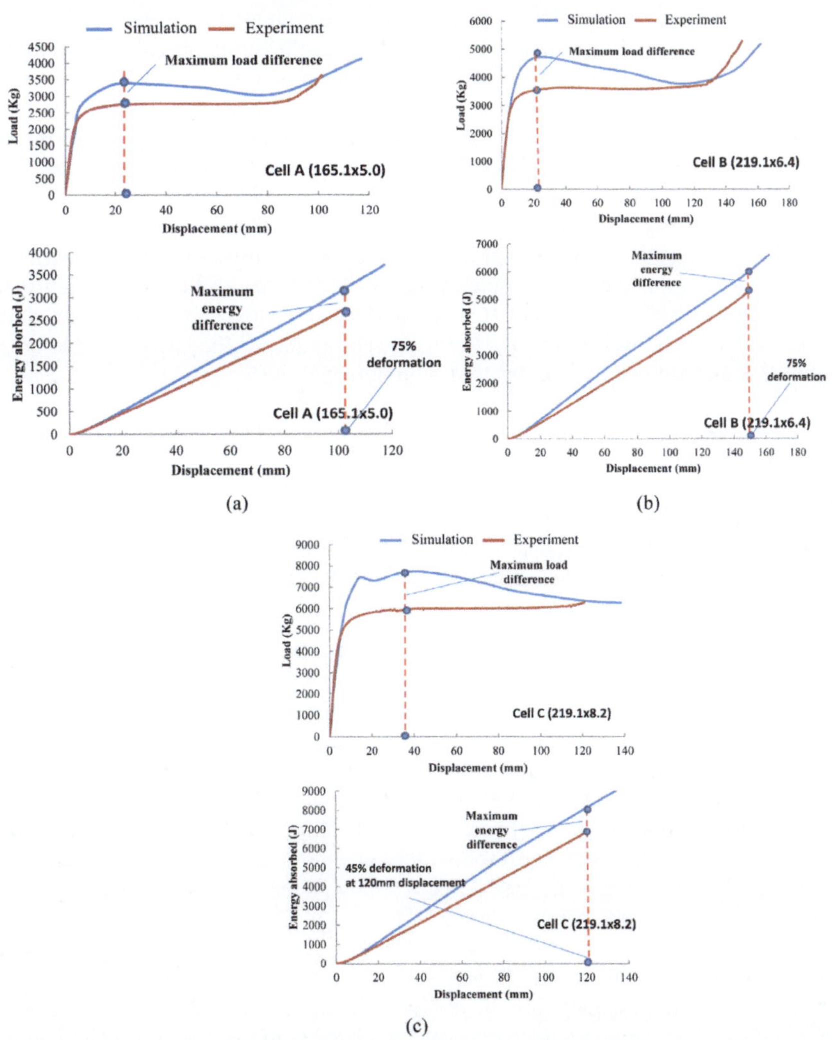


**Figure 7.** Load-displacement(top) and energy absorption (bottom) results for (a) Cell A, (b) Cell B, (c) Cell C.

(push) plate was estimated by integrating area under the load-displacement curve. The energy was estimated and plotted until 75% of the cell's maximum deformation level. Results show that the overall trend of the load-displacement and energy absorption is relatively similar. At early stage, the magnitude of experimental and simulation load to deform the cell is almost same. However, after initial elastic deformation of about 5 - 10 mm, simulation load is higher than experimental, and this pretty much remains constant until the cell reaches its maximum deformation. On the other hand, the cumulative energy absorption capacity increases linearly with the deformation of the cell.

For instance, as can be seen in **Figure 7(a)**, for Cell A, the maximum error between experimental and simulated load was found to be 17.70% at about 22 mm displacement. The maximum error in energy absorption obtained from experiment and simulation was estimated to be 22.97% at 75% of the cell's maximum deformation (at about 113 mm displacement). As shown in **Figure 7(b)**, for Cell B, the maximum error between experimental and simulated load was found to be 24.35% at about 24.75mm displacement. The maximum error in energy absorption obtained from experiment and simulation was estimated to be 11.4% at 75% of the cell's maximum deformation (at about 151 mm displacement). For Cell C, the maximum error between experimental and simulated load was found to be 30% at about 36.11 mm displacement, as can be seen in **Figure 7(c)**. The maximum error in energy absorption obtained from experiment and simulation was estimated to be 15.72% at 45% of the cell's maximum deformation, as the simulation was stopped at about 120 mm displacement.

The difference in the cell strength obtained from experiment and simulation is believed to be the results of pre-stressed state of the cell used in experiments caused by the manufacturing processes. Material properties assigned to the cell in the FE simulation model may not exactly match with the actual mild steel properties of the cell itself. In addition, the contact between the cell and the top and bottom plates was assumed to be an ideal weld joint in simulation. However, a broken weld spot was observed during experiments (see **Figure 6**).

**Figure 8** shows the comparison of load-displacement and energy-displacement among three cell geometries. It can be seen that Cell C has the highest strength among all. This means that one needs a larger amount of load to deform the cell so that the energy is absorbed. This further indicates that Cell C is too hard and rigid. This is generally not desired in an energy absorbing buffer system as the main objective of a buffer is to absorb the impact energy through deformation of the cells as quickly as possible so that a smooth and comfortable ride down deceleration is realized by the vehicle's occupant. On the other hand, Cell A and Cell B show almost the similar level of deformation and indicate more flexibility and are ready to absorb the energy. Therefore, in this study, we use Cell A and Cell B as base cell geometries to design potential buffer systems, which are discussed in the following section.

## 6.2. Buffer Configurations

### 6.2.1. Buffer Designs for 80 km/h Speed

We assume that the mass of a vehicle, $m_{vehicle}$ of 1600 kg [30], travelling at a speed, $v$ of 80 km/h. Using Equation (2) and Equation (3), total kinetic energy of the vehicle is 395 kJ, and the energy to be absorbed by the buffer system would become 197.5 kJ. Given that the amount of energy a cell can absorb, $K.E._{buffer}$ and the number of cells in a row, $m$, the number of rows, $N$ and the number of cells, $m_{total}$ that will form the size of the buffer can be estimated using Equations (4)-(6). The main aim is to design possible buffer sizes that will be relatively shorter in length as compared to the commercially available buffer systems, and be able to use in different roadways to absorb the required energy for the vehicle travelling at 80 km/h,.

**Table 2** summarizes the result of potential buffer designs for 80 km/h speed zone. For the given cell geometry, we vary the number of cells in a row to find different size of the buffer. **Figure 9(a)** shows the configuration of the designed buffers. **Figure 9(b)** shows load-displacement and energy absorption of the buffers at 75% of their maximum capacity, assuming that the buffer will follow the same deformation of a single cell. The deformation and energy absorption of the buffer was estimated by performing an extrapolation on experimental deformation data for a single cell. It is seen that the increase in the cell size and the number of cells in a row increases the width and decreases the length of the buffer. This similarly influences the load-displacement and energy absorption data curves. For example, as can be seen in **Figure 9**, the buffers with Cell A, $m = 4$ and Cell B, $m = 3$ show almost the same energy absorption capacity, but the buffer with larger cell, Cell B, $m = 3$ contains significantly less number of cells. This will potentially reduce the cost associated with the materials and manufacturing of the cells. This further indicates that the buffer with relatively large cells is cost effective, while it withstands high impact energy.

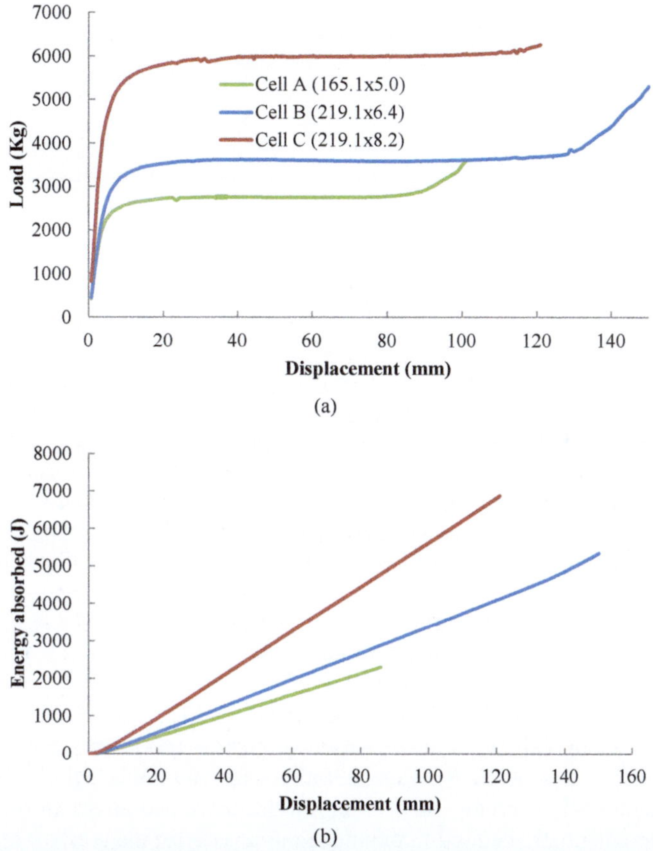


**Figure 8.** Comparison of (a) load-displacement, (b) energy absorption results for Cells A, B, and C.

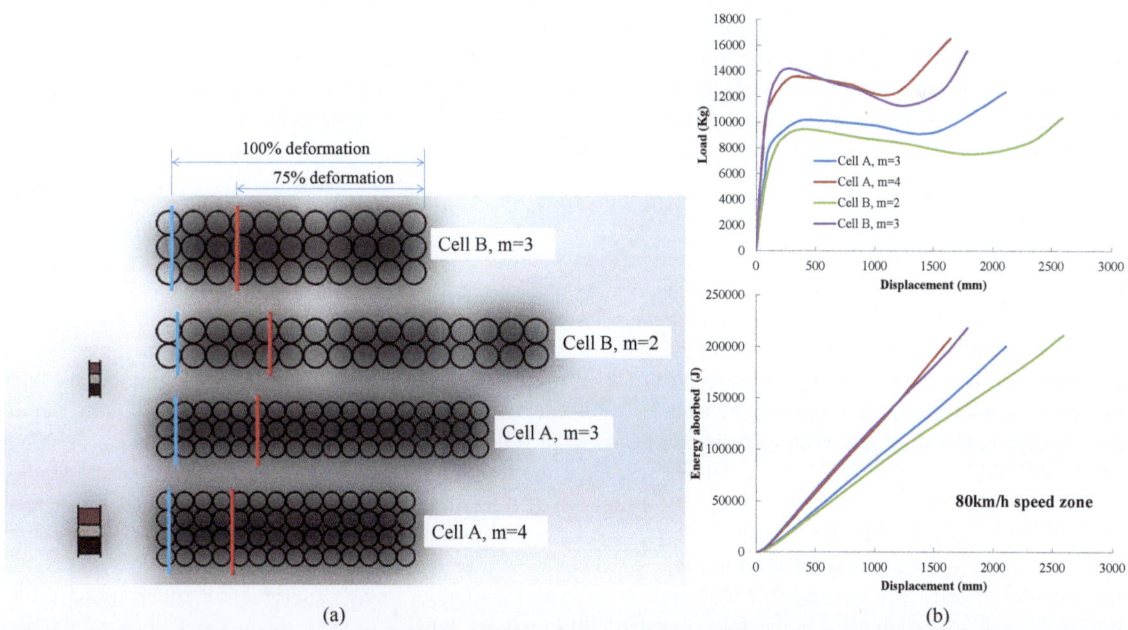


**Figure 9.** (a) Buffer designs, (b) load and energy absorption results for corresponding designs for 80 km/h speed.

**Table 2.** Buffer designs for 80 km/h speed.

| Buffer design | Energy absorbed by buffer, $K.E._{buffer}$ (kJ) | Energy absorbed by buffer, $K.E._{cell}$ (kJ) | No of rows, $N$ | No of cells in a buffer, $m_{total}$ | Width of buffer, $W$ (mm) | Length of buffer, $L$ (mm) |
|---|---|---|---|---|---|---|
| Cell A, $m = 3$ | 197.5 | 3.7 | 18 | 54 | 495.3 | 2971 |
| Cell A, $m = 4$ | 197.5 | 3.7 | 14 | 56 | 660.4 | 2311 |
| Cell B, $m = 2$ | 197.5 | 6.25 | 16 | 32 | 438.2 | 3505 |
| Cell B, $m = 3$ | 197.5 | 6.25 | 11 | 33 | 657.3 | 2401 |

For all the buffer designs, the width and the length were found to be quite less than 1000 mm and 4000 mm, respectively, which are acceptable. The sizes of the buffers were significantly smaller than the available buffer sizes of Quad Guard, e.g. 4910 mm - 8290 mm in length for speed zone of 80 km/h [23].

### 6.2.2. Buffer Designs for 100 km/h Speed

Following the same procedure for 80 km/h speed as described in earlier section, potential buffers for 100 km/h speed zone have been designed. **Table 3** summarizes the results of buffer design for 100 km/h. The number of cells in a row (m) was varied to find the acceptable sizes of the buffers that will absorb the required impact energy. **Figure 10(a)** shows configurations of potential buffers. **Figure 10(b)** illustrates the extrapolated load-displacement and energy absorption of the buffers at 75% of their maximum capacity.

It can be observed from figures that the arrangement of the cells influences the size, deformation, and energy absorption capacity of the buffers in a similar fashion of those for 80 km/h speed. Larger cells with a large number of cells in a row increase energy absorption and decrease the size of the buffers. This also holds the benefit of reduced number of cells in a buffer, which results in significant cost reduction. For all buffer configurations for 100 km/h, the width and length of the buffers are below 1000 mm and 6000 mm, respectively. The sizes are lower than those of commercially available buffers for 100 km/h, e.g. Absorb360 Crash Cushion of 7750 - 9700 mm in length [24]. Therefore, depending on the space and environment of roadways, these proposed buffer designs can potentially be used to minimize the severity of the crash of high speed vehicles hitting the fixed objects.

There are a number of limitations that may hinder the performance of the FE model, and hence, impact the results and analysis. In this study, a quasi-static single cell deformation was tested using both simulation and experiment. However, dynamic experimental crash testing is required to evaluate and justify the deformation and energy absorption capacity of the buffers. In accordance with Manual for Assessing Safety Hardware proposed in [33], occupant risk value and occupant ride-down acceleration ($g$-force) would be estimated to assess the potential injury impact onto vehicle occupants. Uniform deformation of the cells under the impact is important to maximize the energy absorption. The slots on the cells provide the flexibility and enable the deformation. Therefore, size, location and orientation of the circular and vertical slots on the cells may need to be optimized so that the required energy is absorbed by the buffers with a comfortable stopping distance of the vehicle after the impact. This will be addressed in our future work.

## 7. Concluding Remarks

This paper studied design and analysis of cost-effective and high energy absorbing buffer systems for high speed roadways. Deformation and energy absorption capacity of the single cell with different diameters and thicknesses were evaluated using quasi-static tests by finite element simulation and experimental technique. Simulation results matched closely with experimental ones with relatively small errors, suggesting that the FE model presented in this paper was able to accurately estimate the deformation of cells under the quasi-static loading. Based on experimental results of single cells, a number of potential buffer systems were designed for 80 km/h and 100 km/h speed roadways. Results indicate that the buffers with larger diameter cells, e.g. Cell B, $m = 3$, Cell A, $m = 5$ are favorable to be used in high speed zones as they reduce the overall size of buffers and contain less number of cells, while being able to absorb the required amount of impact energy at the given speed. All the buffer designs are relatively shorter than commercially available buffers used in roadways. Consequently, they will potentially result in a reduced cost associated with materials, fabrication, installation and maintenance. In

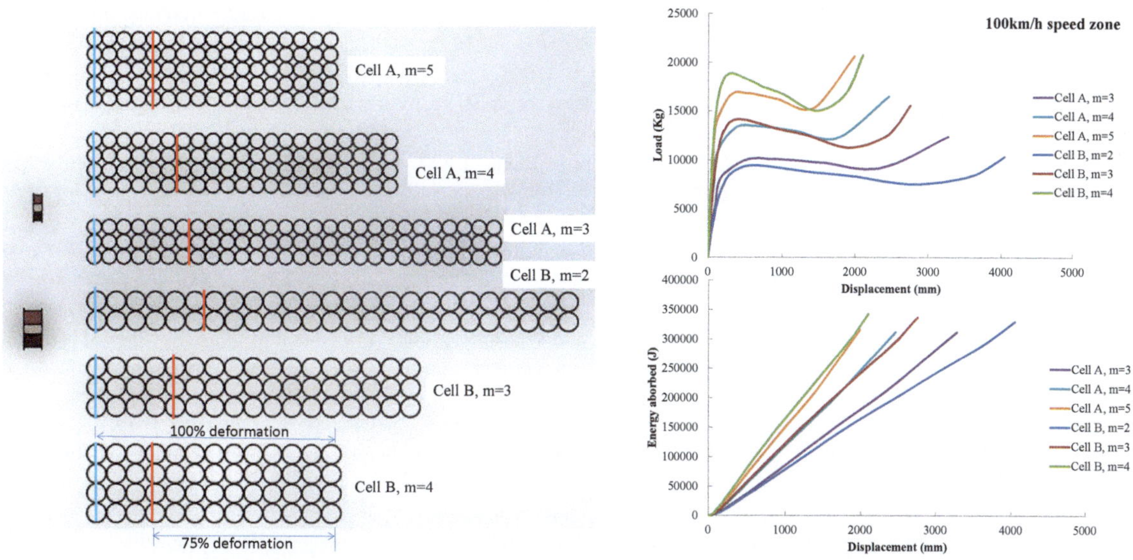


(a) (b)

**Figure 10.** (a) Buffer designs, (b) load and energy absorption results for corresponding designs for 100 km/h speed.

**Table 3.** Buffer designs for 100 km/h speed.

| Buffer design | Energy absorbed by buffer, $K.E._{buffer}$ (kJ) | Energy absorbed by buffer, $K.E._{cell}$ (kJ) | No of rows, $N$ | No of cells in a buffer, $m_{total}$ | Width of buffer, $W$ (mm) | Length of buffer, $L$ (mm) |
|---|---|---|---|---|---|---|
| Cell A, $m = 3$ | 308.6 | 3.7 | 28 | 84 | 495.3 | 4623 |
| Cell A, $m = 4$ | 308.6 | 3.7 | 21 | 84 | 660.4 | 3467 |
| Cell A, $m = 5$ | 308.6 | 3.7 | 17 | 85 | 825.5 | 2807 |
| Cell B, $m = 2$ | 308.6 | 6.25 | 25 | 50 | 438.2 | 3505 |
| Cell B, $m = 3$ | 308.6 | 6.25 | 17 | 51 | 657.3 | 2401 |
| Cell B, $m = 4$ | 308.6 | 6.25 | 13 | 52 | 876.4 | 3725 |

addition, depending on the nature and available space of the roadways, the designed buffers can potentially be installed in a space limited hazardous road environment to reduce the vehicle crash with the fixed objects.

## References

[1] BITRE (2000) Cost of Road Crashes in Australia. Bureau of Infrastructure, Transport and Regional Economics, Canbera, Australia.

[2] BITRE (2011) Cost of Road Crashes in Australia. Bureau of Infrastructure, Transport and Regional Economics, Canbera, Australia.

[3] IIHS (2011) Fatality Facts: Collisions with Fixed Objects and Animals. Insurance Institute for Highway Safety.

[4] Berg, A. and Ahlgrimm, J. (2010) Tree Impacts—Still One of the Most Important Focal Points of Road Deaths. Berichte Bundesanst, Fuer Strassenwes, Unterreihe Fahrzeugtechnik, No. 77.

[5] Zivkovic G. (2008) Energy Absorbing Pole/Tree Buffer. Automotive Safety Engineering Pty Ltd., Adelaide, Australia.

[6] Lake, G. (2010) Local Government and Road Safety. *Journal of the Australasian College of Road Safety*, **21**, 17-19.

[7] Wang, Y.G., Chen, K.M., Ci, Y.S. and Hu, L.W. (2011) Safety Performance Audit for Roadside and Median Barriers Using Freeway Crash Records: Case Study in Jiangxi, China. *Scientia Iranica*, **18**, 1222-1230. http://dx.doi.org/10.1016/j.scient.2011.11.020

[8] Lu, G. and Yu, T.X. (2003) Energy Absorption of Structures and Materials. Woodhead Publishing Limited, Cambridge, UK.

[9] Njuguna, J. (2011) The Application of Energy Absorbing Structures on Side Impact Protection Systems. *International Journal of Computer Applications in Technology*, **40**, 208-207. http://dx.doi.org/10.1504/IJCAT.2011.041657

[10] Baker, B.C., Nolan, J.M., O'Neill, B. and Genetos, A.P. (2008) Crash Compatibility between Cars and Light Trucks: Benefits of Lowering Front-End Energy-Absorbing Structure in SUVs and Pickups. *Accident Analysis & Prevention*, **40**, 116-125. http://dx.doi.org/10.1016/j.aap.2007.04.008

[11] Tay, Y.Y., Lim, C.S. and Lankarani, H.M. (2014) A Finite Element Analysis of High-Energy Absorption Cellular Materials in Enhancing Passive Safety of Road Vehicles in Side-Impact Accidents. *International Journal of Crashworthiness*, **19**, 288-300. http://dx.doi.org/10.1080/13588265.2014.893789

[12] Xie, S., Liang, X. and Zhou, H. (2015) Design and Analysis of a Composite Energy-Absorbing Structure for Use on Railway Vehicles. *Proceedings of the Institution of Mechanical Engineers, Part F: Journal of Rail and Rapid Transit*, 0954409714566058.

[13] Otte, D., Sferco, R., Schaefer, R., Eis, V., Thomas, P. and Welsh, R. (2009) Assessment of Injury Severity of Nearside Occupants in Pole Impacts to Side of Passenger Cars in European Traffic Accidents—Analysis of German and UK In-Depth Data.

[14] Lacy, K., Srinivasan, R., Zegeer, C., Pfefer, R., Neuman, T., Slack, K. and Hardy, K. (2004) NCHRP Report 500: Guidance for Implementation of the AASHTO Strategic Highway Safety Plan—Volume 8: A Guide for Reducing Collisions Involving Utility Poles. Transportation Research Board of the National Academies, Washington DC.

[15] Elvik, R. (1995) The Safety Value of Guardrails and Crash Cushions: A Meta-Analysis of Evidence from Evaluation Studies. *Accident Analysis & Prevention*, **27**, 523-549. http://dx.doi.org/10.1016/0001-4575(95)00003-I

[16] Pang, T.Y. and Tristian, H. (2014) Experimental and Finite Element Nonlinear Dynamics Analysis of Formula SAE Impact Attenuator. In: Jazar, R.N. and Dai, L., Eds., *Nonlinear Approaches in Engineering Applications* 2, Springer, New York, 213-229. http://dx.doi.org/10.1007/978-1-4614-6877-6_7

[17] Thiyahuddin, M.I., Gu, Y.T., Thambiratnam, D.P. and Thilakarathna, H.M. (2014) Impact and Energy Absorption of Portable Water-Filled Road Safety Barrier System Fitted with Foam. *International Journal of Impact Engineering*, **72**, 26-39. http://dx.doi.org/10.1016/j.ijimpeng.2014.04.008

[18] Roberts, M. (2014) Development of a Finite Element Model for Predicting the Impact Energy Absorbing Performance of a Composite Structure. Masters Theses Project Reports, California Polytechnic State University, San Luis Obispo, CA, USA.

[19] Boria, S. and Pettinari, S. (2014) Mathematical Design of Electric Vehicle Impact Attenuators: Metallic vs Composite Material. *Composite Structures*, **115**, 51-59. http://dx.doi.org/10.1016/j.compstruct.2014.04.010

[20] Wilson, A.H. (1979) Roadside Tree/Pole Crash Barrier Field Test. Jet Propulsion Laboratory, California Institute of Technology, Pasadena.

[21] Krage, W.G. and Denman, O.S. (1988) Collapsible Highway Barrier. US4784515 A.

[22] Nabors, D.T. and Hossain, M. (1998) Development of a Low-Cost Crash Cushion Using Recycled Automobile Tires.

[23] Michalec, C.R. (2005) Development of an Optimal Impact Energy Absorber for Highway Crash Cushions. Texas A&M University, College Station.

[24] Whitehead, M. (2014) Road Safety Barrier Systems, End Treatments and Other Related Road Safety Devices. Department of Transport and Main Roads, Queensland.

[25] Saferoads (2015) Rubber Speed Crash Cushion. Saferoads, Australia.

[26] Australian Construction Product (2015) Universal TAU-II Crash Cushion. Australian Construction Product (ACP), Australia.

[27] Bolyan (2015) Quadguard Cushion System. Bolyah Group, Australia.

[28] Road Management Solutions (2015) Absorb 360 Crash Cushion. Road Management Solutions, Australia.

[29] Zivkovic, G. and Day, J. (2012) Safety Barrier Product Information Report. Automotive Safety Engineering Pty Ltd., Adelaide.

[30] Zivkovic, G. and Winfield, M. (2010) A Study into the Lateral Compression of a 115 mm Diameter Steel Tube. Automotive Safety Engineering Pty Ltd., Adelaide.

[31] Ansys (2015) Finite Element Analysis Tool. Ansys Inc. http://www.ansys.com

[32] Fildes, B.N., Lane, J.C., Lenard, J. and Vulcan, A.P. (1991) Passenger Cars and Occupant Injury. Monash University Accident Research Centre.

[33] Sicking, D., Mak, K., Rohde, J. and Reid, J. (2009) Manual for Assessing Safety Hardware. American Association of State Highway Transportation Officials, Washington DC.

# 22

# Flame Spread over Binary Mixed Liquid Fuels—Super-Flash/Super-Flash Mixed Condition

**Koichiro Sasaki, Akihiko Ito, Hiroyuki Torikai**

Department of Intelligent Machines and System Engineering, Hirosaki University, Hirosaki, Japan
Email: h14ms611@hirosaki-u.ac.jp

## Abstract

**In this study, we investigate the flame spread characteristics of binary mixed liquid fuels that are both in super-flash condition. We propose a method for estimating flashpoint and flame spread rate for binary mixed liquid fuels from binary vapor concentration distribution and compare these estimated values with experimental results. In addition, we measure the flame height and center of the flame leading edge for binary mixed liquid fuels. The results show that, when the flashpoint of each single fuel is known, the flashpoint of the binary mixed liquid fuel can be calculated. Moreover, the fuel that most influence combustion changes near the intersection of vapor concentration distribution in accordance with Raoult's law. Photographs of the flame indicate discontinuity. We can calculate flame spread rate by using the ratio of each diffusion coefficient to judge equivalently. Finally, it is found that the flame height and center of flame leading edge are proportional to total vapor concentration of the binary mixed liquid fuels.**

## Keywords

**Flame Spread, Binary Mixed Liquid Fuels, Raoult's Law, Flashpoint, Super-Flash Condition**

## 1. Introduction

Recently, serious disasters related to flame spread over liquid fuel such as tsunami fires happened during the Great East Japan earthquake in 2011. With this background, the study of flames spread over liquid fuel has drawn current interest, because of its relevance to fire safety and our basic curiosity about its complex mechanism. A series of studies are performed to understand the mechanism on alcohols [1]-[13]. But there are few studies of flame spread over binary mixed liquid fuels. This flame spread phenomenon is more similar to the real situation. So this study is needed to prevent unexpected damage.

The flame spread characteristics of mixed liquid fuel depend on the distribution of vapor concentration by each liquid component that is formed. However, the relationship between vapor concentration distribution and flame spread rate is not clear. The purpose of this study is to reveal the relationship between vapor concentration distribution of mixed liquid fuel and flame spread rate. In this study, we try ethanol-methanol combinations which are both in super-flash condition. We measure distribution of vapor concentration and flame spread rate of binary mixed liquid fuels under these conditions and consider the relationship.

## 2. Estimation of Flashpoint and Flame Spread Rate for Binary Mixed Liquid Fuels

### 2.1. Flashpoint of Binary Mixed Liquid Fuels

To derive the flashpoint of binary mixed liquid fuels, we measured the vapor concentration of binary mixed liquid fuels in accordance with Raoult's law [14]. The flashpoint of binary mixed liquid fuels depends on the composition of each liquid composition. So it is very hard to measure the flashpoint for all liquid compositions. With this background, we examined prediction methods of the flashpoint of binary mixed liquid fuels using the flashpoint of simple ingredients and Raoult's law. In this study, we used vapor concentration at 1mm from the liquid surface and regarded it as a quasi-steady condition. At first, we measured the flashpoint of each simple ingredient and defined A which has lower flashpoint and $B$ which has higher flashpoint. And each flash point is determined as $T_A$ $T_B$. Secondly, we calculated vapor concentration of each ingredient of the binary mixed liquid fuels. To keep the lower limit of the different flammable materials constant, we used a correction value. We multiplied the correction value $F$ by the vapor concentration of fuel $B$ $C_B$ and summed up $FC_B$ and vapor concentration of fuel $A$. The sum is $c_m$. The equations are below:

$$F = c_A / c_B \tag{1}$$

$$c_m = c_A + FC_B \tag{2}$$

At each liquid phase composition, if the vapor concentration is higher than the flammable material's lower limit $c_m$, the temperature at this time is defined as the flashpoint.

### 2.2. Flame Spread Rate of Binary Mixed Liquid Fuels

The flame spread rate of binary mixed liquid fuels depends on the fuel vapor concentration which is generated by each liquid composition. Therefore, we tried to calculate by converting total vapor concentration into single component fuel. Equation (2) indicates the concentration conversion equation

$$C_t' = X \cdot C_t \tag{3}$$

$C_t'$ is vapor concentration after conversion. $C_t$ is vapor concentration before conversion. The basic characteristics of mixed fuels under super-flash conditions are pre-mixed flame and burning velocity. To judge the vapor concentration of binary mixed liquid fuels by the burning velocity equivalently, we used the conversion value $X$ for single component fuel and composition of maximum burning velocity with an equivalent ratio of 1. We calculated the maximum burning velocity of methanol as 55 cm/s [15] and ethanol as 47 cm/s [16].

Secondly, flame spread over mixed liquid fuels when both of them are in super-flash condition depends on total fuel vapor concentration. Therefore, we used Equation (3) to calculate flame spread rate over binary mixed liquid fuels

$$V_a = c_t V_i \tag{4}$$

$V_a$ is the flame spread rate of mixed fuels when both of them are in super-flash condition. $V_i$ is defined as flame spread rate per 1 vol% by flame spread rate of the single component fuel divided by vapor concentration. If liquid temperature is determined, vapor concentration is also determined and if vapor concentration distribution is determined, flame spread rate is also determined. $c_t$ is the total vapor concentration of mixed fuels in each liquid fuel composition.

## 3. Experimental Apparatus and Methods

### 3.1. Measurement of Flashpoint of Binary Mixed Liquid Fuels

**Figure 1** shows the experimental apparatus to measure flashpoint. We performed the flashpoint experiment

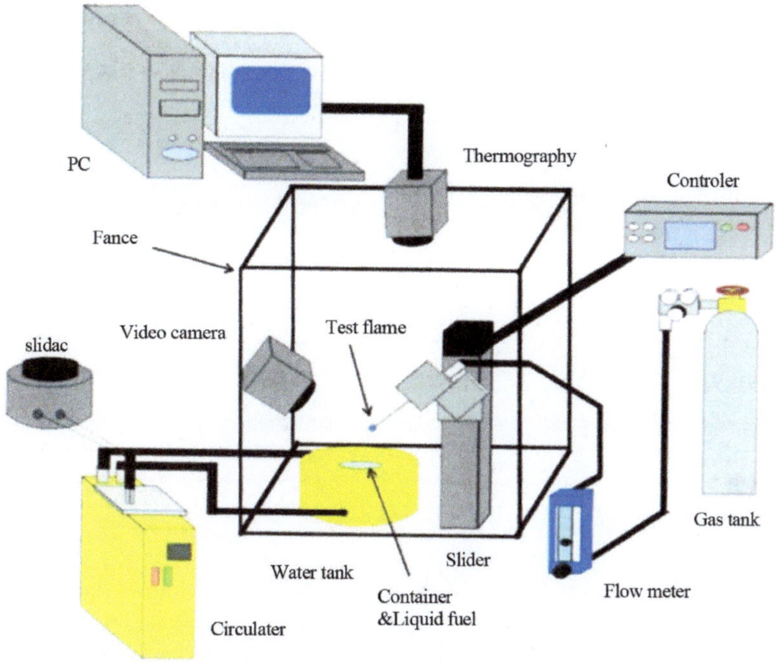


**Figure 1.** Experimental apparatus for measurement of flashpoint.

according to [JISK2256]. We poured the fuel 10 mm from edge of the fuel container to the fuel surface. When the test flame approached the fuel, the distance from the top of the test flame nozzle at the lowest to the fuel surface was 9.3 mm. The fuel container was kept cool by cooling circulation water from a circulator to keep the fuel container temperature less than 10˚C from the predicted flashpoint. When the fuel container temperature neared the flashpoint, we opened the cover and move the apparatus up and down. After that, we judged ignition by observation. We raised the liquid temperature 1˚C/1 min and measured for 30 s. Finally, we performed the experiment five times for each condition and we used methanol and ethanol. The flashpoint of methanol is 15.6˚C and ethanol is 18˚C in an open cup [17].

## 3.2. Measurement of Flame Spread Rate over Binary Mixed Liquid Fuels

**Figure 2** shows the experimental apparatus to measure flame spread rate over binary mixed liquid fuels. The fuel container is made of aluminum and the size is 20 mm wide 250 mm long and 3 mm thickness. We performed the experiment in a wind tunnel with a counter current of 0.1 m/s. The cross-sectional size was 70 mm × 70 mm and the length was 300 mm. To keep the initial fuel temperature constant, we set aluminum piping under the fuel container with diameter 10 mm × length 270 mm and circulated water. We used nichrome wire to ignite and filmed by video camera (CASIO EX-F1). To measure liquid temperature, we used a thermocouple which was attached on the edge of the fuel container. We performed the experiment five times and the initial liquid temperatures were 18˚C, 20˚C, 24˚C and 28˚C. The fuels were ethanol and methanol the same as the flashpoint measurement—both of them were in super-flash condition.

## 3.3. Measurement of Flame Shape on Binary Mixed Liquid Fuels

The spreading flame shape depends on vapor concentration distribution near the liquid surface. In this study, judging from the flame shape, it was a so-called “triple flame” structure. In this structure, the vapor concentration is rich near the surface and the flame shape changes toward the direction of height-vapor concentration is lean and vapor concentration seems to be a theoretical mixture ratio at the apex of flame curvature. We defined the maximum height from the liquid surface to the apex of flame curvature as flame center because the theoretical mixture of each liquid phase composition was not known. This is the inherent value for each liquid fuel. Thus we performed this experiment because we seemed to reveal the predominating liquid fuel of each liquid

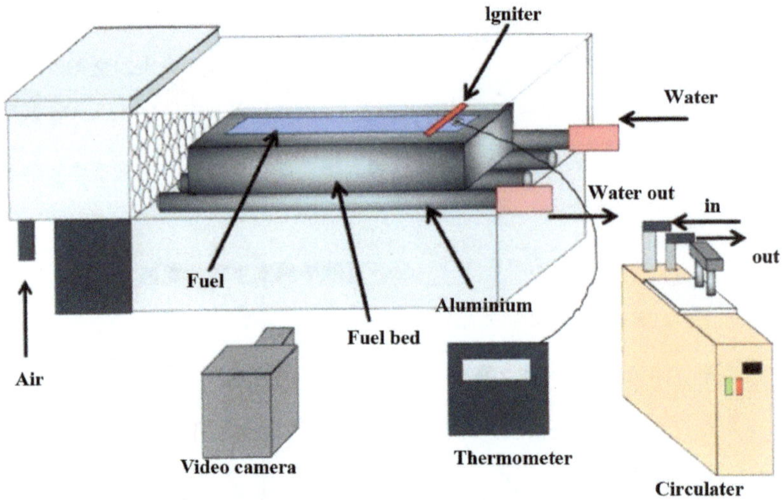


**Figure 2.** Experimental apparatus for measurement of flame spread rate.

phase composition by investigating the theoretical mixture ratio. **Figure 3** shows the measurement method of flame center. We measured the maximum height of flame curvature from the liquid surface on the photographs which we filmed to measure flame spread rate. We indicated the relationship between measured flame height and vapor concentration distribution. Vapor concentration distribution is calculated by Equation (4) [14] and (5)

$$c_i^*/c_s = 1 - \text{erferf}\left(y/\sqrt{4Dt}\right) \tag{5}$$

$$c_i^* = \alpha k c_s \tag{6}$$

where $k$ is the infrared absorption coefficient [14]. $c_s$ is saturated vapor concentration calculated by the literature [18] [19]. $D$ is the diffusion coefficient which depends on temperature and pressure. $D$ can be calculated by equation (6) [20] and using the temperature in the laboratory $T_0$ and pressure $P$

$$D = D_0\left(T_0/273\right)^2\left(P_0/P\right) \tag{7}$$

where $D_0$ is the diffusion coefficient at 0˚C 1 atm (standard conditions) and methanol is 0.135 $cm^2/s$, ethanol is 0.1020 $cm^2/s$ [21]. $P_0$ is atmospheric pressure.

### 3.4. Measurement of Flame Height on Binary Mixed Liquid Fuels

We measured flame height and flame center at the same time. **Figure 4** shows the measurement method for flame height. We investigated how the flame shape changes at each liquid phase composition. In the photographs, we defined the maximum height of blue flame by eyesight from the liquid surface as flame height. We indicated the relationship between flame height and vapor concentration distribution calculated by Equation (5) and Equation (6).

## 4. Experimental Results and Examination

### 4.1. Appearance of Flame Spread on Binary Mixed Liquid Fuels

**Figure 5** shows the appearance of flame spread on methanol/ethanol mixed fuels at 20˚C. The appearances are different due to the component of binary mixed liquid fuels. Based on these photographs, we considered the relationship between vapor concentration distribution and flashpoint and flame spread rate.

### 4.2. The Relationship between Flashpoint and Liquid Fuel Composition

**Figure 6** shows the calculation result of flashpoint of methanol/ethanol mixed liquid fuels. The more the compo-

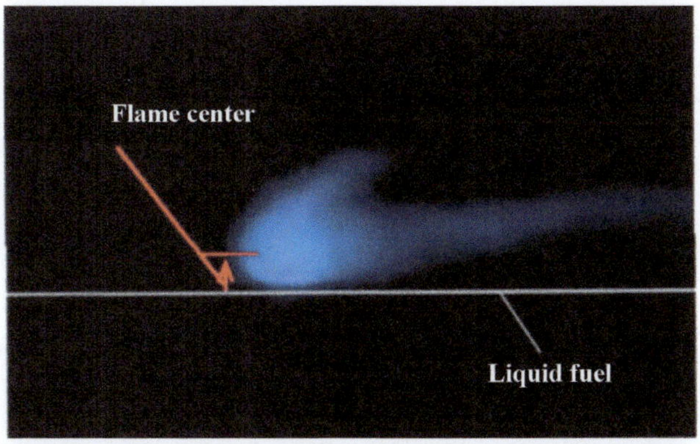


**Figure 3.** Measurement of flame center.

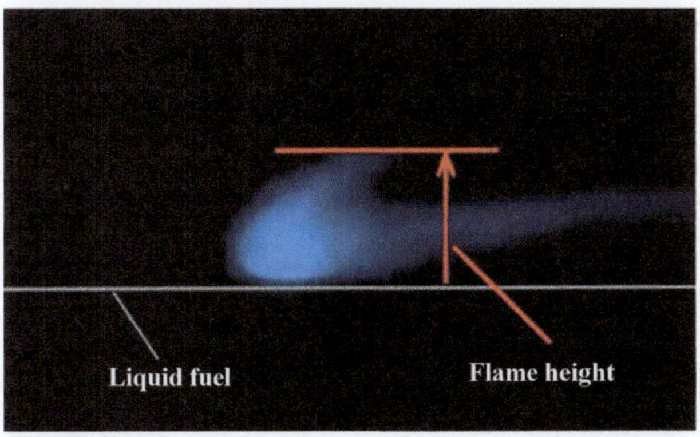


**Figure 4.** Measurement of flame height.

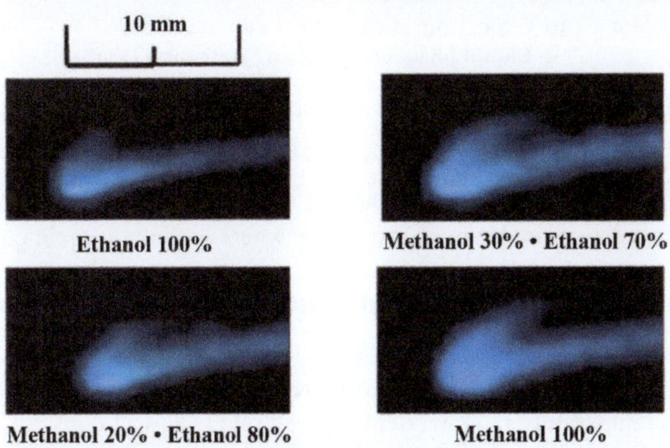


**Figure 5.** Photographs of flame ($T = 20^{\circ}C$).

sition of ethanol increases, the less the flashpoint decreases. This is because the vapor of methanol increases near the liquid surface and it reaches the lower limit concentration. **Figure 7** shows a comparison of experimental results and estimated results using Equation (1) and Equation (2).

This is the better method to predict flashpoint because the estimated result and experimental result match well.

Considering that methanol vapor tends to increase, in case a gap is generated, it needs to be a lower value than real one. But considering that the flashpoint is measured under higher temperature than the estimated one, it is thought that when methanol vapor concentration increases, ethanol vapor concentration has an influence on the

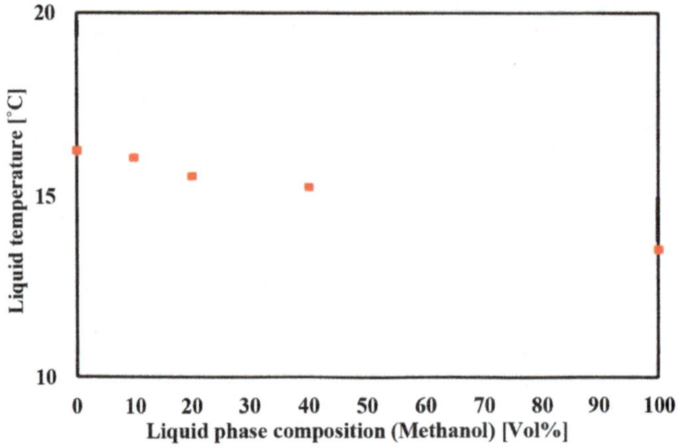


**Figure 6.** Flashpoint of binary liquid fuels.

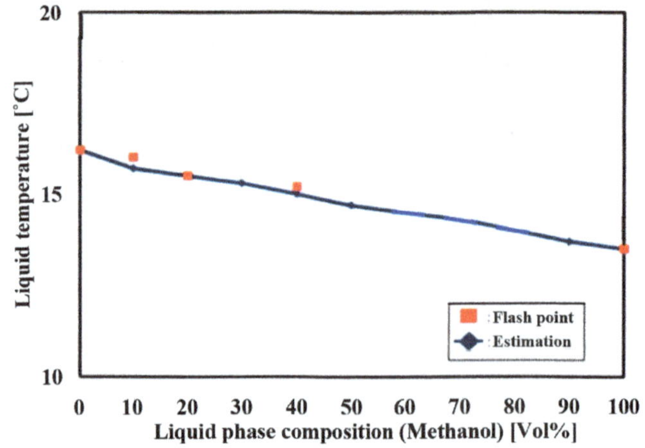


**Figure 7.** Comparison of experimental result and estimated value of flashpoint.

flashpoint. The reason why we couldn't measure flashpoint under 60 - 90 methanol composition, is because the gap between flashpoint and burning point is small in the case where methanol liquid composition increases.

## 4.3. Relationship between Flame Spread Rate and Distribution of Vapor Concentration

**Figure 8** shows the relationship between liquid composition and flame spread rate under each liquid component. Flame spread rate under super-flash condition tends to increase at the same time the methanol composition increases. But when the liquid temperature is 28°C, flame spread rate is almost constant. This is because methanol and ethanol reach stoichiometric ratio. In addition, the flame spread rate is discontinuous at 20°C with 20% - 30% methanol composition and 24°C with 30% - 40% methanol composition as shown in **Figure 8**. It is thought that the dominant liquid fuel changes. **Figure 9** shows the relationship between vapor concentration and methanol liquid phase. In **Figure 8**, the dominant liquid fuel is changed by around 20%. From Equation (2), we changed ethanol to methanol vapor and methanol to ethanol vapor. The chain line in **Figure 9** shows the converse value which is mainly fuel vapor concentration. **Figure 10** shows the comparison with calculated flame spread rate using Equation (3) with equivalent concentration and experimental results. Estimated values of flame spread rate correlate with experimental results well. **Figure 5** shows fuel temperature of 20°C. We focused on the circle in **Figure 8**. To examine the cause of discontinuity, we compared the flame around the discontinuity point. Compared with the images of 20% methanol composition and 30% methanol composition, the flame shape of 20% methanol composition is similar to the flame of ethanol 100% ratio. On the other hand, the flame of methanol

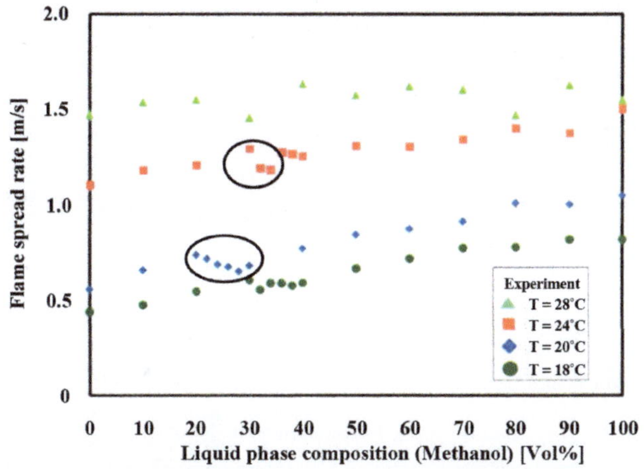


**Figure 8.** Flame spread rate on binary mixed liquid fuels.

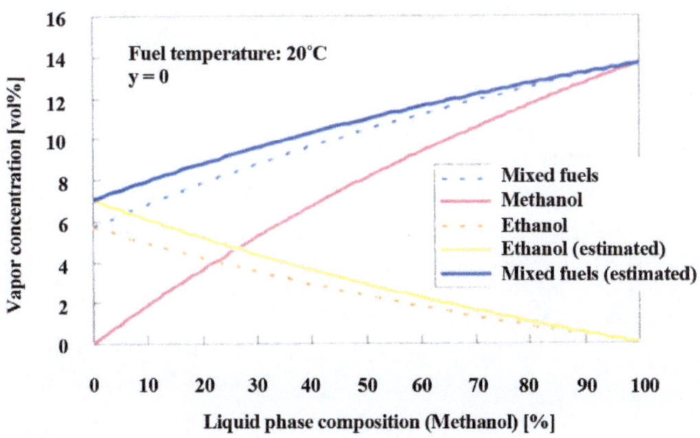


**Figure 9.** Vapor concentration of binary mixed liquid fuels.

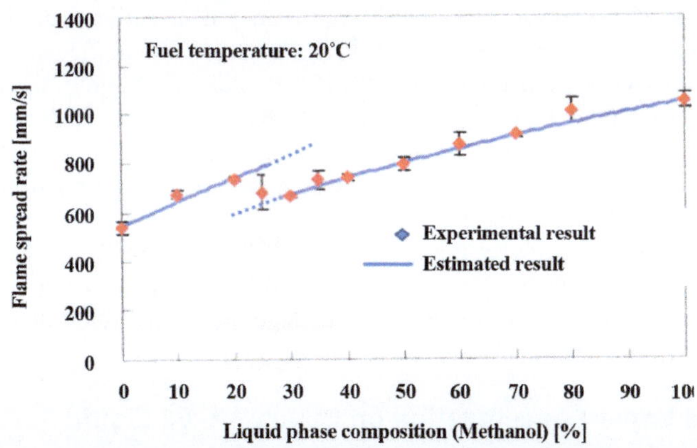


**Figure 10.** Estimation of flame spread rate on binary mixed liquid fuels.

30% is similar to the flame of methanol 100%. Therefore, the cause of decline in flame spread rate is substituted of mainly fuel vapor. In short, binary mixed liquid fuels which are both in super-flash condition have boundary values for which the dominant fuel changes. By the way, **Figure 11** shows a comparison of experimental results and estimated values of flame spread rate. When liquid temperature is low, a discontinuity point appears. But at 28°C, the discontinuity point doesn't appear and the flame spread rate is almost constant. The reason seems to be

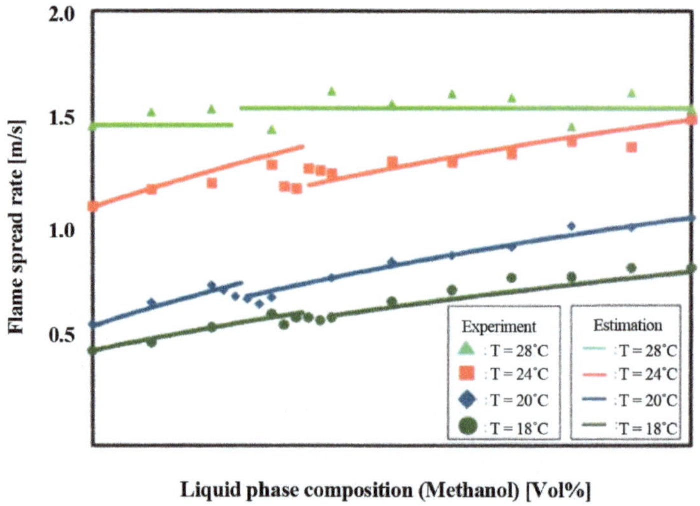


**Figure 11.** Comparison of experimental results and estimated values.

that flammable vapor on the liquid surface reaches the theoretical mixture ratio. But by the influence of the rise of liquid fuel, a gap is generated between experimental value and estimated value. So we need to consider equivalent concentration. As for flame spread over binary mixed liquid fuel, the predominating liquid fuel changes at the cross section of vapor concentration distribution in accordance with Raoult's law. To calculate flame spread rate over binary mixed liquid fuels by vapor concentration, one liquid fuel must be converted to the other' by Equation (2). The calculation equation is indicated below:

$$c_t = c_b + c_a X \tag{8}$$

$$V_i = V_b / c_b \tag{9}$$

where $V_a$ is flame spread rate over binary mixed liquid fuels, $V_i$ is flame spread rate per 1 vol% of predominated liquid fuel and $c_t$ is total vapor concentration at each liquid phase composition. We defined the extinguish distance as 1 mm from the liquid surface. Raoult's law is adopted on the liquid surface, but actually a fuel vapor distribution exists. So we considered this. Vapor concentration distribution depends on the kinds of liquid fuel. So we used a conversion coefficient $X$ which is the ratio of diffusion coefficient. Thus to calculate flame spread rate over binary liquid fuels, we use Equation (9). If the value at extinguish distance 1 mm reaches the theoretical mixture ratio of each fuel, the flame spread rate is constant. From the experiment of measurement of vapor concentration and flame shape, if the liquid phase composition of methanol is 0% - 25%, we regarded the predominating liquid fuel to be ethanol. If the liquid phase composition of methanol is 26% - 100%, we regarded the predominating liquid fuel to be methanol. From **Figure 9** and **Figure 10**, the experimental results and estimated results match well. In the case that liquid temperature is 28°C, the flame spread rate is constant because ethanol and methanol reach the theoretical mixture ratio. We changed the predominating liquid fuel to 20% because the cross section of Raoult's law is near 20%. However we found the result that the cross section didn't change, but the discontinuity point was displaced.

## 4.4. Flame Center and Vapor Concentration on Binary Mixed Liquid Fuels

**Figure 12** shows the relationship between the flame center of vapor concentration distribution and each liquid phase composition at each liquid temperature. Vapor concentration is calculated by the vapor concentration distribution of the diffusion equation. Flame center and vapor concentration distribution rise proportionally with the increase of methanol composition. From this result, the lean limit of binary mixed liquid fuels seems to expand by increasing ethanol composition. But some calculation results which the flame center makes corresponding to vapor concentration distribution sometimes indicate more than the theoretical mixture ratio. So we need to consider a more accurate calculation method.

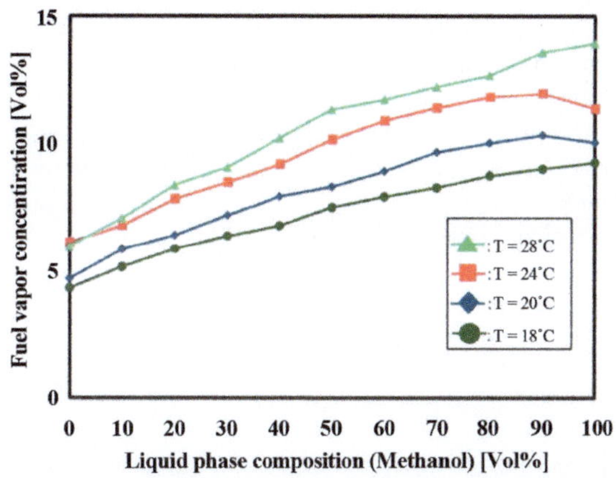


**Figure 12.** Flame center on binary mixed liquid fuels.

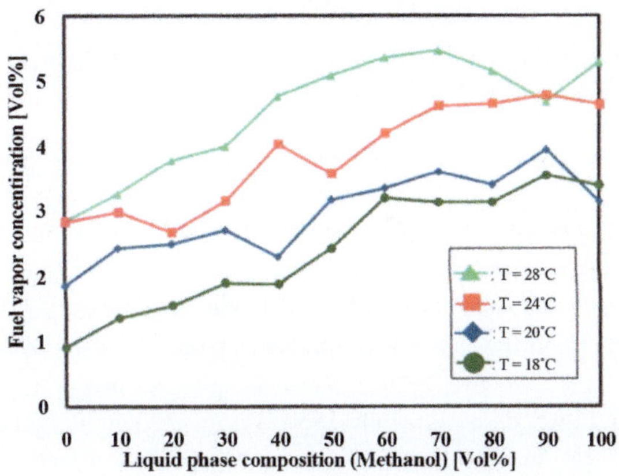


**Figure 13.** Flame height on binary mixed liquid fuels.

## 4.5. Flame Height and Vapor Concentration on Binary Mixed Liquid Fuels

To calculate the flashpoint of binary mixed liquid fuels, if the flashpoint of a single liquid fuel is known, we can calculate one liquid fuel's flammable lower limit by re-compensating the other using Raoult's law and a correction value. **Figure 13** shows the relationship between flame height of vapor concentration distribution and each liquid phase composition at each liquid temperature. Flame height and vapor concentration rise proportionally to the rise of methanol composition in the same way as the flame center. Thus the lean limit of binary mixed liquid fuels also expands by increasing ethanol composition. But the judgement of flame height is by eyesight and the blue flame height changes at each place. So we need to consider more precise method.

# 5. Conclusions

We aim to establish a prediction method by using the mixed liquid fuels methanol and ethanol in super-flash condition and Raoult's law to calculate vapor concentration distribution. The conclusions are follows:

1) To calculate the flashpoint of binary mixed liquid fuels, if the flashpoint of a single liquid fuel is known, we can calculate one liquid fuel's flammable lower limit by re-compensating the other using Raoult's law and a correction value.

2) For the flame spread rate over binary mixed liquid fuels, the predominating liquid fuel changes near the cross section of vapor concentration distribution in accordance with Raoult's law.

3) For the discontinuity of flame spread rate over binary mixed liquid fuels, we confirm the decrease of flame spread rate by experiment. In addition, flame shape changes near the discontinuity point. From these results, we find discontinuity as a phenomenon.

4) When the liquid temperature is 28°C, the flame spread rate is constant. This is why flammable vapor formed near the liquid surface reaches theoretical mixture ratio.

5) In the case that we calculate flame spread rate over binary mixed liquid fuel by vapor concentration distribution from flame spread rate over single liquid fuel and Raoult's law, we cannot calculate it by the total amount of binary mixed liquid fuel vapor. So we can calculate the flame spread rate by using a conversion coefficient to judge the liquid fuels equivalently.

6) The flame center of binary mixed liquid fuels is proportional to the rise of total vapor concentration of binary mixed liquid fuels.

7) The flame height on binary mixed liquid fuels is proportional to the rise of total vapor concentration of binary mixed liquid fuels.

## Acknowledgements

We would like to thank Kenji Takimoto working in Shigeru Corporation, Ltd and Kota Mikami working in Toshiba Plant Systems and Services Corporation for experimental supports and discussions.

## References

[1] Burgoyne, J.H., Roberts, A.F. and Quinton, P.G. (1968) The Spread of Flame Across a Liquid Surface. I. The Induction Period. *Proceedings of the Royal Society*, **A308**, 39-53. http://dx.doi.org/10.1098/rspa.1968.0206

[2] Burgoyne, J.H., Roberts, A.F. and Quinton, P.G. (1968) The Spread of Flame Across a Liquid Surface. II. Steady-State Conditions. *Proceedings of the Royal Society*, **A308**, 55-68. http://dx.doi.org/10.1098/rspa.1968.0207

[3] Burgoyne, J.H., Roberts, A.F. and Quinton, P.G. (1968) The Spread of Flame Across a Liquid Surface. III. A Theoretical Model. *Proceedings of the Royal Society*, **A308**, 69-79. http://dx.doi.org/10.1098/rspa.1968.0208

[4] Glassman, I., Hansel, J. and Eklund, T. (1969) Hydrodynamic Effects in the Flame Spreading, Ignitability and Steady Burning of Liquid Fuels. *Combustion and Flame*, **13**, 99-101. http://dx.doi.org/10.1016/0010-2180(69)90037-6

[5] Kikaino-Kenkyu, I.H. (1971) Hyomenwokanetsusaretaekitai no ryudo. *Kikai no Kenkyu*, **23**, 1505-1510.

[6] Akita, K. and, Fujiwara, O. (1971) Pulsating Flame Spread along the Surface of Liquid Fuels. *Combustion and Flame*, **17**, 268-269. http://dx.doi.org/10.1016/S0010-2180(71)80172-4

[7] Akita, K. (1972) Some Problems of Flame Spread along a Liquid Surface. *Symposium (International) on Combustion*, **14**, 1075-1083. http://dx.doi.org/10.1016/S0082-0784(73)80097-9

[8] Newman, J.S. (1979) Lase Droplet Velocimeter Measurements of the Gas and Liquid Flow Fields Induced by Flame Propagation over a Liquid Surface. MSE Thesis, Princeton University, Princeton.

[9] Hirano, T., Suzuki, T., Mashiko, I. and Tanabe, N. (1980) Gas Movements in Front of Flames Propagating across Methanol. *Combustion Science and Technology*, **22**, 83-91. http://dx.doi.org/10.1080/00102208008952373

[10] Mackinven, R., Hansel, F. and Glassman, I. (1970) Influence of Laboratory Parameters on Flame Spread across Liquid Fuels. *Combustion Science and Technology*, **1**, 293-306. http://dx.doi.org/10.1080/00102206908952209

[11] Matsumoto, Y. and Saito, T. (1980) Ekimennenshodenpanikansurukenkyu. *JSEM*, **46**, 996-1006. (in Japanese)

[12] Matsumoto, Y. and Saito, T. (1981) Flame Propagation over Liquid Fuel at Sub-Flash Temperature. *Bulletin of JSEM*, **24**, 123-138.

[13] Takahashi, K., Ito, A., Kudo, Y., Konishi, T. and Saito, K. (2008) Scaling Analysis on Pulsating Flame Spread over Liquids. *International Journal of Chemical Engineering*, **2008**, Article ID: 178292. http://dx.doi.org/10.1155/2008/178292

[14] Takimoto, K., *et al.* (2008) Speech Papers Correction, Combustion Symposium Part 46, 286-287.

[15] Hirano, T. (1986) Nenshogaku-Nenshogensho to sonoseigyo-, Kaibun-do, 49-63. (In Japanese)

[16] Gulber, O.L. (1982) Laminar Burning Velocities of Methanol, Ethanol and Isooctane-Air Mixtures. *Proceedings of the Combustion Institute*, **19**, 275-281.

[17] Drysdale, D. (1999) An Introduction to Fire Dynamics. 2nd Edition, Wiley, Hoboken, 203.

[18] Ueda, M. (Year) Shitsudo to Johatsu—Kiso to Keisokugijutsu Made. Colona Co., 2-4. (In Japanese)

[19] Kuratera, T., Terasaki, K., Nagai, N. and Ito, K. (1982) Kogyo netsurikigaku. Asakurashoten, Tokyo, 96-98. (In Japanese)

[20] Katayama, K., *et al.* (1986) JSME Data Book: Heat Transfer. 4th Edition, JSME, Tokyo, 114-115.

[21] Kato, Y. (1964) Heat Dennetsu Gairon. Yokendo, Tokyo, 222-223. (In Japanese)

## Nomenclature

$C_a$: Under limiting concentration of single component of liquid fuel [vol%]
$C_b$: Under limiting concentration of single component of liquid fuel [vol%]
$C_A$: Vapor concentration of fuel A [vol%]
$C_B$: Vapor concentration of fuel B [vol%]
$C_m$: Under limiting concentration of binary mixed liquid fuels [vol%]
$F$: Correction value [-]
$C_t$: Total vapor concentration at each liquid phase composition [vol%]
$X$: Vapor conversion coefficient [-]
$C_t'$ : Vapor concentration after conversion [vol%]
$V_a$: Flame spread rate over binary mixed liquid fuels [m/s]
$V_i$: Flame spread rate over mainly liquid fuel per 1 vol% [m/s]
$Ci$: Vapor concentration of ingredient $i$ of binary mixed liquid fuels [vol%]
$C_i^*$ : Vapor concentration of ingredient ionly [vol%]
$C_s$: Saturated vapor concentration [vol%]
$y$: The height from liquid surface [mm]
$T$: Temperature [°C]
$T_0$: Temperature in laboratory [°C]
$T_A$: Flashpoint of fuel which has lower flashpoint [°C]
$T_B$: Flashpoint of fuel which has higher flashpoint [°C]
$t$: Time [s]
$P$: Pressure [Pa]
$D$: Diffusion coefficient [$cm^2/s$]
$D_0$: Diffusion coefficient at 0°C, 1 atm [$cm^2/s$]
$\alpha$: Inherent value of each liquid fuel [-]
$k$: Radiation absorption coefficient [/m]

# Permissions

All chapters in this book were first published in OJSST, by Scientific Research Publishing; hereby published with permission under the Creative Commons Attribution License or equivalent. Every chapter published in this book has been scrutinized by our experts. Their significance has been extensively debated. The topics covered herein carry significant findings which will fuel the growth of the discipline. They may even be implemented as practical applications or may be referred to as a beginning point for another development.

The contributors of this book come from diverse backgrounds, making this book a truly international effort. This book will bring forth new frontiers with its revolutionizing research information and detailed analysis of the nascent developments around the world.

We would like to thank all the contributing authors for lending their expertise to make the book truly unique. They have played a crucial role in the development of this book. Without their invaluable contributions this book wouldn't have been possible. They have made vital efforts to compile up to date information on the varied aspects of this subject to make this book a valuable addition to the collection of many professionals and students.

This book was conceptualized with the vision of imparting up-to-date information and advanced data in this field. To ensure the same, a matchless editorial board was set up. Every individual on the board went through rigorous rounds of assessment to prove their worth. After which they invested a large part of their time researching and compiling the most relevant data for our readers.

The editorial board has been involved in producing this book since its inception. They have spent rigorous hours researching and exploring the diverse topics which have resulted in the successful publishing of this book. They have passed on their knowledge of decades through this book. To expedite this challenging task, the publisher supported the team at every step. A small team of assistant editors was also appointed to further simplify the editing procedure and attain best results for the readers.

Apart from the editorial board, the designing team has also invested a significant amount of their time in understanding the subject and creating the most relevant covers. They scrutinized every image to scout for the most suitable representation of the subject and create an appropriate cover for the book.

The publishing team has been an ardent support to the editorial, designing and production team. Their endless efforts to recruit the best for this project, has resulted in the accomplishment of this book. They are a veteran in the field of academics and their pool of knowledge is as vast as their experience in printing. Their expertise and guidance has proved useful at every step. Their uncompromising quality standards have made this book an exceptional effort. Their encouragement from time to time has been an inspiration for everyone.

The publisher and the editorial board hope that this book will prove to be a valuable piece of knowledge for researchers, students, practitioners and scholars across the globe.

# List of Contributors

**Mehrzad Ebrahemzadih**
Department of Occupational Health, Faculty of Health, Kurdistan University of Medical Sciences, Sanandaj, Iran
Kurdistan Environmental Health Research Center, Kurdistan University of Medical Sciences, Sanandaj, Iran

**G. H. Halvani**
Departments of Occupational Health, Faculty of Health, Shahid Sadoghi University of Medical Sciences, Yazd, Iran

**Behzad Shahmoradi**
Kurdistan Environmental Health Research Center, Kurdistan University of Medical Sciences, Sanandaj, Iran

**Omid Giahi**
Department of Occupational Health, Faculty of Health, Kurdistan University of Medical Sciences, Sanandaj, Iran
Kurdistan Environmental Health Research Center, Kurdistan University of Medical Sciences, Sanandaj, Iran

**Raqibull Hasan, Mohammad Monirujjaman Khan, Asaduzzaman Ashek, Israt Jahan Rumpa**
Department of Electrical and Electronic Engineering, Primeasia University, Dhaka, Bangladesh

**Joseph M. Makaya Alfred S. Traoré and Nicolas Barro**
Research Center for Biological, Food and Nutritional Sciences, Research and Training Unit, Life
Earth Sciences, University of Ouagadougou, Ouagadougou, Burkina Faso

**Serge Aho**
Epidemiology and Infection Control Unit, Public Hospital of Dijon, Dijon, France

**Joseph Wethé**
Pan African Agency, Water and Sanitation for Africa, Ouagadougou, Burkina Faso

**Dayéri Dianou**
National Center for Scientific and Technological Researches, Institute for Health Sciences Research, Ouagadougou, Burkina Faso

**Mehrzad Ebrahemzadih and Ebrahim Darvishi**
Kurdistan Environmental Health Research Center, Kurdistan University of Medical Sciences, Sanandaj, Iran
Department of Occupational Safety & Health, School of Health, Kurdistan University of Medical Sciences, Sanandaj, Iran

**Mehdi Meimanat Abadi**
Children's Cancer Charity, Kurdistan University of Medical Sciences, Sanandaj, Iran

**Saeed Dehestaniathar and Afshin Maleki**
Kurdistan Environmental Health Research Center, Kurdistan University of Medical Sciences, Sanandaj, Iran

**Yuichi Otsuka**
Department of System safety, Nagaoka University of Technology, Nagaoka, Japan

**Akihiko Hidaka, Yuichiro Izato and Atsumi Miyake**
Graduate School of Environment and Information Sciences, Yokohama National University, Yokohama, Japan

**Wen-Fa Zhu, Xiao-Dong Chai and Shu-Bin Zhen**
College of Urban Railway Transportation, Shanghai University of Engineering Science, Shanghai, China

**Hui-Zhen Ma**
Library, Shanghai University of Engineering Science, Shanghai, China

**Yan-Ling Qiu**
School of Civil Engineering, Southwest Jiaotong University, Chengdu, China

**Ling-Kan Yao**
School of Civil Engineering, Southwest Jiaotong University, Chengdu, China
Road and Railway Engineering Research Institute, Sichuan Key Laboratory of Aseismic Engineering and Technology, Chengdu, China

**Quan Yuan**
Xingshu Company, Sichuan Provincial Transport Department, Chengdu, China

**John Nsikan Efiok**
Faculty of Business Administration, University of Uyo, Uyo, Nigeria

**Opeoluwa Oluseye**
Warsash Maritime Academy, Southampton Solent University, Southampton, UK

**Tommy Uduak**
Faculty of Management Sciences, University of Calabar, Calabar, Nigeria

**Raimi Olalekan**
Department of Environmental Health Management, University of Uyo, Uyo, Nigeria

**Akihiko Hidaka and Atsumi Miyake**
Graduate School of Environment and Information Sciences, Yokohama National University, Yokohama, Japan

**Yukinobu Yoshida, Akihiko Ito and Hiroyuki Torikai**
Graduate School of Science and Technology, Hirosaki University, Hirosakicity, Japan

**Wanda D. Minnick and Jonathan D. Cunkelman**
Safety Sciences Department, Indiana University of Pennsylvania, Indiana, USA

**Stephen J. Mitchual, Mark Donkoh and Francis Bih**
Department of Construction and Wood Technology Education, University of Education, Winneba, Kumasi Campus, Kumasi, Ghana

**Christopher Allen Janicak**
Department of Safety Sciences, Indiana University of Pennsylvania, Indiana, USA

**Yusuke Koshiba**
Department of Materials Science and Chemical Engineering, Yokohama National University, Yokohama, Japan

**Hideo Ohtani**
Department of Safety Management, Yokohama National University, Yokohama, Japan

**Samuel A. Lawoyin and Ding-Yu Fei, Ou Bai**
Department of Biomedical Engineering, Virginia Commonwealth University, Richmond, VA, USA

**Helmut Paschold**
Department of Safety Sciences, College of Health and Human Services, Indiana University of Pennsylvania, Indiana, PA, USA

**Ahmad Bahoo Toroody**
Faculty of Engineering, Kar Higher Education Institute of Qazvin, Qazvin, Iran

**Mohammad Mahdi Abaiee**
Faculty of Marine Technology, Amirkabir University of Technology, Tehran, Iran

**Reza Gholamnia**
School of Health Safety and Environment, Shahid Beheshti University of Medical Science, Tehran, Iran

**Mohammad Bahoo Torody**
Faculty of Engineering, Azad University of Zanjan, Zanjan, Iran

**Nastaran Hekmat Nejad**
Faculty of Engineering, Kar Higher Education Institute of Qazvin, Qazvin, Iran

**Eugie Kabwe and Yiming Wang**
Department of Mining and Mineral Resources Engineering, University of Science and Technology Beijing, Beijing, China

**Fatih Yilmaz and Selcuk Alp**
Department of Industrial Engineering, Yildiz Technical University, Istanbul, Turkey

**Mohammad Sharif Uddin**
School of Engineering, University of South Australia, Mawson Lakes, Australia

**Jacob Quintel**
School of Engineering, University of South Australia, Mawson Lakes, Australia
Automotive Safety Engineering Pty Ltd., Lonsdale, Australia

**Grad Zivkovic**
Automotive Safety Engineering Pty Ltd., Lonsdale, Australia

**Koichiro Sasaki, Akihiko Ito and Hiroyuki Torikai**
Department of Intelligent Machines and System Engineering, Hirosaki University, Hirosaki, Japan
Midori-ku, Yokohama 226-8503, Japan

# Index